A. S. Marfunin

Physics of Minerals and Inorganic Materials

An Introduction

Translated by N.G. Egorova and A.G. Mishchenko

With 138 Figures

Springer-Verlag
Berlin Heidelberg New York 1979

Dr. Arnold S. Marfunin
IGEM Academy of Sciences
Staromonetniy 35
Moscow 109017/USSR

The cover motif has been combined from parts of figures 62, 74, 76.

The original Russian edition was published by Nedra, Moscow, in 1974.

ISBN 3-540-08982-9 Springer-Verlag Berlin Heidelberg New York
ISBN 0-387-08982-9 Springer-Verlag New York Heidelberg Berlin

Library of Congress Cataloging in Publication Data. Marfunin, Arnold Sergeevich. Physics of minerals and inorganic materials. Translation of Vvedenie v fiziku mineralov. Bibliography: p. Includes index. 1. Mineralogical chemistry. 2. Chemistry, Physical and theoretical. 3. Chemistry, Inorganic. I. Title. QE371.M3713 549 78-23186.

This work is subject to copyright. All rights are reserved, whether the whole or part of the material is concerned, specifically those of translation, reprinting, re-use of illustrations, broadcasting, reproduction by photocopying machine or similar means, and storage in data banks.
Under § 54 of the German Copyright Law where copies are made for other than private use, a fee is payable to the publisher, the amount of the fee to be determined by agreement with the publisher.
© by Springer-Verlag Berlin Heidelberg 1979.
Printed in Germany.

The use of registered names, trademarks, etc. in this publication does not imply, even in the absence of a specific statement, that such names are exempt from the relevant protective laws and regulations and therefore free for general use.

Typesetting, printing, and bookbinding by Brühlsche Universitätsdruckerei, Lahn-Gießen
2132/3130-543210

Preface

The physics of minerals in a broad sense implies the fundamental aspects of understanding mineral matter: the electronic structure of atoms related to their behavior in geochemical processes; the atomic and electronic structures of minerals; the properties of minerals, with their genetic, geophysical, and technical significance, and their pressure and temperature dependence; the mechanisms of phenomena and reactions in mineral formation and transformation processes; the physical methods applied in mineralogical, geochemical and petrological studies, and to a great extent in geological surveys and prospecting.

In a narrower sense, it is a branch lying in the border area between mineralogy and solid-state physics, dealing with those aspects of mineralogy which require, for their understanding and investigation, special knowledge in contemporary physics and chemistry of solids.

The physics of minerals accounts for the third crucial change within this century in the conceptual foundations of mineralogy: after physicochemical mineralogy, from experimental studies of phase relations to paragenetic analyses, and crystal chemistry of minerals, there followed solid-state physics, which has evolved to its present state over the past 25 years.

The task of mineralogy has expanded greatly. In addition to the identification and description of minerals, it is becoming necessary to establish the relationships between structure, composition and properties of minerals and their genesis, their distribution within geological regions, magmatic, metamorphic and sedimentary formations and types of ore deposits. The development of new methods of investigation requires an understanding of the physical meaning of the parameters under evaluation.

At present an enormous amount of empirical material has been accumulated and is rapidly growing, due to the continuous development of new methods, techniques and instruments; but further accumulation appears to be ineffective without an understanding of the causal and nonempirical relationships.

The scope of application of minerals has also widened. As a result of the synthesis of large single crystals, minerals have become available for use as lasers, optical crystals, and piezocrystals; as probing systems they are utilized in studying luminophors, scintillators, semi-conductors and superconductors, ferromagnetics and ferroelectrics, thermoluminescent dosimeters, molecular sieves, nonlinear and acoustic optics crystals.

Minerals such as scheelite, fluorite, corundum, garnet, spinel, sphalerite, diamond, galena, pyrrhotite, pyrite, chalcopyrite, quartz, beryl, anhydrite, apatite, the zeolites, and many others represent not only structural types, but also whole classes of compounds that possess special properties and have long been model systems in many physical and applied studies. This fact explains the close contact between the physics of minerals and the steadily growing science of inorganic materials.

The investigation of lunar samples has proved impressively that modern mineralogy and geochemistry are a source of high potential for various techniques, and has thus also shown the need for adequate knowledge of all new methods.

The physics of minerals refers to a system of ideas, models, special theories, formalisms, methods of calculation and experimental parameters describing the nature of minerals. It is based on crystal field theory, molecular orbital theory and energy band theory related to structure, spectroscopy and chemical bonds in minerals[1]. These theories gave rise to new prerequisites in mineralogy and a new geochemical outlook to replace the ideas introduced in the works of Vernadsky, Goldschmidt, and Fersman, which have governed the studies of terrestrial matter over the past decades. This means that the conceptions of atoms with invariable properties (e.g., atomic and ionic radii, ionization potentials, electronegativities) must be replaced in principle by theories of the self-consistent atoms in minerals, with atomic properties changing in each compound. Attempts to determine the behavior of atoms in the earth's crust directly from these invariable properties and the periodic table of elements must be replaced by measurements and calculations in specific mineral systems. In place of the most general physical laws, the most sophisticated and up-to-date specialized theories of individual properties and particular phenomena thus become indispensable.

This book is primarily intended to provide a general introduction to all sections of the physics of minerals, and hence comprises mainly the solid-state and chemical bond theories with special reference to minerals.

Of course, the advanced side of the theories of solids has been thoroughly treated in special texts and reviews so far published, but this book (1) provides a full account of the topics selected for mineralogists: (2) it treats them as part of mineralogy, in fact the most essential part of concepts and methods in mineralogy, and (3) it is adapted to the geological style of thinking, training and manner of formulating questions.

The number of new concepts and theories brought into mineralogy by the physics of solids is extremely large, probably larger than that introduced into mineralogy by any other related science. It was therefore difficult to expound these theories and concepts because of the necessity

---

[1] Further chapters of physics of minerals are contained in A. S. Marfunin: Spectroscopy, Luminescence and Radiation Centers in Minerals. Berlin, Heidelberg, New York: Springer 1979.

Preface VII

to begin with the basic, most profound and most complicated concepts, and at the same time to cover the latest highly sophisticated treatments; mineralogists need results that make use of final data for minerals, and not only explanations of general laws.

Throughout the book the author endeavors to make these new and complicated subjects intelligible and accessible for mineralogists and material researchers. As experience shows, difficulty arises when dealing with the most general concepts. Therefore, in the preliminary chapters of the book the principles of quantum theory and atomic structure are discussed in greater detail than in the works of Fyfe, Grigoriev, Lebedev, Povarennikh, in the geochemistry texts, and in the early works of Fersman and Goldschmidt, to provide a first emphasis on understanding the meanings of parameters, concepts and calculations.

Thus organized, the material reveals the great research potential that lies in nearly every section of this system. Of great importance is that it is the physics of minerals which for the first time in the history of mineralogy provides an understanding of the properties of minerals, mineralogical phenomena and methods.

The references contain only bibliography selected mainly over the last 10–15 years, but include all key references. They are compiled as a comprehensive guide to experimental data, special monographs and reviews.

The author wishes to express his sincere thanks to all his co-workers for their helpful discussions of many problems treated in this book: Drs. L. V. Bershov, A. N. Platonov, A. N. Tarashchan, L. N. Vyalsov, Y. M. Nussick, A. M. Bondar, R. M. Mineeva, A. R. Mkrtchan, V. O. Martirosan, and A. V. Speransky.

The author is deeply grateful to Professors S. A. Altshuler, V. M. Vinokurov, M. M. Zaripov, I. N. Penkov, V. I. Nefedov for valuable suggestions, and remembers with gratitude the valuable discussions of chemical bond problems with Professors Y. K. Sirkin and M. E. Dyatkina.

Moscow, January 1979   A. S. MARFUNIN

Contents

1	**Quantum Theory and the Structure of Atoms**	
1.1	Geochemistry – History of Self-Consistent Atoms	1
1.2	The Beginnings of Quantum Theory	3
1.2.1	Rutherford-Bohr Model of the Hydrogen Atom and Three Postulates of the "Old Quantum Theory"	3
1.2.2	Calculation of the Radius and Energy of Hydrogen Atom Orbits	5
1.2.3	Atomic Structure and Spectra. The Calculation of Line Spectrum of the Hydrogen Atom	6
1.2.4	Fine Structure of Spectra and Sommerfeld's Development of the Bohr Theory; Quantum Numbers	9
1.3	The Schrödinger Equation as the Basic Equation of Quantum Mechanics	14
1.3.1	The Physical Bases	14
1.3.2	Derivation of the Schrödinger Equation	20
1.4	Atomic Orbitals (Solutions of the Schrödinger Equation)	27
1.4.1	Physical Meaning of the Schrödinger Equation Solutions	27
1.4.2	The Physical Meaning of the Atomic Orbitals	28
1.4.3	The s, p, d, f Systematics of Atomic Orbitals	33
1.5	Orbital Radii (Solutions of the Schrödinger Equation)	35
1.5.1	Concerning the Calculations of Electronic Structure of Many-Electron Atoms	35
1.5.2	Orbital Radii and the Wave Functions of Atoms	38
1.6	Electron Spin	40
1.7	Electronic Configurations and the Periodic System of Elements	42
1.8	Term Symbols and Atomic States	46
1.8.1	Description of Atomic States	46
1.8.2	Term Symbols :	49
1.8.3	Term Derivation from Electron Configurations	49
1.8.4	Free Atom Energy Levels and Hamiltonian	52
1.8.5	Atomic Spectroscopy and Spectrochemical Analysis of Minerals, Rocks, and Ores; Spectroscopy – Cosmochemistry – Astrophysics	54
2	**Crystal Field Theory**	
2.1	The Actions of the Crystal Field upon the Atomic Orbitals and Terms	57

2.1.1	Symmetry of Atomic Orbitals in the Crystal Field; the Concepts of Characters and Irreducible Representations	57
2.1.2	Correlation Tables for the Symmetry Types in Various Point Groups	69
2.1.3	Selection Rules Related to Symmetry Types	70
2.2	Three Types of Ion Behavior in Crystal Fields: Weak, Medium, and Strong Crystalline Fields	75
2.3	Iron Group: Term Splitting by Crystal Field	77
2.3.1	Electron Configurations, Terms, Cubic Field Splitting (Qualitative Schemes)	77
2.3.2	Crystal Field Parameters; Tanabe-Sugano Diagrams	81
2.3.3	Splitting by Spin-Orbit Interaction, Jahn-Teller Effect, and Lowering of Symmetry	87

3 Molecular Orbital Theory

3.1	Introduction	94
3.2	General Theory of the Chemical Bond; Molecular Orbital Method; Valence Bond Method	98
3.2.1	Description and Systematics of Molecular Orbitals	98
3.2.2	Molecular Orbital Energy and Coefficient Calculations (the H_2^+ Molecule Ion Example)	108
3.2.3	Molecular Orbital Calculation for Octahedral and Tetrahedral Complexes of Transition Metal and Nontransition Element Ions	116
3.2.4	Valence Bond Method; Hybrid Atomic Orbitals	130
3.3	Analysis of the MO Scheme: Information Obtained from MO and Basic Concepts of the Theory of the Chemical Bond	134
3.3.1	Coulomb Integrals H_{AA} in the MO Method-Ionization Potentials-VSIE; Deep Meaning of the Self-Consistency; Electronegativity	136
3.3.2	LCAO Coefficients c_i and Electronic Population Analysis; Ionicity – Covalency of Chemical Bonding and Effective Charge; Valence and Charge	143
3.4	Further Development of Molecular Orbital Methods	152
3.4.1	About the Methods of the MO Calculations for Isolated Clusters	152
3.4.2	Molecular Orbitals for the Larger Clusters	154
3.4.3	Bond Orbital Model	154

4 Energy Band Theory and Reflectance Spectra of Minerals

4.1	Basic Principles and Methods of the Energy Band Theory	156
4.1.1	Wave Vector k in the Free-Electron Case	156
4.1.2	Two Approximations of the Energy Band Theory: Nearly Free Electrons and Tight Binding Models	158
4.1.3	Concept of k-Space and Brillouin Zones	159

4.1.4	Classification of Orbitals in Crystals with Respect to Symmetry Types 162
4.1.5	Energy Band Structure Schemes 164
4.1.6	Band Occupation; Densities of States; Fermi Surface . . . 166
4.1.7	The Methods of Band Structure Calculation 167
4.2	Analysis of the Band Schemes and Reflectance Spectra of Minerals 168
4.2.1	Intrinsic Absorption and Reflectance Spectra. Measured and Calculated Parameters 170
4.2.2	Structure Type of NaCl–MgO–PbS 171
4.2.3	Structure Type of Sphalerite (Cubic ZnS) 176
4.2.4	Structure Type of Wurtzite (ZnS Hexagonal) 176
4.2.5	Data for Other Minerals 177

5 Spectroscopy and the Chemical Bond

5.1	General Outline and Parameters of Solid State Spectroscopy 178
5.2	Principal Concepts and Parameters of the Chemical Bond from the Standpoint of Spectroscopy 183

6 Optical Absorption Spectra and Nature of Colors of Minerals

6.1	Parameters of Optical Absorption Spectra 190
6.1.1	Units of Measurement of Optical Transition Energies . . . 191
6.1.2	Intensity of Absorption 192
6.1.3	Diffuse Reflectance Spectra 197
6.2	Types of Optical Absorption Spectra and Selection Rules . . 198
6.3	Analysis and Experimental Survey of Transition Metal Ions Spectra. 202
6.4	The Nature of Colors of Minerals 236
6.4.1	Types of Colors of Minerals 236

7 Structure and the Chemical Bond

7.1	Contemporary Methods of Description and Calculations of the Chemical Bond in Solids 239
7.1.1	Extension of the Bond Orbital Methods for Cristobalite and Quartz Structures 242
7.2	Lattice Energy of Ionic Crystals 246
7.3	Lattice Sums, Crystal Field Parameters, Spectroscopical Parameters, and Intracrystalline Distribution 254
7.4	Atomic and Ionic, Orbital, and Mean Radii 263
7.4.1	Ionic Radii and Molecular Orbitals 264
7.4.2	Systems of Additive Ionic and Atomic Radii 268
7.4.3	Appraisal of the Systems of Additive Radii 270
7.4.4	Orbital Radii 270
7.4.5	Experimental X-Ray and Electron Diffraction Determinations of Atomic Sizes 271

8 Chemical Bond in Some Classes and Groups of Minerals

8.1 Diversity of the Aspects of a Complex Phenomenon of Chemical Bond in Solids 274
8.2 The Chemical Bond in Silicates 277
8.2.1 Description of the Chemical Bond in SiO_4^{4-} in Terms of the Calculated Molecular Orbital Diagram 277
8.2.2 Molecular Orbital Diagram for the SiO_4^{4-} According to X-Ray and ESCA Spectra 279
8.2.3 Effective Charges of Si and Al in the Silicates and Aluminosilicates 279
8.2.4 Silica Polymorphs: Energy Band Schemes; Bond Orbital Model and Calculations of the Electronic Structure and Properties 280
8.2.5 Cation Polyhedra in Crystal Structures of Silicates . . . 281
8.2.6 Degree of Ionicity-Covalency in Cation Polyhedra According to Superfine Structure of Electron Paramagnetic Resonance (EPR) Spectra 282
8.2.7 Energies of the Structural Sites, Energies of Stabilization, and Intracrystalline Fields in Silicates 283
8.2.8 Mössbauer Characteristics of the Bonding of Iron and of the Site Population in Silicate Minerals 285
8.2.9 Crystal-Chemical Meaning of the Nuclear Magnetic Resonance (NMR) Parameters in Silicates 285
8.2.10 Bond Length and Angle Variations; Bridging and Non-bridging Oxygens 286
8.2.11 Interlayer Bonding and Surface Energy Calculations in Sheet Silicates 288
8.2.12 Mantle Properties, High Pressure Spectroscopy, and Electronic Structure of Silicates 288
8.3 The Chemical Bond in Sulfides and Related Compounds . . 289
8.3.1 Diversity of the Aspects of the Chemical Bond in Sulfide and Related Compounds and the Theoretical Schemes 290
8.3.2 Energy Gaps in Sulfides, Types of Crystal and Types of Optical Transition; Ionicity and Band Scheme 292
8.3.3 Interactions M–M and M–S–M in Transition Metal Sulfides and Their Relation to Properties and Structures 294
8.3.4 States of Iron in Sulfides According to Mössbauer Spectra Parameters 300
8.3.5 Polarity and Donor-Acceptor Bonds in Sulfides and Sulfosalts of As, Sb, Bi According to NQR Data 301
8.3.6 Structural Features of Sulfides and Related Compounds from the Standpoint of the Electronic Structure 303
8.3.7 Survey of Data on the Chemical Bond in Sulfides 307
8.4 Features of Chemical Bonding in Other Classes of Mineral 311

References 316
Subject Index 335

1. Quantum Theory and the Structure of Atoms

1.1 Geochemistry – History of Self-Consistent Atoms

The essence of the definition of geochemistry as a science by Vernadsky [12], Fersman [3] and Goldschmidt [6] dealt not only with the accumulation of analytical data for geological matter and their interpretation, but also stressed acceptance of the atom as an object of investigation. Geochemistry is, by their definitions, a natural history of the atoms of the earth [8–11]. The distribution of atoms in the earth's crust, their associations (paragenesis of elements), and their migration in geochemical processes, dispersion or concentration in ore deposits, were connected with the structure of atoms and especially with their outer electron shells. In such discussions, one considered, for example, electronic configurations of atoms, ionic radii and volumes and ionization potentials. The position of the atoms in the Mendeleev periodic table was used to systematize the distribution of elements between different types of geological processes. Thus geochemical tables of elements were proposed, with special reference to the chemistry of rock and the mineralogy of ore deposits [14]. The electronic structure of atoms begins usually by considering the geochemistry of individual elements (see [4, 5, 7–11]). Basic ideas about mineral matter in geochemistry and mineralogy depend on the general concepts of the structure of matter.

However, it was historically so that mineralogy and economic geology developed many centuries before natural scientists attempted to understand the basic laws of the structure of matter; geochemistry also arose in the prequantum period (*The Data of Geochemistry* of Clark goes back to 1908, while the beginnings of the Bohr "old quantum theory" emerged in 1912). Geochemistry was already undergoing intensive development before 1926, the time of the emergence of quantum mechanics; its framework as a branch of science was formed in the decade 1920–1930 by the surveys of Goldschmidt, Vernadsky and Fersman, at the time when quantum-mechanical concepts still played no great role in the understanding of matter, before properties of multielectron atoms were calculated, and before quantum theories of atomic behavior in solids had been developed.

It is thus to be expected that in these foundation years of geochemistry, information on the structure of atoms was confined to the primary Bohr planetary model; most important, specific or general problems of the chemistry of the earth's crust were considered in terms of fixed properties of atoms, without taking into account changes of these properties when atoms enter into various solids. The idea of the additivity of properties of free atoms was widely applied to solids.

Elaborated in the decade 1950–1960, quantum theories describing the behavior of atoms in solids demonstrated the inadequacy of this proposition. Data on atomic properties represent only the beginning of a long path towards the understanding of solids, which must include chemical bond theories (electronic structure of solids), molecular orbitals, crystal fields, energy bands and experimental methods (mostly those of solid-state spectroscopy) and investigations of the electronic structure of real compounds and crystals.

If one tries to identify the main feature in contemporary solid-state theories, with all their sophisticated calculations and experimental aspects, one would probably select the self-consistency of the atom's electronic properties in a compound, with one another and with the overall structure of the compound. It follows from this that geochemistry is now defined as a history of self-consistent atoms in minerals (solutions and melts) in the earth (see. p. 137).

Thus, an understanding of atoms in geochemistry is connected with an understanding of atoms in minerals, i.e., with quantum theories describing changes of atomic properties in solids. For this, however, it is necessary to present in sufficient detail the properties of free atoms as a starting point.

Quantum mechanics is the mechanics of electron motion. As in classical mechanics, it considers velocity, mass, position and other characteristics of moving particles; however, unlike movement of macroscopic bodies, either terrestrial or heavenly, the movement of electron and other elementary particles is governed by quantum laws and is considered not per se, but as cause and mechanism of phenomena determining all physical and chemical properties of matter.

Classical mechanics deals with varying objects, while quantum mechanics always deals with the same electron having constant mass and charge and distinguishable only by its position with respect to the nucleus and other electrons.

Processes in microcosm are beyond the realm of direct perception. Quantum-mechanical concepts are alien to everyday experience. The mechanics of electron motion differ fundamentally from the motion of tangible bodies observed by experience and described by classical mechanics. This causes difficulties in understanding the physical meaning of the basic concepts of quantum mechanics. Another difficulty lies in the fact that the description of electron motion is reduced to the solution of problems of applied mathematics. Thus understanding the principles of quantum mechanics means first of all understanding its mathematical apparatus. This, however, is not all. In the course of the development of quantum mechanics, a mathematical apparatus was first formulated, and only subsequently was this interpreted and translated into physical concepts. Mathematical formalism is also relatively alien to geological thinking, although one encounters this approach in descriptions of crystal optics, with observations conveyed by means of the optical indicatrix, or in the presentation of electron paramagnetic resonance spectra by means of spin Hamiltonian parameters.

It should definitely be stated that knowledge of the mathematical apparatus of quantum mechanics is unavoidable. The Schrödinger equation has the same fundamental significance for quantum mechanics as the mathematical expressions of the first and second basic laws for thermodynamics. The methods of approximative solution of the Schrödinger equation lead to the principal methods of quantum chemistry, molecular orbital and valence bonds. Group theory represents the basis of crystal field theory. The Hamiltonian operator describes electron absorption spectra.

Not only can the basic concepts of quantum mechanics not be understood without their mathematical expressions, but contemporary results of solid-state physics, including many data on minerals and their synthetic analogs, would be inaccessible without the application of mathematics; results of solid state spectroscopy could not be interpreted without it.

We shall try, however, to reduce the use of mathematics to a minimum and to restrict it by showing how basic physical concepts are expressed mathematically, and by tracing general features of calculations leading to certain final expressions necessary for treatment and understanding of the experimental data.

It is helpful to be aware of the causes of difficulties in understanding the mathematical aspects of quantum mechanics. Most of which are usually associated with the simplest mathematical concepts and with notions that are either forgotten or not included in university textbooks. Furthermore, for each of these concepts it is necessary to reflect on its physical meaning. One must also "grow accustomed" to these concepts and be able to consider them in different contexts for different examples. The fundamentals of mathematics are thus similar to those of languages: to know them one needs to use them.

1.2 The Beginnings of Quantum Theory

1.2.1 Rutherford-Bohr Model of the Hydrogen Atom and Three Postulates of the "Old Quantum Theory"

In Rutherford's planetary depiction of atomic structure, the electron moves around the nucleus, obeying the classical laws of Newtonian mechanics (Fig. 1). The angular momentum of the electron moving in a circular orbit is expressed by mvr, where m and v are respectively the mass and velocity of the electron, and r is the radius of the orbit.

Since the atom is characterized by electrostatic forces, as distinct from a planetary system, which is governed by gravitational attractions, the interaction between an electron and nucleus is expressed by Coulomb's law as a product of the charges divided by the square of the distance (i.e., by the square of the radius of the orbit). Since the charge of the nucleus is equal to $+Ze$ and the charge of the electron $-e$, the force of the interaction between them will be $-Ze^2/r^2$. This centripetal force must be balanced by the centrifugal force mv^2/r:

$$\frac{Ze^2}{r^2} = \frac{mv^2}{r}. \tag{1}$$

This model however, is unstable from the standpoint of classical electrodynamics, and it fails to explain the line spectrum of atoms. Bohr (1913) eliminated these contradictions by assuming that the existence in atoms of discrete stationary states, and the ability to emit and absorb energy only in certain discrete portions, which cannot be explained by classical mechanics, should be taken as

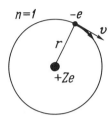

Fig. 1. Bohr model of the hydrogen atom. A single electron with mass m and charge $-e$ moves around the nucleus with charge $+Ze(Z=1)$ in orbit with the principal quantum number $n=1$ with velocity v; r is the first orbit radius

postulates consistent with the quantum energy concept introduced by Planck (1900).

According to this concept, the absorption and emission of electromagnetic radiation occur in discrete portions or quanta with the energy ε proportional to the frequency of radiation v:

$$\varepsilon = hv$$

where Planck's constant h has the value of $\sim 6.625 \cdot 10^{-27}$ erg·s.

Planck's constant has the dimensions of action (energy·time) and is often called the quantum of action.

Since frequency v is measured in s^{-1} (or cycles per s or Hz), the product h (erg·s)·v (s^{-1}) is expressed in units of energy (erg).

If the frequency v corresponds to the optical region of the spectrum, hv is called the quantum of light, or photon.

For the radio-frequency region (for instance, in nuclear magnetic resonance spectroscopy) this value is the radio-frequency quantum; for the superhigh-frequency region (electron paramagnetic resonance) it is the microwave quantum; for other parts of spectrum it is respectively the X-ray, UV or IR quanta.

The value $\varepsilon = hv$ is a quantum of energy, i.e., the energy absorbed or emitted by an atom when transitions occur from one state to another.

In many calculations instead of h the value $\hbar$ is used: $\hbar = h/2\pi = 1.05 \cdot 10^{-27}$ erg·s (since $2\pi = 6.2832$ and $h \approx 6.625 \cdot 10^{-27}$ the value $\hbar$ is easy to remember as approximately $1 \cdot 10^{-27}$ erg·s). In this case the energy quantum is $\varepsilon = \hbar\omega$, where ω is an angular frequency equal to $2\pi v$.

Planck's constant is one of the fundamental constants of nature; it is the universal measure of microcosm mechanics.

The principles of the "old quantum theory" (as distinct from modern quantum mechanics which owes its origin to the Schrödinger and Heisenberg equations) are given by the three Bohr's postulates:

1. The Stationary States Postulate. The electron rotates round the nucleus without radiating energy in certain orbits called stationary. In the atom there is a discrete set of such "quantized" stationary orbits. One of these with minimum energy corresponds to the normal or ground state. Other stationary orbits, empty in the ground state, are populated by electrons in excited states of the atom.

2. The Postulate of Transition Frequencies (Bohr's "Frequency Condition"). Emission of radiation occurs not when an electron moves in the orbit, but when it

Calculation of the Radius and Energy of Hydrogen Atom Orbits

jumps from one stationary state to another. Radiation energy proportional to the frequency of transition is equal to the energy difference of these states: $h\nu = E_2 - E_1$.

3. The Angular Momentum Quantization Postulate. The stationary orbits are those for which the angular momentum of an electron mvr is quantized and is an integral multiple of $h/2\pi$, that is:

$$mvr = n \cdot h/2\pi = n\hbar, \tag{2}$$

where n is an integer corresponding to the orbit's "number" and is termed the principal quantum number.

By means of these postulates Bohr was able to calculate the mechanics of hydrogen atom electron motion and to interpret the hydrogen atom line spectrum.

1.2.2 Calculation of the Radius and Energy of Hydrogen Atom Orbits

Calculations of most simple atoms (and primarily of the hydrogen atom) are thought to have a kind of "didactic" significance, i.e., they help to retrace the course of all calculations of atomic properties using a few fundamental constants that describe the electron properties. Afterwards one can conceive the course of calculations for much more complex systems such as atoms in crystals too difficult to be considered here in detail.

Using only two values, the mass and charge of the electron, and Planck's constant (a measure for atomic system mechanics), and with the help of the simplest relations of classical mechanics complemented with Borh's postulates, it is possible to perform all calculations for the hydrogen atom.

The radii of the electron orbits can be readily determined by the comparison of the Eqs. (1) and (2). From Eq. (1) of equilibrium between electrostatic attraction and centrifugal force we obtain $v^2 = e^2/mr$ (for the hydrogen atom $Z=1$). From Eq. (2) of quantization of stationary orbits $mvr = n\hbar$, or $m^2v^2r^2 = n^2\hbar^2$ and on substituting the value v^2 from Eq. (1) we obtain:

$$r = \frac{n^2\hbar^2}{me^2} \quad \text{or} \quad r = \frac{n^2 h^2}{4\pi^2 me^2}. \tag{3}$$

The radius of the first ($n=1$) circular orbit of the hydrogen atom is:

$$r_a = \frac{(1)^2 \cdot (6.62491 \cdot 10^{-27} \text{ erg} \cdot \text{s})^2}{4(3.1416)^2 \cdot (9.1091 \cdot 10^{-28} \text{ g})(4.8030 \cdot 10^{-10} \text{ ESU})^2}$$
$$= 0.529 \cdot 10^{-8} \text{ cm} = 0.529 \text{ Å}.$$

The radius of the second orbit ($n=2$) is equal to $4r_a$, the radius of the third orbit ($n=3$) $9r_a$ and so on ($r_n = n^2 r_a$).

The energy of an electron in a particular orbit is determined from the equations of classical mechanics in which we substitute the expression for the radius obtained above.

The total energy E is equal to the sum of the kinetic energy, T, and the potential energy V. $T = mv^2/2$. From Eq. (1) $v^2 = e^2/mr$; then $T = e^2/2r$. $V = -Ze^2/r = -e^2/r$ i.e., the kinetic energy is equal to one half the potential energy. The total energy of the electron (for the hydrogen atom) is:

$$E = T + V = \frac{e^2}{2r} - \frac{e^2}{r} = -\frac{e^2}{2r}. \tag{4}$$

Substituting the value of the Bohr orbit radius from Eq. (3)

$$E = -\frac{e^2 m e^2}{2n^2 h^2} = -\frac{me^4}{2n^2 h^2} = -\frac{2\pi^2 me^4}{n^2 h^2}. \tag{5}$$

Since the radius of the first Bohr orbit equalling $r_a = (1)^2 h^2/(4\pi^2 me^2) = 0.529$ Å, is taken for the atomic unit of length, the electron energy for the hydrogen atom is frequently written as:

$$E = -\frac{Z^2 e^2}{2r_a n^2}. \tag{6}$$

1.2.3 Atomic Structure and Spectra. The Calculation of Line Spectrum of the Hydrogen Atom

If the energy of an electron in orbit n_1 is equal to E_1, and in orbit n_2 to E_2, the emission energy $E = hv$ (when transition occurs from one orbit to another) is calculated according to Bohr's second postulate: $E_{1-2} = hv = E_2 - E_1$. Substituting in this expression the value of energy from Eq. (5) we obtain:

$$E_{1-2} = hv = \frac{2\pi^2 me^4}{n_2^2 h^2} - \left(-\frac{2\pi^2 me^4}{n_1^2 h^2}\right) = \frac{2\pi^2 me^4}{h^2}\left(\frac{1}{n_1^2} - \frac{1}{n_2^2}\right). \tag{7}$$

From here the transition frequency (in s^{-1}) $v = E/h$ is:

$$v = \frac{2\pi^2 me^4}{h^3}\left(\frac{1}{n_1^2} - \frac{1}{n_2^2}\right). \tag{8}$$

Or, if we express frequency in wavenumbers cm^{-1},

$$v(cm^{-1}) = \frac{v(s^{-1})}{c(cm \cdot s^{-1})}:$$

$$v = \frac{2\pi^2 me^4}{h^3 c}\left(\frac{1}{n_1^2} - \frac{1}{n_2^2}\right) \tag{9}$$

or

$$v = R\left(\frac{1}{n_1^2} - \frac{1}{n_2^2}\right) \tag{10}$$

where

$$R = \frac{2\pi^2 m e^4}{h^3 c} = \frac{2\cdot(3.1416)^2 \cdot 9.1091\cdot 10^{-28} \cdot (4.8030\cdot 10^{-10})^4}{(6.62491)^3 \cdot 2.9979\cdot 10^{10}}$$
$$= 109{,}737 \text{ cm}^{-1}.$$

The calculated value R (the Rydberg constant) is in good agreement with the experimental value $R = 109{,}677$ cm^{-1} obtained very accurately from the measurements of the atomic hydrogen spectrum and Eq. (10) coincides with that derived empirically from the observations of the hydrogen atom spectrum before the introduction of Bohr's theory.

The electron in the ground state moves in one stable orbit with $n = 1$. When an atom is excited (for example, by electric arc during spectral analysis) electrons are "showered" on different excited orbits with $n > 1$. From these the electron returns to its normal ground state (through several intermediate states) by emitting radiation. A transition between every two states gives a spectral line; a system of spectral lines composes the emission spectrum of the atom.

The hydrogen atom spectrum (Fig. 2a) ranges from vacuum-UV to near-IR regions. The lines in the spectrum are grouped into five series which are named after their discoverers (Fig. 2b). Lines of each series occur as a result of transitions to one and the same orbit from different higher-lying orbits. Line sequences within a series are governed by Eq. (10).

It should be noted that there is a direct relation between the spectrum lines and the atomic structure: each line with a definite frequency and, consequently, energy $E = hv$, corresponds to a transition between two orbits. Thus, the spectra are a clue to understanding the atomic structure.

Energy Level Diagram for the Hydrogen Atom. Electron energies in different orbits (in different states) are usually represented as energy level diagrams (Fig. 2c). Transitions from higher to lower-lying levels correspond to emission of radiation, while transitions in the opposite direction correspond to absorption of radiation. The requencies of the transitions are determined by the energy difference of these levels.

Note that since the radii of the orbits are directly proportional and the electron energies of these orbits are inversely proportional to n^2 [see Eqs. (3) and (5)], the difference between the radii of subsequent orbits increases (Fig. 2b) and the distances between subsequent energy levels rapidly decrease (Fig. 2c). This corresponds to a rapid decrease in the line separation within each series towards shorter wavelengths (greater energies). A limit of each series represented by the most short-wave line corresponds to $n_2 \to \infty$ and to the frequency $v = R/n_1^2$.

Note the two scales in the energy level diagram Figure 2c.

On the right-hand scale the lowest level corresponds to the normal inexcited ground state of an atom. This ground level is taken for the zero state as is usual in spectroscopy. Then the right-hand scale in Figure 2c shows the positive energy values required for the transitions from the ground level (from the first orbit) to one of the excited states (to one of the higher lying orbits), i.e., it is the energy difference $E_n - E_1$ or the excitation potentials for the transitions to one of the

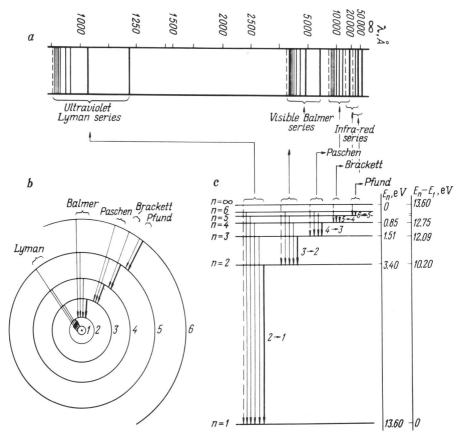

Fig. 2a–c. Hydrogen atom structure and spectrum. **a** Emission spectrum of the hydrogen atom. **b** Transitions between circular orbits giving rise to the spectral lines grouped in the series: Lyman series (transitions to the orbit with $n=1$ from all higher-lying orbits), Balmer series (to the orbit with $n=2$); Paschen series ($n=3$); Brackett ($n=4$); Pfund ($n=5$). **c** Energy level diagram of the hydrogen atom corresponding to the transitions in **b** and explaining the spectrogram in **a**. The levels are shown corresponding to circular orbits only (i.e., to principal quantum number n only) (see also Figs. 3 and 4)

excited orbits. The uppermost level meaning the complete removal of an electron (i.e., the transition to the "orbit" with $n_2 = \infty$) corresponds to the ionization potential (i.e., the ionization energy divided by the electron charge) and is calculated from Eq. (7) with $n_1 = 1$ and $n_2 = \infty$:

$$E_{1-2} = \frac{2\pi^2 me^4}{h^2}\left(\frac{1}{n_1^2} - \frac{1}{n_2^2}\right) = \frac{2\pi^2 me^4}{h^2} = 2.180 \cdot 10^{-11}\,\text{erg} = 13.60\,\text{eV}$$

$(1\,\text{erg} = 6.2419 \cdot 10^{-4}\,\text{eV})$.

On the left-hand scale (Fig. 2c) the ionization energy of the atom is taken as the zero level. The negative energy values correspond here to the energies required for removal of an electron from this level. For the first (ground) level this value is represented by the negative ionization energy.

The energy scale is presented not in erg, but in eV; in spectroscopy, however, instead of this energy scale in eV, the proportional frequency scales v (s^{-1}) or wave-number scale v (cm^{-1}) are frequently used, since $E = h$ (erg·s)·v (s^{-1}) and $E = h \cdot v \cdot c = h$ (erg·s)·v(cm^{-1})·c(cm·s^{-1}). From here we derive a spectral term $T = E/hc$(cm^{-1}). Diagrams thus constructed are spectral term diagrams but are often referred to as energy level diagrams.

The difference of terms gives immediately the frequency of the corresponding transition. The unit of s^{-1} (Hz) is more convenient in radio-frequency regions, and the cm^{-1} unit in optical regions.

1.2.4 Fine Structure of Spectra and Sommerfeld's Development of the Bohr Theory; Quantum Numbers

Higher-resolution spectrograms show that the hydrogen lines observed in low-resolution spectrograms as single lines in fact represent the closely spaced pairs of lines or doublets forming the fine structure of the spectrum.

Observation of the fine structure has revealed a more complicated electron structure of the atom and indicated the existence in it of complementary quantized states (Fig. 3).

The principal quantum number n takes into account only circular (Bohr's) orbits. The transition 3→2 gives only one line (Fig. 2 and left part of Fig. 3). Note that the distance between the levels $n=3$ and $n=2$ is extremely large (15,420 cm^{-1}) as compared to the distances between the split levels (of the order of $0.03 \div 0.3$ cm^{-1}). This shows that the principal quantum number n determines almost completely the total energy of the electron state.

The orbital quantum number l was introduced as a result of Sommerfeld's refinements of Bohr's picture of the moving electron in atoms: it includes not only circular but also elliptical orbits. The motion of the electron in elliptical orbits already required two quantum numbers for the allowed states, hence the principal quantum number n was represented as a sum of the radial n_r and azimuthal n_φ quantum numbers $n = n_r + n_\varphi$ (in accordance with the radius vector r and azimuth φ in the elliptical model). Since these quantum numbers are interdependent, it is worth mentioning only two of them, the principal n and azimuthal $k = n_\varphi$.

In quantum mechanics the azimuthal quantum number k of the Bohr-Sommerfeld theory was replaced by the azimuthal or orbital angular quantum number $l = k - 1$. The orbital angular momentum l^* is related to the orbital quantum number l as $l^* = \sqrt{l(l+1)} \cdot \hbar$ where l assumes integral values from 0 to $(n-1)$.

The states with different l are designated by spectroscopic symbols, $s, p, d, f...$ based originally on the spectral series nomenclature: sharp, principal, diffuse, fundamental series in atomic spectra of alkaline metals:

$l = 0, \quad 1, \quad 2, \quad 3, \quad 4, \quad 5$
$ s \quad p \quad d \quad f \quad g \quad h.$

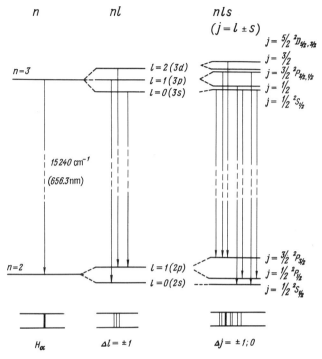

Fig. 3. Fine structure of the H_α line (656.3 nm; $n=3 \rightarrow n=2$), first most intense line in the Balmer series of the hydrogen spectrum. *Left:* Circular orbits only are taken into account; only one line occurs arising from the electron transition from the orbit with principal quantum number $n=3$ to the orbit with $n=2$ (cf. Fig. 2). *Middle:* Elliptic orbits with the orbital quantum numbers $l=0,1,2$ and the selection rule $\Delta l = \pm 1$ are taken into account; instead of one H_α line one observes (with higher instrumental resolution) three closely spaced lines. *Right:* Electron spin and spin-orbit coupling (the inner quantum number $j=l+s$) are taken into account; further splitting of the H_α line in seven lines can be observed with higher resolution (see also Fig. 4)

The electron states are then described thus:

if $n=1$ $l=0$ (1s state),

$n=2$ $l=0.1$ (2s, 2p),

$n=3$ $l=0, 1, 2$ (3s, 3p, 3d),

$n=4$ $l=0, 1, 2, 3$ (4s, 4p, 4d, 4f states) etc.

In the energy level diagram (Fig. 4) the levels with the same n are split in accordance with the l quantum number. Thus instead of one energy level E_3 (Fig. 3) three levels with somewhat different energy values appear: $E_{3,0}(E_{3s})$, $E_{3,1}(E_{3p})$, $E_{3,2}(E_{3d})$; instead of E_2 there are $E_{2,0}(E_{2s})$ and $E_{2,1}(E_{2p})$.

However, the number of lines observed in the spectra is less than the number of possible transitions between these levels. Therefore, empirically selection rules were introduced to restrict allowed transitions by the condition $\Delta l = \pm 1$ (i.e.,

Fig. 4a and b. Energy level diagram of the hydrogen atom taking into account orbital interaction (see Figs. 2 and 3). **a** In the form of usual energy level diagrams. **b** According to Grotrian-type diagrams common in atomic spectroscopy. In accordance with the selection rule $\Delta l = \pm 1$, the series in the hydrogen spectrum are interpreted as follows: Lyman series $(n \to 1)$, $np \to 1s$ transitions; Balmer series $(n \to 2)$, $ns \to 2p$, $np \to 2s$, and $nd \to 2p$ transitions; Paschen series $(n \to 3)$, $nf \to 3d$ transitions

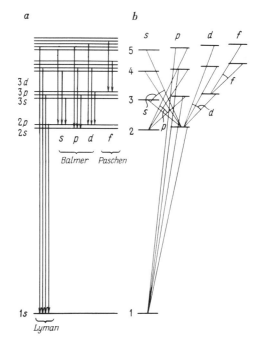

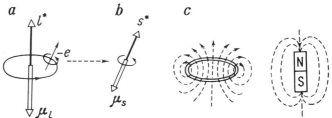

Fig. 5a–c. Mutual orientation of **a** angular orbital moment (l^*) and magnetic orbital momentum (μ_l); **b** spin (s^*) and spin magnetic momentum (μ_s) and **c** their analogy with magnetic field (magnetic dipole) arising from circular current loop

transitions are allowed only when l changes by ± 1). Other transitions are forbidden by the selection rule (Fig. 3).

The motion of the electron in an orbit may be treated as a circular current loop creating a magnetic field (Fig. 5). Here, the orbital angular momentum l^* of the electron may be related to the orbital magnetic moment μ_l (with the orbital quantum number m_l) which is oriented in a direction opposite to that of the angular momentum (due to the negative charge of the electron).

To account for the doublet splitting in alkaline metal spectra Goudsmith and Uhlenbeck (1925) suggested that the electron possesses the intrinsic or the spin angular momentum associated with the spinning of the electron about its own axis (see also Chap. 1.6). The spin quantum numbers (or simply spin) take for one electron the single value $s = 1/2$.

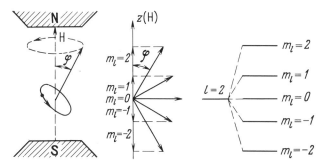

Fig. 6. Precession of orbital momentum l^* around the magnetic field H direction

The spin angular momentum s^* is equal: $s^* = \sqrt{s(s+1)} \cdot \hbar = \sqrt{1/2(1/2+1)} \cdot \hbar = \sqrt{3/2} \cdot \hbar = 0.866\hbar$.

The spin is accompanied by the spin magnetic momentum, i.e., the electron could be considered as a magnetic dipole (Fig. 5).

The interaction between the spin and orbit momenta (the spin-orbit or magnetic dipole–dipole coupling), like all other interactions in atomic systems, is governed by quantum laws. The total angular momentum of the electron j^* obtained by the vector addition is $j^* = \sqrt{j(j+1)} \cdot \hbar$, where j is the inner quantum number equal to $l \pm s$. For all values except $l=0$, the energy levels split into two sublevels. Thus for $l=2$ there are two sublevels with $j=3/2$ and $j=5/2$, for $l=1$ the sublevels with $j=1/2$ and $j=3/2$. Only for the s-states with $l=0$ is there one value $j=1/2$ and so the s-levels do not split (see Figs. 3 and 4). The selection rule for j is: $\Delta j = \pm 1, 0$.

The magnetic quantum numbers (m_l and m_s) are not connected with the size or shape or energy of the orbits which are completely determined by the principal, orbital and spin quantum numbers, but m_l and m_s indicate the orientation of axes of the orbital and spin rotation with respect to some special directions, thus determining the spatial quantization of the electron orbits. Since this is usually the external magnetic field direction, these quantum numbers are called the magnetic orbital and magnetic spin quantum numbers (m_l and m_s).

Once the atom is in the magnetic field (Fig. 6), the direction of the orbital rotation axis, i.e., the vector of the orbital angular momentum (and antiparallel vector of the orbital magnetic moment), because of its magnetic dipole properties, tends to assume parallel or antiparallel orientation in relation to the applied field (depending on the direction of electron motion in the orbit). This is prevented, however, by the motion of the electron in the orbit. In consequence, there is a double motion of the electron in the orbit and of its orbital spinning axis about the direction of the magnetic field, i.e., the precession of orbital momentum of the electron known as the Larmor precession. This occurs in such a manner that the angle between the magnetic field direction (z axis) and the axis of orbital motion is quantized, i.e., it assumes only discrete values determined by the magnetic quantum number m_l. The number m_l governs allowed projections of the electron orbital angular momentum l^* on the z axis (direction of the magnetic field). The

Fig. 7. Spin momentum quantization in magnetic field

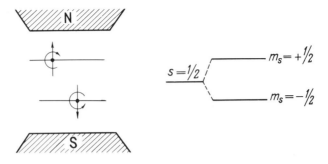

quantum condition for the angle φ is given by $\cos\varphi = m_l/l$. The m_l quantum number assumes values $+l, l-1, \ldots -l$, in all $(2l+1)$ values, i.e., with

$l = 0 \quad m_l = 0$,
$l = 1 \quad m_l = 1, 0, -1$,
$l = 2 \quad m_l = 2, 1, 0, -1, -2$, etc.

The spin angular momentum of the electron assumes only two values in relation to the direction of magnetic field, parallel and antiparallel (Fig. 7). Correspondingly, the spin magnetic quantum number m_s determining the allowed spin projection on the z axis assumes only two values: $m_s = +1/2$ and $-1/2$.

The space quantization of the electron total angular momentum j^* being a vector sum of $l^* + s^* = j^*$ is governed by the magnetic inner quantum number m_j, which assumes the values of $+j, j-1, \ldots -j$, in all $(2j+1)$ values. For example, if $j = 3/2$ then $m_j = 3/2, 1/2, -1/2, -3/2$.

The electron states characterized by different values of magnetic quantum numbers, i.e., different orientations of orbital, spin and total angular momenta in relation to the magnetic field direction, have slightly different energy values. Consequently, a further splitting of the energy levels occurs. The transition from the state with one orientation to another causes emission or absorption of energy. This energy is extremely small and the transitions correspond to the cm wavelength region and are observed, for example, in the electron spin resonance spectra.

Transitions from these magnetic sublevels of one orbital level to the magnetic sublevels of the other orbital level cause splitting of atomic spectral lines: the single lines are split into groups of closely spaced lines in a strong magnetic field (Zeeman effect).

In the absence of the external magnetic field, these electron states (with different m_l, m_s, m_j) correspond to one and the same energy level and are said to be degenerate states, which are not distinguishable in the absence of a magnetic field and acquire different energy values in the imposed magnetic field. It is said in the latter case that imposing an external magnetic field removes the degeneracy.

The Relation of the Bohr Theory to Quantum Mechanics. Bohr's theory and its development opened a great epoch in atomic physics. The theory was most successful in interpreting atomic spectra (optical and X-ray) and the effects exerted on these spectra by magnetic and electric fields, in explanation of the Mendeleev's periodic system of elements, in some approaches to the chemical bond, to magnetic and other properties of atoms.

At the same time it should be emphasized that the planetary model of the atom has been abandoned by contemporary physics. The Bohr model of the atom already lost its simplicity when the orbital quantum number l, which can take a zero value and correspond thus to the straight line orbit, was first introduced. How far this picture differs from the contemporary one can be seen from comparison of the circular and elliptical orbits of the old quantum theory with the quantum mechanical pictures of s, p, d orbitals (see Fig. 10).

Even the conception of electron behavior in the atom has been changed: the planetary electrons of an atom were thought of as moving in circular and elliptical orbits, while in wave mechanics the electrons are regarded as having dual wave-particle properties which are expressed by wave functions.

1.3 The Schrödinger Equation as the Basic Equation of Quantum Mechanics

1.3.1 The Physical Bases

It was not the Bohr quantum theory development that gave a rise to quantum mechanics, but the fundamental discovery by de Broglie of the wave-like nature of matter, and the establishment of the uncertainty principle by Heisenberg promoted the development of new physical concepts on the electron and its behavior. The mathematical representation of these concepts in the form of the Schrödinger equation, providing the description and means of calculation of electron behavior, led to quantum mechanics.

Wave-like Properties of the Electron. The quantum approach revealed the dual nature of light, possessing properties of both waves and particles (the latter being photons of varying energy corresponding to the radiation in different parts of the spectrum). From the Einstein relation $E = mc^2$, establishing the coordination of energy and mass, and the Plank relation $E = hv$, it follows that

$$mc^2 = hv \quad \text{or} \quad mc = \frac{h}{\lambda}, \quad \text{since} \quad \frac{v(\text{s}^{-1})}{c(\text{cm}\cdot\text{s}^{-1})} = \frac{1}{\lambda}(\text{cm}).$$

De Broglie (1924) suggested that the same relationship exists for electrons if the mass of photon m is replaced by the mass of electron, and the velocity of light c by the velocity of electron v:

$$mv = \frac{h}{\lambda}.$$

In this expression the left term represents the electron behavior as a particle with the mass m, the velocity v and impulse $p=mv$, the right gives the wave-like behavior of electrons with the wavelength λ ("de Broglie wavelength").

$$\lambda = \frac{h}{mv} = \frac{h}{p}.$$

This suggestion was soon confirmed experimentally by the observations of Davisson and Germers (1927) on the diffraction of electrons by crystals, and was the first application of the electron diffraction method (like X-ray diffraction) to studies of the structure of matter. Thus, the dual nature of electrons was established, exhibiting the properties of both particles and waves.

An electron has a mass of approximately $9 \cdot 10^{-28}$ g and velocity of approximately $6 \cdot 10^7$ to $6 \cdot 10^8$ cm s^{-1}, thus the wavelength is of an order of 10^{-8} cm ($\lambda = h/mv = 6.6 \cdot 10^{-27}/9 \cdot 10^{-28} \cdot 6 \cdot 10^8 \approx 1.2 \cdot 10^{-8}$ cm $= 1.2$ Å), i.e., the order of the atom sizes.

The wave behavior was established for other microparticles such as neutrons, protons, atoms, and molecules, and their diffraction also applies to structure studies (neutronography, atomic and molecular beams). For the macroscopic particles, for example with mass 1 g and velocity 1 cm s^{-1}, the wavelength (of an order of $6.6 \cdot 10^{-27}$ cm) appears too small to be detected.

Probability and Statistic Nature of the Electron Behavior and the Heisenberg Uncertainty Principle. The establishing of the dual wave particle existence of the electron leads inevitably to another important property of its behavior consisting of a probability character of the description of electron motion. This property is revealed when the wave theory is applied to a description of behavior of the microparticle (electron).

Let us first compare the wave and quantum description of light waves (photons).

In the distribution of light intensity over a diffraction pattern when a beam of light passes through a narrow slit, one can observe on the screen alternating bright and dark fringes correlating, by ordinary theory, the distribution of light intensity to the square of wave amplitude. On the other hand, according to the quantum theory (photon theory of light) the intensity distribution in the diffraction pattern is proportional to the number of photons which hit different parts of the screen.

If we consider an individual photon, the exact position of its collision with a screen is uncertain within the whole diffraction pattern; a photon may hit any point of the screen. However, the probability of a photon striking different points differs and corresponds to the diffraction bands predicted by the wave diffraction theory.

The same features characterize the diffraction of the waves connected with electrons (de Broglie waves): (1) uncertainty of the electron position on a screen being hit by electrons; (2) the different hit probabilities describing the distribution of electron collisions with a screen; (3) the statistical character of the phenomenon; (4) the correspondance between hit probability and electron wave intensity; (5) proportionality of the hit probability to the square of the electron wave amplitude.

The uncertainty of the space distribution, the "spreading" of the electron, and a probability character of this distribution explain the finite width of lines in spectra and in diffraction patterns, and serve as a reason for the "charge clouds" models of atoms.

In a general form this is formulated as Heisenberg's uncertainty principle:

$$\Delta x \cdot \Delta p_x \geq h,$$

where Δx is the uncertainty in position of an electron (uncertainty in determination of coordinate x after the electron has passed the diffraction slit in the above observation), Δp_x is the uncertainty in impulse $p_x = mv$ (uncertainty of determination of the velocity component v_x after the electron passes through the diffraction slit).

The Planck constant h determines a minimum error when simultaneously geometric (position in space) and dynamic (impulse, angular momentum, etc.) values are measured, i.e., the two conjugate values that describe the atomic system.

By means of the uncertainty relation it can be shown that: (1) the inaccuracy in measuring x for the fixed values of p_x leads to the error in x that excludes any certainty of this value, (2) the product of $\Delta x \cdot \Delta p_x$ (with the magnitude of these values being sufficient for a simultaneous determination of the position and impulse of the electron in an atom) will be less than the h value, and thus is an impossible combination of these values.

For the electron with the mass $9 \cdot 10^{-28}$ g ($\sim 10^{-27}$ g) and velocity of about $2 \cdot 10^8$ cm s^{-1} (i.e., electron velocity in the first Bohr orbit which can be easily calculated from the third Bohr postulate) the error in the velocity measurement 0.1% is $\Delta v = 2 \cdot 10^5$ cm s^{-1}. Then the error in the coordinate measurement is:

$$\Delta x = \frac{h}{\Delta p_x} = \frac{h}{m \cdot \Delta v_x} \approx \frac{6.6 \cdot 10^{-27}}{10^{-27} \cdot 2 \cdot 10^5} = 3.3 \cdot 10^{-5} \text{ cm}.$$

This error in determination of the electron position in an atom is 3300 Å, which exceeds by 1000 times the size of the atom proper.

If, on the other hand, we assume the error in the position measurement to be at least $\Delta x = 10^{-8}$ (i.e., of the order of a size of the atom) then:

$$\Delta x \cdot \Delta p_x = \Delta x \cdot (m \cdot \Delta v) \approx 10^{-8} \cdot (10^{-27} \cdot 2 \cdot 10^5) = 2 \cdot 10^{-30} \ll 6.6 \cdot 10^{-27},$$

i.e., this product appears to be less than Planck's constant, which is in contradiction to the uncertainty principle.

Let us note that in the uncertainty relation, the product $\Delta x \cdot \Delta p_x = \Delta x (m \cdot \Delta v)$ has the same dimension (cm·g·cm·s^{-1} = g·cm^2·s^{-1}) as has the quantum constant h (erg·s = dyn·cm·s = g·cm·s^{-1}·cm·s = g·cm^2·s^{-1}).

Thus, the product of uncertainty in position and uncertainty of impulse $\Delta x \cdot \Delta p_x$ cannot be less than the quantum constant h. If the electron position is exactly known ($\Delta x = \Delta y = \Delta z = 0$), then $p = \infty$, i.e., all the values of the impulse p

are equally probable. Moreover, on the contrary, when $\Delta p = 0$, all the positions of the electron are equally probable ($\Delta x = \infty$).

It also follows from the uncertainty principle that an electron state can be determined by the specification either of its position only or of its impulse.

For macroscopic objects with greater mass values, the product $\Delta x \cdot \Delta p$ greatly exceeds the value h; here, the error, whose value is of one order of h, is practically negligible.

Wave Function and Probability Distribution. Owing to the conclusion drawn on the wave nature of the electron and its behavior described in statistical and probability terms, the old concept of the orbit vanishes completely. The electron describes an arbitrary path moving predominantly toward and away from the nucleus, not in an orbit plane but in all directions, its velocity being variable. It cannot be determined at the same time by the definite position and impulse, i.e., the uncertainty of these variables is admitted. The distance at which the electron can be found does not correspond to the radii of the Bohr orbits (though these are the most probable distances), but it can be in any part of the whole volume in a diameter of about 2 Å.

Instead of the concept of an orbit incarnating the most essential features of the old quantum theory, the concept of a wave function (or state function, eigenfunction) has been introduced by quantum mechanics to describe the wave, statistical and probability behavior of the electron. From the wave function concept such pictorial approaches were derived (replacing the orbital motion model) as charge (electronic) cloud, boundary surfaces and atomic orbitals.

Let us consider now the physical meaning of the following concepts (Fig. 8, see also Chap. 1.4):

ψ = wave function,

$\psi^2 dv$ = probability distribution function,

$\psi^2 4\pi r^2$ = radial probability distribution function.

Each of these functions is of great importance and cannot be replaced by others.

Wave function ψ is a function, the square of which is a measure of the probability of finding the electron in a certain volume element around the nucleus. Note that it is only the square of the absolute magnitude of a wave function ψ^2 that has the physical meaning.

To understand the significance of the wave function and why its square determines the probability distribution of an electron in the atom, one has to consider the wave equation which includes this function. This equation gives mathematical expression to all types of wave motion, such as, for example, light and sound waves, elastic waves, string vibrations, electromagnetic waves, etc.

It is well known, for example, from crystal optics, that the equation for sine waves is:

$\psi = a \cdot \sin 2\pi (vt - x/\lambda)$,

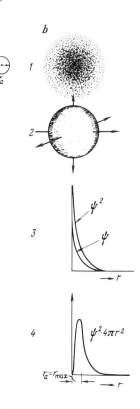

Fig. 8a and b. Comparison of Bohr and quantum-mechanical models of atom. **a** First ($n=1$) circular ($l=0$) Bohr orbit. **b** 1s atomic orbital in quantum-mechanical presentation: *1*, charge cloud; *2*, boundary surface (95% of the charge cloud); *3*, electron density distribution; *4*, radial probability distribution function (r_{max} coincides with the first Bohr orbit radius r_a)

where ψ is the wave amplitude, a and x determine the maximum amplitude and the coordinate, ν and λ are frequency and wavelength. For the standing sine wave this equation is written as

$$\psi = a \cdot \sin 2\pi \cdot x/\lambda.$$

In this time-independent equation the wave amplitude is expressed as a function of the coordinate x.

A double differentiation of the equation of the standing wave with respect to x gives:

$$\frac{\partial^2 \psi}{\partial x^2} = -\frac{4\pi^2}{\lambda} \cdot a \cdot \sin 2\pi \cdot x/\lambda = -\frac{4\pi^2}{\lambda^2} \psi,$$

where the amplitude ψ is the wave function of coordinates x, y, z, which does not depend on time and thus corresponds to the standing wave.

It is easy now to pass to the three-dimensional equation (see below) from which the Schrödinger electron wave equation is derived.

It is also well known from crystal optics that the light intensity is proportional to the square of amplitude. However, as was shown in considering the diffraction pattern, the light intensity is proportional to the number of photons, and thus determines photon distribution.

Since the electron possesses wave properties, its motion should also be described by the wave equation. Then the wave function of the electron ψ is equivalent to the amplitude and its square ψ^2 corresponds to the light intensity. Although in all observations we deal mainly with ψ^2 and not with ψ, the latter enters into the wave equation and is used in all calculations. It is a quite concrete notion and its value has been calculated and tabulated for the majority of atoms (see Chap. 1.5).

Probability distribution function $\psi^2 dv$. There are two aspects in its physical interpretation: (1) it represents probability of finding an electron in an elementary volume dv that surrounds the point with coordinates of x, y, z, i.e., $\psi^2 \cdot dx \cdot dy \cdot dz = \psi^2 dv$, (2) it is a measure of electron density in the elementary volume dv; the electron is smeared over a whole space around the nucleus in the form of a "charge cloud" or "electron cloud"; an electron density equalling, for example, 0.2 indicates that the electron moving rapidly for 0.2 instants of time occurs in this part of atom volume or, in other words, 0.2 of the charge and mass of the electron is concentrated in the given volume of an atom.

The latter interpretation based on the electron density is more pictorial, but it becomes understandable and more easy to correlate with the more rigorous first interpretation, if the results of statistical observations are taken into account. In fact, a single electron cannot be distributed over the whole space occupied by an atom (as in the case of a diffraction pattern, where a single electron or photon cannot from a diffraction fringe). However, just as the position of each electron or photon is uncertain within a range of a diffraction pattern and the probability of their hitting the different parts of this pattern is determined by the square of the amplitude, so the electron position in the atom is uncertain but is characterized by the probability function $\psi^2 dv$. When a great number of atoms are observed (as is usual), or when an electron moves rapidly in an atom, the statistical pattern will reflect the distribution of electron density within the charge cloud.

Radial probability distribution function $\psi^2 4\pi r^2$ determines the probability of finding the electron (electron density) in concentric shells of the thickness dr thick which occur at a distance r from the nucleus. The volume of these thinnest spherical shells is $4\pi r^2 dr$. The probability of finding the electron in such a shell is equal to the product of $\psi^2 dv$ (probabilities of finding the electron in the elementary volume dv at a distance r) and $4\pi r^2 dr$ (the volume of the shell).

Note that the dependence of the function $4\pi r^2 \psi^2$ on the radius differs from that of the ψ^2 function. The ψ^2 probability of finding the electron in the dv volume is maximal in the immediate neighborhood of the nucleus and decreases with increasing distance from the nucleus. The radial probability distribution changes, however, from zero at $r=0$, passes the maximum and gradually decreases (Fig. 8). The curve of this dependence is determined by the fact that the first term $4\pi r^2$ in the product $4\pi r^2 \cdot \psi^2$ is rapidly increasing, starting from a zero value (with $r=0$), whereas the second term ψ^2 is decreasing rapidly as the radius is increased. At first the $4\pi r^2$ value increases more rapidly than the ψ^2 decreases, thus leading to an increase of their product; then the ψ^2 starts decreasing more rapidly than the $4\pi r^2$ increases, and their product passes the maximum and shows a further gradual decrease.

The maximum of the radial probability distribution corresponds to the atomic radius. It is evident that this maximum is not coincident with the boundary of the charge cloud. The radial density maximum in the hydrogen atom corresponds to the radius of the first Bohr orbit with $n=1$ equalling 0.528 Å; the difference of their physical meaning however, is apparent.

1.3.2 Derivation of the Schrödinger Equation

Derivation from the Wave Equation. After the discussion of the concept of the wave function and its analogy with the amplitude function in the general wave equation, it seems very simple to derive the electron wave equation (the Schrödinger equation) in the following way:

1. Let us make use of the three-dimensional time-independent equation (the equation of the standing wave) which is common for all wave motions:

$$\frac{\partial^2 \psi}{\partial x^2} + \frac{\partial^2 \psi}{\partial y^2} + \frac{\partial^2 \psi}{\partial z^2} = -\frac{4\pi^2}{\lambda^2} \cdot \psi, \qquad (11)$$

where ψ is the amplitude function which here represents the electron wave function; x, y, z are space coordinates, and λ is wavelength.

2. For the electron the λ value is the de Broglie wavelength of the particle equalling $\lambda = h/p = h/mv$. Substituting this de Broglie wavelength value of the electron in the general wave equation, we obtain the electron wave equation.

3. There is another characteristic feature of the quantum-mechanic description of electron behavior. Since the velocity in the de Broglie expression $\lambda = h/mv$ cannot be observed experimentally, we may express it through the energy value, because the spectral lines measured in the experiment are connected with energy levels.

From the expression for the kinetic energy $T = mv^2/2$ and from the energy conservation law $T = E - V$ we obtain:

$$v^2 = \frac{2T}{m} = \frac{2(E-V)}{m},$$

then:

$$\lambda^2 = \frac{h^2}{m^2 v^2} = \frac{h^2}{m^2} \cdot \frac{m}{2(E-V)} = \frac{h^2}{2m(E-V)}.$$

4. Substituting this value λ in the general wave equation we obtain the Schrödinger equation:

$$\frac{\partial^2 \psi}{\partial x^2} + \frac{\partial^2 \psi}{\partial y^2} + \frac{\partial^2 \psi}{\partial z^2} = -\frac{8\pi^2 m}{h^2}(E-V)\psi. \qquad (12)$$

Derivation of the Schrödinger Equation

The left-hand term of the equation can be written as the Laplacian (Laplace operator), denoted by symbol $\nabla^2\psi$; sometimes instead of $\nabla^2\psi$ the symbol $\Delta\psi$ is used:

$$\nabla^2\psi = -\frac{8\pi^2 m}{h^2}(E-V)\psi.$$

Instead of $h^2/4\pi^2$ we may also use $\hbar^2$ (since $\hbar = h/2\pi$):

$$\nabla^2\psi = -\frac{2m}{\hbar^2}(E-V)\psi \quad \text{or} \quad -\frac{\hbar^2}{2m}\nabla^2\psi + V\psi = E\psi. \tag{13}$$

However, in order to "read" the Schrödinger equation and to analyze its physical meaning and mathematical properties obtain it in somewhat different forms, namely: (a) in the operator form (in terms of the Hamiltonian) which is convenient for derivation of the energy level value; (b) in polar (spherical) coordinates which are commonly used to separate the angular part of the wave function (from which the atomic orbitals are derived) and the radial part, which is determined in the atomic structure calculations.

Derivation with the Help of the Hamiltonian Operator

First Preface to the Derivation: the Notion of Operators. Since the electron state is described by the wave function, all the mathematical operations in quantum mechanics, unlike classical mechanics, are carried out on functions, namely, on the electron wave functions ψ. All physical values describing the electron state, such as velocity, coordinate, impulse, angular momentum, energy, etc. must be associated with the wave function.

However, since these physical values are related to the functions, they are used in quantum mechanics not by themselves, but are replaced by operators which denote corresponding operations performed over the function. Thus, for physical values of velocity, coordinate, impulse, angular momentum, energy, there are corresponding operators of velocity, coordinate, impulse, momentum and energy.

The operators are often designated by the same symbol as the corresponding value, but are distinguished from the latter by superscript (for example, the impulse p and the corresponding impulse operator $\hat{p}$, the potential energy V and the corresponding potential energy operator $\hat{V}$, etc.) or are designated by the same letter in a different print. Since the operators by themselves do not have any physical meaning and are used only to designate the operation performed over the function, their symbols are always followed by the symbols of the function. Hence one does not often use any special designation for operators, but when for example p or V are followed by the function ψ (i.e., $p\psi$ or $V\psi$), they are understood to be the operators of the momentum and potential energy, respectively.

The Second Preface: the Notion of the Hamiltonian Operator. In classical mechanics the Hamiltonian H represents the means of expressing the total energy E through a sum of kinetic energy T, which is expressed in terms of the impulse p, and of the potential energy V expressed in terms of a space coordinate $q(x, y, z)$:

$$E = H(p, q) = T(p) + V(x, y, z).$$

In quantum mechanics one uses the Hamiltonian operator $H(p,q)$, representing the sum of the kinetic energy operator $T(p)$ and the potential energy operator $V(x,y,z)$. However, since, according to the uncertainty principle ($\Delta p \cdot \Delta q \geq h$), the impulse and coordinate in atoms cannot be accurately determined simultaneously, the Hamilton operator should be treated as a single operator and its eigenvalues (see below) do not represent the sum of the T and V eigenvalues.

Derivation of the Schrödinger equation is reduced in this case to writing the energy conservation law ($T+V=E$ or $H=E$) in the operator form:

$$\hat{H}\psi = \hat{E}\psi.$$

In order to identify this expression with the wave equation we must:
1. Decompose the Hamiltonian operator expression:

$$H\psi = (T+V)\psi.$$

2. Express the kinetic energy $T = mv^2/2$ through the impulse

$$p = mv; \quad v^2 = \frac{p^2}{m^2}; \quad T = \frac{p^2}{2m}.$$

3. Replace the impulse p components by operators

$$p_x \to \hat{p}_x = \frac{\hbar}{i} \cdot \frac{\partial}{\partial x}; \quad p_y \to \hat{p}_y = \frac{\hbar}{i} \cdot \frac{\partial}{\partial y}; \quad p_z \to \hat{p}_z = \frac{\hbar}{i} \cdot \frac{\partial}{\partial z}.$$

Note once more that the operators (in this case $\hbar/i \cdot \partial/\partial x$ and so on) do not represent physical value. They must be applied to the function, which in the given case is function ψ, in order to obtain another value (i.e., another function) $\hbar/i \cdot \partial \psi/\partial x$.

By squaring and introducing the Laplacian designation we obtain:

$$\hat{p}^2 \psi = -\hbar \left(\frac{\partial^2}{\partial x^2} + \frac{\partial^2}{\partial y^2} + \frac{\partial^2}{\partial z^2} \right) \psi = -\hbar^2 \nabla^2 \psi.$$

4. Then the kinetic energy operator is:

$$\hat{T}\psi = -\frac{\hbar^2}{2m} \nabla^2 \psi.$$

5. Finally we obtain the Schrödinger equation in the operator form:

$$(\hat{T} + \hat{V})\psi = \hat{E}\psi \quad \text{or} \quad -\frac{\hbar^2}{2m} \nabla^2 \psi + V\psi = E\psi.$$

On dividing all members of this equation by $\hbar/2m$, it is easy to arrive at the conventional form of the wave equation.

Derivation in Polar Coordinate Form.

The solution of many problems is simplified if the Schrödinger equation is represented in spherical polar coordinates. This form is especially convenient for understanding the physical meaning of the solutions of the equation: atomic orbitals, their form, quantum numbers, and orbital radii.

To change from the Cartesian system of coordinates (in which the position of an electron is determined by the x, y, z coordinates) into polar coordinates that determine the electron position by the radial distance r, polar angle ϑ (or colatitude) and azimuthal angle φ, we have to make use of simple relations (Fig. 9):
$x = r \cdot \sin\vartheta \cdot \cos\varphi$, $y = r \cdot \sin\vartheta \cdot \sin\varphi$, $z = r \cdot \cos\vartheta$, $r = \sqrt{x^2 + y^2 + z^2}$.

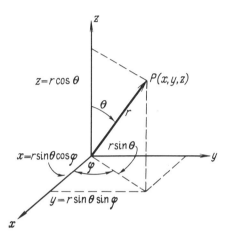

Fig. 9. Relation between rectangular and polar (spherical) coordinates

The transformation of the Schrödinger equation from the Cartesian into the spherical polar system of coordinates is accomplished by means of a standard mathematical operation consisting of replacing the Laplacian in the Cartesian system of coordinates

$$\nabla^2(x, y, z) = \frac{\partial^2}{\partial x^2} + \frac{\partial^2}{\partial y^2} + \frac{\partial^2}{\partial z^2}$$

into the Laplacian in the spherical system of coordinates

$$\nabla^2(r, \vartheta, \varphi) = \frac{1}{r^2} \cdot \frac{\partial}{\partial r}\left(r^2 \frac{\partial}{\partial r}\right) + \frac{1}{r^2 \sin\vartheta} \cdot \frac{\partial}{\partial \vartheta}\left(\sin\vartheta \frac{\partial}{\partial \vartheta}\right) + \frac{1}{r^2 \sin^2\vartheta} \cdot \frac{\partial^2}{\partial \varphi^2}.$$

Substitute also the potential energy value $V = -e^2/r$ for the hydrogen atom.
The Schrödinger equation then instead of forms used earlier Eqs. (2) and (9):

$$\nabla^2 \psi(x, y, z) + \frac{8\pi^2 m}{h^2}(E - V)\psi(x, y, z) = 0$$

will become

$$\nabla^2 \psi(r, \vartheta, \varphi) + \frac{8\pi^2 m}{h^2}\left(E + \frac{e^2}{r}\right)\psi(r, \vartheta, \varphi) = 0.$$

By decomposing the Laplacian expression $\nabla^2 \psi(r, \varphi)$, we obtain the Schrödinger equation in polar coordinates:

$$\frac{1}{r^2}\frac{\partial}{\partial r}\left(r^2 \frac{\partial \psi}{\partial r}\right) + \frac{1}{r^2 \sin\vartheta}\frac{\partial}{\partial \vartheta}\left(\sin\vartheta \frac{\partial \psi}{\partial \vartheta}\right) + \frac{1}{r^2 \sin^2\vartheta}\frac{\partial \psi}{\partial \varphi^2} + \frac{8\pi^2 m}{h^2}\left(E + \frac{e^2}{r}\right)\varphi = 0.$$

Note that this complicated expression is no more than the same expression as in Eq. (2), (9) and (16).

The reason for representing the Schrödinger equation in polar coordinates lies in the fact that it can be transformed further into a form which enables its separation into three more simple equations, each containing only one variable: r, or ϑ, or φ. Each of these three equations can be solved separately. The solutions of these equations are the three functions:

$R(r)$, a radial function of r only,
$\Theta(\vartheta)$, an angular function of ϑ only,
$\Phi(\varphi)$, an angular function of φ only.

The total wave function is the product of these functions:

$$\Psi(r, \vartheta, \varphi) = R(r) \cdot \Theta(\vartheta) \cdot \Phi(\varphi).$$

Substituting $\Psi = R \cdot \Theta \cdot \Phi$ [after abbreviating $R(r)$ to R etc.] into the above form of the Schrödinger equation, and multiplying throughout by $r^2 \sin^2\vartheta / (R \cdot \Theta \cdot \Phi)$, we obtain the equation in which the separate terms involve one variable only:

$$\frac{\sin^2 \vartheta}{R}\frac{d}{dr}\left(r^2 \frac{dR}{dr}\right) + \frac{\sin\vartheta}{\Theta}\frac{d}{d\vartheta}\left(\sin\vartheta \frac{d\Theta}{d\vartheta}\right) + \frac{1}{\Phi}\frac{d^2\Phi}{d\varphi^2}$$
$$+ r^2 \sin^2 \vartheta \frac{8\pi^2 m}{h^2}\left(E + \frac{e^2}{r}\right) = 0.$$

(Since here each function depends on one variable only, the partial derivative sign ∂ is replaced by that of the ordinary derivative d.)

Let us obtain three separate equations from Eq. (17). The third term of Eq. (17), $1/\Phi \cdot d^2\Phi/d\varphi^2$, depends on φ only and other terms depend on r and ϑ only. Hence the sum of these terms can be equal to zero for all the φ values only when the term $1/\Phi \cdot d^2\Phi/d\varphi^2$ is equal to a constant. Denoting this constant as $-m_l^2$, we can obtain the first equation resulting from the separation of the Schrödinger equation in polar coordinates, namely the equation involving φ only:

$$\frac{1}{\Phi}\frac{d^2\Phi}{d\varphi^2} = \text{const} = -m_l^2.$$

Substituting $-m_l^2 = 1/\Phi \cdot d^2\Phi/d\varphi^2$ and dividing throughout by $\sin^2\vartheta$ we obtain:

$$\frac{1}{R}\frac{d}{dr}\left(r^2\frac{dR}{dr}\right) + \frac{r^2 8\pi^2 m}{h^2}\left(E + \frac{e^2}{r}\right) = -\frac{1}{\Theta \sin\vartheta}\frac{d}{d\vartheta}\left(\sin\vartheta\frac{d\Theta}{d\vartheta}\right) + \frac{m_l^2}{\sin^2\vartheta}.$$

Here the left-hand side depends on r only, but the right hand side on ϑ only. Hence it follows that the two sides can be equated to the same constant, which we shall call β.

Thus we obtain the second equation involving r only:

$$\frac{1}{R}\frac{d}{dr}\left(r^2\frac{dR}{dr}\right) + \frac{r^2 8\pi^2 m}{h^2}\left(E + \frac{e^2}{r}\right) = -\beta$$

and the third equation involving ϑ only:

$$\frac{1}{\sin\vartheta}\frac{d}{d\vartheta}\left(\sin\vartheta\frac{d\Theta}{d\vartheta}\right) - \frac{m_l^2}{\sin^2\vartheta} = -\beta.$$

The Solution of the Φ Equation; the Magnetic Quantum Number m_l. The wave functions $\Phi = N\sin(m_l\varphi)$ and $\Phi = N\cdot\cos(m_l\varphi)$ are the solution of $d^2\Theta/d\varphi^2 + m_l^2\Phi = 0$ (taken without derivation). For Φ to be a single-valued function of φ (having one unique probability distribution value everywhere and changing from 0 to 2π), it should be a periodic function with integer m_l values (then changes of φ by 2π would repeat the same value of Φ).

The m_l value corresponds to the magnetic quantum number thus arising as the immediate consequence of the wave equation solution. The Φ function must be normalized to unity. This means that the probability of finding the electron in the φ interval between 0 and 2π must be equated to unity.

Thus from this condition the normalizing constant N can be fixed by integration over the entire region from 0 to 2π and by then equating this integral to unity:

$$\int_0^{2\pi} \Phi^2 d\varphi = \int_0^{2\pi} N^2 \sin^2(m_l\varphi) d\varphi = 1,$$

but $\sin^2(m_l\varphi)d\varphi = \frac{1}{2}(1-\cos(2m_l\varphi)d\varphi)$, then

$$\int_0^{2\pi} \Phi^2 d\varphi = N^2\left\{\int_0^{2\pi}\frac{d\Phi}{2} - \int_0^{2\pi}\frac{1}{2}\cos(2m_l\varphi)d\varphi\right\} = N^2\left\{\frac{2\pi}{2} - 0\right\} = 1; \quad N = \frac{1}{\sqrt{\pi}};$$

hence

$$\Phi_{m_l}(\varphi) = \frac{1}{\sqrt{\pi}}\sin(m_l\varphi),$$

where $m_l = 0, \pm 1, \pm 2 \ldots$

Table 1. Complex and real function Φ_{m_l} values

m_l		Real
0	$1/\sqrt{2\pi}\cdot e$	$1/\sqrt{\pi}\cdot 1$
1	$1/\sqrt{2\pi}\cdot e^{i\varphi}$	$1/\sqrt{\pi}\cdot \cos\varphi$
−1	$1/\sqrt{2\pi}\cdot e^{-i\varphi}$	$1/\sqrt{\pi}\cdot \sin\varphi$
2	$1/\sqrt{2\pi}\cdot e^{2i\varphi}$	$1/\sqrt{\pi}\cdot \cos 2\varphi$
−2	$1/\sqrt{2\pi}\cdot e^{2i\varphi}$	$1/\sqrt{\pi}\cdot \sin 2\varphi$

Table 2. Function $\Theta_{lm_l}(\vartheta)$ values

l	m_l	$\Theta_{lm_l}\vartheta$
0	0	$\sqrt{2}/2 \cdot 1$
1	0	$\sqrt{3}/2 \cdot \cos\vartheta$
1	± 1	$\sqrt{3}/2 \cdot \sin\vartheta$
2	0	$\sqrt{5}/2 \cdot (3\cos^2\vartheta - 1)$
2	± 1	$\sqrt{15}/2 \cdot \sin\vartheta \cdot \cos\vartheta$
2	± 2	$\sqrt{15}/4 \cdot \sin^2\vartheta$

The $\Phi_{m_l}(\varphi)$ function written above as real function can be expressed also in the complex form:

$$\Phi_{m_l}(\varphi) = N \cdot e^{im_l\varphi}.$$

The probability distribution function $\psi^2(\varphi)$ is the product of ψ and its complex conjugate ψ^*, then

$$\int_0^{2\pi} \Phi_{m_l} \cdot \Phi_{m_l}^* \cdot d\varphi = N^2 \int e^{im_l\varphi} = N^2 \cdot 2\pi = 1;$$

hence $N = \dfrac{1}{\sqrt{2\pi}}$ and $\Phi_{m_l}(\varphi) = \dfrac{1}{\sqrt{2\pi}} e^{im_l\varphi}$.

(Note the different N values in real and complex forms of the Φ_{m_l} function.)

The expressions for the normalized $\Phi_{m_l}(\varphi)$ functions are given in Table 1 in both the complex and the real form for the m_l values from 0 to 2.

The Solutions of the Θ and R Equations; the Quantum Numbers n and l. These solutions, though reduced to standard mathematical operations such as in the preceding case, are solved by a rather elaborate procedure and for this reason we shall directly consider their results (for their solutions see, for example [41]).

From the solution of the Θ equation one obtains the orbital quantum number l: its acceptable solutions are those where $\beta = l(l+1)$, with $l = 0, 1, 2 \ldots (n-1)$; $l \geq m_l$; $m_l = l, l-1 \ldots -l-1, -l$. The Θ function is determined besides l by the m_l value, i.e., it is $\Theta_{lm_l}(\vartheta)$ function (Table 2).

From the solution of the R equation one obtains the principal quantum number n. The R function is determined by the n and l values: $R_{nl}(r)$. The R function can be written in terms of the associated Laguerre polynomials and the Θ function in terms of the associated Legendre polynomials. Their general expressions are complicated enough (for Legendre polynomials see p. 255) to be considered here, however since the quantum numbers are small integer (0, 1, 2, 3) the polynomials transform into very simple expressions for these concrete quantum number values (Table 2).

In Table 2 the numerical coefficients are the normalizing constants N (obtained in the same manner as for the Φ function); further terms showing angular dependence from the ϑ angle value are written.

The R function values are given below in Chap. 1.4.

1.4 Atomic Orbitals (Solutions of the Schrödinger Equation)

1.4.1 Physical Meaning of the Schrödinger Equation Solutions

Let us consider the three components composing the Schrödinger equation $H\psi_{nlm_l} = E\psi_{nlm_l}$ i.e., ψ_{nlm_l}, E, H.

1. The solutions to the Schrödinger equation represent the wave functions ψ_{nlm_l} (or eigenfunctions), referred to as one-electron functions called atomic orbitals and described by the quantum numbers n, l, m_l.

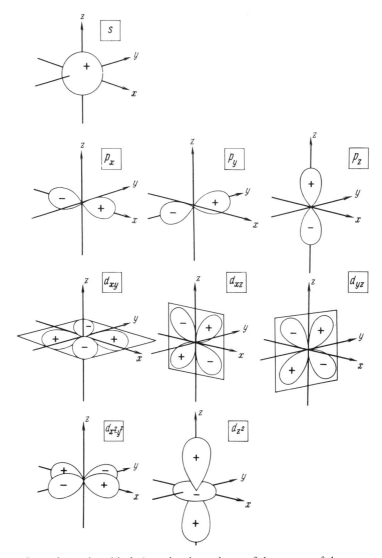

Fig. 10. The forms of s, p, d atomic orbitals (angular dependence of the square of the wave functions)

These atomic orbitals describe the form, size and electron-density distribution in charge clouds corresponding to the state with the given values of the quantum numbers (Fig. 10).

2. The acceptable solutions to the Schrödinger equation exist only for certain values of the energy E_n (determined by the principal quantum number n only) corresponding to the wave functions ψ_{nlm_l} (being thus degenerated with respect to the l and m_l quantum numbers), i.e., the energy E_n is restricted to a discrete set of values called eigenvalues. The corresponding energy levels determine the transitions between the states described by the different eigenfunctions ψ_{nlm_l} and having the energy values E_n.

3. The particular form of the Schrödinger equation is determined by the Hamiltonian operator H. For the case of the hydrogen atom, the H operator in front of the wave function ψ_{nlm_l} decomposes into the sum of the kinetic energy and the Coulomb interaction between the hydrogen atom electron and the nucleus. For more complex atomic systems the Hamiltonian operator contains additional terms which can reflect any interaction in these systems.

1.4.2 The Physical Meaning of the Atomic Orbitals

Atomic Orbital: Names and Notations. The notion characterizing the space distribution and motion of an electron in an atom evolved from the Bohr circular and elliptical orbit model to the wave description of the electron and finally to the eigenfunctions representing the Schrödinger equation solutions, and permitting the separation of the angular functions (describing the forms of the atomic orbitals) and of the radial function (describing the electron density distribution in the charge clouds).

Since the electron motion in atoms is described by means of these functions, they are called the atomic orbitals or simply orbitals, retaining this old quantum theory name for quite another physical and mathematical concept. Similarly the notation of the atomic orbitals is also retained:

1. The orbitals with l quantum numbers are denoted by the letters (see p. 9):

$l=0,\quad 1,\quad 2,\quad 3,\quad 4,\quad 5\ldots$
$s,\quad p,\quad d,\quad f,\quad g,\quad h\ldots$

2. The principal quantum number n is placed before these letters; for example, the 1s orbital has $n=1$, $l=0$, the 2p-orbital has $n=2$, $l=1$, the 3d-orbital $n=3$, $l=2$, the 4f orbital $n=4$, $l=3$ etc.

3. The magnetic quantum number m_l (or substituting letters x, y, z and xy, xz etc., indicating the orientation of the orbitals; see below) is placed as subscript: $2p_0$ (or $2p_z$), $3d_2$ (or $3d_{xy}$) etc.

The List of Atomic Orbitals. The quantum numbers can assume the following values (see also Chap. 1.3):

$n=1, 2, 3, 4, 5, 6 \ldots,$
$l=(n-1), \ldots 0,$
$m_l=l, l-1, \ldots 0, -(l-1), -l.$

Table 3. Wave functions ψ_{nlm_l}

n	l	nl	m_l	Atomic orbital
1	0	1s	0	1s
2	0	2s	0	2s
	1	2p	1, 0, −1	$2p_x, 2p_y, 2p_z$
3	0	3s	0	3s
	1	3p	1, 0, −1	$3p_x, 3p_y, 3p_z$
	2	3d	2, 1, 0, −1, −2	$3d_{xy}, 3d_{xz}, 3d_{yz}, 3d_{x^2-y^2}, 3d_{z^2}$
4	0	4s	0	4s
	1	4p	1, 0, −1	$4p_x, 4p_y, 4p_z$
	2	4d	2, 1, 0, −1, −2	$4d_{xy}, 4d_{xz}, 4d_{yz}, 4d_{x^2-y^2}, 4d_{z^2}$
	3	4f	3, 2, 1, 0, −1, −2, −3	$4f_{xyz}, 4f_{x^3}, 4f_{y^3}, 4f_{z^3},$ $4f_{xz^2}, 4f_{yz^2}, 4f_{z(x^2-y^2)}$

The possible combinations of these quantum numbers lead to the list of all possible one-electron wave functions (Table 3).

Three terms which determine the wave function are

$$\psi_{nlm_l}(r, \vartheta, \varphi) = N_{nlm_l} \cdot R_{nl}(r) \cdot Y_{lm_l}(\vartheta, \varphi).$$

a) The normalizing constant N_{nlm_l} (i.e., written with the subscript $_{nlm_l}$) for the wave function ψ_{nlm_l} is obtained as the product of the normalizing constants for the R_{nl}, Θ_{lm_l} and Φ_{m_l} functions (see Chap. 1.3). The normalizing constant takes into account the geometry of a space in which the probability of finding the electron ψ^2 is equal to unity, i.e., it reflects the AO form and enters as an indispensable term into expressions for any wave functions including linear combinations of atomic orbital molecular orbitals (see Chap. 3.2) etc. Usually N contains a square root of some value, since it is ψ^2 which is normalized.

b) $R_{nl}(r)$ is the radial part of the $\psi_{nlm_l}(r, \vartheta, \varphi)$ wave function representing its dependence on the radius r only and describing the values and the distribution of the electron density in AO. Note that in the designation $R(r)$, as in a designation $f(x)$, the sign R corresponds to f (the sign for function) and r corresponds to x (variable) as in $f(x)$.

The radial function depends on the quantum numbers n and l only and does not depend on m_l, i.e., for all the three 3p functions ($3p_x, 3p_y, 3p_z$) there is the same radial distribution of electron density. Radial functions (Table 4) are depicted (Figs. 11–13) in three types of plots showing the radial dependence (1) of the R function (obtained as a result of the partial solution of the Schrödinger equation), or (2) of the R^2 function (which is in fact of real significance as a probability of finding the electron) or (3) of the $R^2 \cdot 4\pi r^2 dr$ function (describing the radial distribution of the electron density and showing the existence of a maximum of the electron density distribution related to the orbital radius; see also Chap. 1.3).

For s orbitals (see Fig. 11) showing no angular dependence, the ψ_{nlm_l} total wave function coincides with the radial functions $R_{nl}(r)$, but from the angular part the

Table 4. Radial part of the wave functions (see Figs. 11–13)

Ψ_{nl}	N	$R_{nl(r)}$
1s	2	e^{-r}
2s	$-1/2\sqrt{2}$	$e^{-r/2} \cdot (2-r)$
2p	$1/2\sqrt{6}$	$e^{-r/2} \cdot r$
3s	$2/81\sqrt{3}$	$e^{-r/3} \cdot (27 - 18r + 2r^2)$
3p	$4/81\sqrt{6}$	$e^{-r/3} \cdot (6r - r^2)$
3d	$4/81\sqrt{30}$	$e^{-r/3} \cdot r^2$

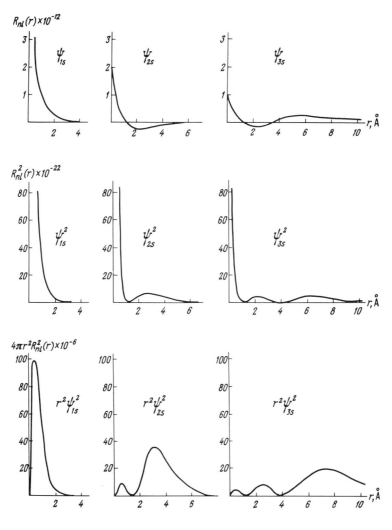

Fig. 11. Plot of the radial distribution function for s orbitals

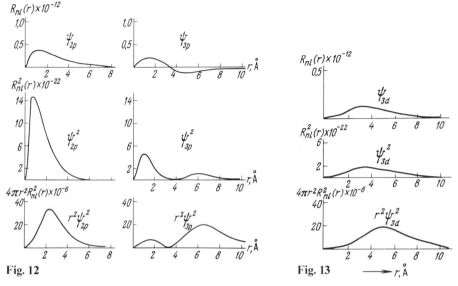

Fig. 12. Plot of the radial distribution function for *p* orbitals

Fig. 13. Plot of the radial distribution function for *d* orbitals

normalizing coefficient only [multiplied by the normalizing coefficient for the $R_{nl}(r)$ function] enters in the ψ_{nlm_l} expression. The radial parts of the wave functions for the hydrogen atom are given in Table 4. The radial distribution for other atoms will be considered in Chap. 1.5.

c) $Y_{lm_l}(\vartheta, \varphi)$ is the angular part of the $\psi_{nlm_l}(r, \vartheta, \varphi)$ wave function showing the dependence on the ϑ and φ angles only (see Table 5) and describing the forms of the AO (Fig. 11).

Since free atoms and ions have spherical symmetry, $Y_{lm_l}(\vartheta, \varphi)$ functions are called spherical functions or spherical harmonics.

The $Y_{lm_l}(\vartheta, \varphi)$ angular function represents the product of the two functions: $\Theta(\vartheta) \cdot \Phi(\varphi)$ and depends on the quantum numbers l and m_l only and not on the principal quantum number n. It means that all the *s* orbitals (or all the *p*, *d*, *f*, orbitals), i.e., orbitals with the same l, have the same form, thus independent of the n value. The magnetic quantum number determines the orbital orientation.

The n quantum number determines the sizes of the atomic orbitals. The angular functions are pictured as boundary surfaces (comprising about 95% of electron density, see Fig. 8).

In Table 5 the normalizing constant N is the product of the normalizing constants for the $\Theta_{lm_l}(\vartheta)$ and $\Phi_{m_l}(\varphi)$ functions.

The $Y_{lm_l}(\vartheta, \varphi)$ function represents the angular dependence in spherical coordinates, the $Y_{lm_l}(x/r, y/r, z/r)$ in Cartesian coordinates. The former represents the angular dependence more pictorially, while the latter is more convenient for the designation of atomic orbital orientations, for the expression of linear combinations of atomic orbitals etc.

Table 5. Angular part of the wave functions (see Fig. 10)

Φ_{lm_l}	N	$Y_{lm_l}(\vartheta, \varphi)$	$Y_{lm_l}(x/r, y/r, z/r)$	Linear combinations
s	$1/\sqrt{2\pi}\cdot\sqrt{1/2}$	—	—	$s = s_0$
p_z	$1/\sqrt{2\pi}\cdot\sqrt{3/2}$	$\cos\varphi$	z/r	$p_z = p_0$
p_x	$1/\sqrt{2\pi}\cdot\sqrt{3/4}$	$\sin\vartheta\cdot\cos\vartheta$	x/r	$p_x = -1/\sqrt{2}(p_1 + p_{-1})$
p_y	$1/\sqrt{2\pi}\cdot\sqrt{3/4}$	$\sin\vartheta\cdot\sin\vartheta$	y/r	$p_y = -1/\sqrt{2}(p_1 - p_{-1})$
d_{xy}	$1/\sqrt{\pi}\sqrt{15/16}$	$\sin^2\vartheta\cdot\sin^2\vartheta$	xy/r^2	$d_{xy} = 1/i\sqrt{2}(d_2 - d_{-2})$
$d_{x^2-y^2}$	$1/\sqrt{\pi}\sqrt{15/16}$	$\sin^2\vartheta\cdot\cos 2\varphi$	$(x^2 - y^2)/r^2$	$d_{x^2-y^2} = 1/\sqrt{2}(d_2 + d_{-2})$
d_{xz}	$1/\sqrt{\pi}\sqrt{15/4}$	$\cos\vartheta\cdot\sin\vartheta\cdot\cos\varphi$	xz/r	$d_{xz} = -1/\sqrt{2}(d_1 - d_{-1})$
d_{yz}	$1/\sqrt{\pi}\sqrt{15/4}$	$\cos\vartheta\cdot\sin\vartheta\cdot\sin\varphi$	yz/r^2	$d_{yz} = -1/i\sqrt{2}(d_1 + d_{-1})$
d_{z^2}	$1/\sqrt{2\pi}\sqrt{5/8}$	$3\cos^2\vartheta - 1$	$(3z^2 - r)/r^2$	$d_{z^2} = d_0$

The designations, for example p_z or d_{xz}, reflect the orientation of these atomic orbitals (along the z axis for p_z or in the xz plane for d_{xz}) and their angular dependence: z/r or xz/r^2 (see Table 5). Using the relation between Cartesian and polar coordinates (Fig. 9) it is easy to pass from z/r to $\cos\varphi$ (since $z = r\cos\varphi$) or from xz/r^2 to $\sin\vartheta\cdot\cos\varphi\cdot\cos\vartheta$ (since $x = r\sin\vartheta\cdot\cos\varphi$ and $z = r\cos\vartheta$).

The $\Theta_{lm_l}(\vartheta)$ and $\Phi_{m_l}(\varphi)$ functions were given earlier (see Tables 1 and 2) for the hydrogen atom for different values of the magnetic quantum number m_l which determines orientations of atomic orbitals with respect to a coordinate axes.

The only functions designated by m_l values which are real are $p_0 = p_z$ and $d_0 = d_{z^2}$, while expressions for other functions (p_1, d_1, d_2 etc.; see Table 1) include the imaginary quantity $i = \sqrt{-1}$, and hence these functions are complex ones. The representation of orbitals as p_0, p_1, p_{-1} and $d_2, d_1, d_0, d_{-1}, d_{-2}$ is more convenient in calculations and in considerations of magnetic properties. In descriptions of chemical bonding (including molecular orbitals) those real functions (p_x, d_{xy} etc.) are used which do not contain i and give real forms of orbitals (Fig. 10).

These real orbitals $d_{xy}, d_{xz}, d_{yz}, d_{z^2}, d_{x^2y^2}$ are derived as linear combinations of the $d_2, d_1, d_0, d_{-1}, d_{-2}$ functions (see Table 5).

Now we can retrace the train of the considerations leading to the atomic orbitals in the form in which they are shown in Figure 10 and in Table 5: (1) write the Schrödinger equation for the hydrogen atom; (2) express it in spherical coordinates; (3) obtain the separate solutions for the Θ and Φ functions (see Tables 1 and 2); (4) write the expression for the $Y_{lm_l}(\vartheta, \varphi)$ functions as the product of $\Theta_{lm_l}(\vartheta)$ (see Table 2) and $\Phi_{m_l}(\varphi)$ (see Table 1): for example, for $p_1(l=1, m_l=1)$ and $p_{-1}(l=1, m_l=-1)$:

$$Y_{p_1}(\vartheta, \varphi) = \sqrt{3/2}\cdot\sin\vartheta\cdot 1/\sqrt{2\pi}\cdot e^{i\varphi} = \sqrt{3/2}\cdot 1/\sqrt{2\pi}\cdot\sin\vartheta\cdot e^{i\varphi},$$

$$Y_{p_{-1}}(\vartheta, \varphi) = \sqrt{3/2}\cdot\sin\vartheta\cdot 1/\sqrt{2\pi}\cdot e^{-i\varphi} = \sqrt{3/2}\cdot 1/\sqrt{2\pi}\cdot\sin\vartheta\cdot e^{-i\varphi};$$

(5) adding these two functions (in order to obtain the linear combination $p_1 + p_{-1}$ and taking into account that $e^{i\varphi} = \cos\varphi + i\sin\varphi$ and $e^{-i\varphi} = \cos\varphi - i\sin\varphi$) gives:

$$p_1 + p_{-1} = 2\cdot\sqrt{3/2}\cdot 1/\sqrt{2\pi}\cdot\sin\vartheta\cdot\cos\varphi,$$

$$p_1 - p_{-1} = 2\cdot\sqrt{3/2}\cdot 1/\sqrt{2\pi}\cdot\sin\vartheta\cdot\sin\varphi.$$

Normalizing these linear combinations to unity, we obtain the expressions for p_x and p_y orbitals (Table 5) in spherical coordinates or (since from Fig. 9 $\sin\vartheta\cdot\cos\varphi = x/r$ and $\sin\vartheta\cdot\sin\varphi = y/r$) in Cartesian coordinates.

As another example consider in the same way the $d_{x^2-y^2}$ orbital which can be derived as the linear combination of the d_2 and d_{-2} wave functions (see Tables 1, 2 and 5):

$$Y_{d_2}(\vartheta,\varphi) = \sqrt{15/16}\cdot 1/\sqrt{2\pi}\cdot\sin^2\vartheta(\cos 2\varphi + i\sin 2\varphi),$$

$$Y_{d_{-2}}(\vartheta,\varphi) = \sqrt{15/16}\cdot 1/\sqrt{2\pi}\cdot\sin^2\vartheta(\cos 2\varphi - i\sin 2\varphi).$$

Since from Figure 9 $\sin^2\vartheta(\cos^2\varphi - \sin^2\varphi) = (x^2 - y^2)r$ we can obtain also the expression in Cartesian coordinates.

1.4.3 The s, p, d, f Systematics of Atomic Orbitals

It is convenient in the description of atomic orbitals to consider separately their form (determined by the angular part of the wave function) and the electron density distribution (determined by the radial part of the wave function).

In such cases as chemical bond and valency problems, crystal field and molecular orbital theories, the descriptions of the form of atomic orbitals, especially their symmetry and directional properties are more important and thus orbitals are pictured without reference to their sizes (which depends on the radial part of the wave function) and then represent not the whole wave functions ψ_{nlm_l} or $\psi^2_{nlm_l}$, but their angular parts $Y_{lm_l}(\vartheta,\varphi)$ or $Y^2_{lm_l}(\vartheta,\varphi)$.

These atomic orbital pictures (the boundary surfaces) are the same for all orbitals with the same l quantum number (designated as s, p, d, f for $l=0, 1, 2, 3$), not only for the single-electron hydrogen atom but for many-electron atoms (considered in one-electron approximation); they are the same for all atoms and ions, for their ground and excited states. For all atoms and states there are moreover only four types of boundary surface: s, p, d, f.

The number of possible orientations of the orbitals is given by the number of possible values of the m_l magnetic quantum number and is equal for the s orbitals to 1, for the p orbitals to 3 for the d orbitals to 5, for the f-orbitals to 7, i.e., to $(2l+1$; see Table 3).

Without any fixed direction, for instance, without fixed crystal field axes (for atoms in crystals), electric or magnetic field directions, all the possible orientations of the orbitals are equivalent and correspond to the same energy value. These orbitals are called degenerate. A degree of degeneracy corresponds to a number of the possible orientations of the orbitals.

s Orbitals. Two peculiarities are notable for the s-orbitals: (1) lack of the angular dependence; the s boundary surface has a spherical form (see Figs. 8, 10), the s wave functions are independent of ϑ and φ angles and depend only on radius r; (2) only the s orbitals have nonzero electron density at the nucleus, while the p, d, f electrons do not contact directly with the nuclei.

Because of the s electron penetration through the nucleus, hyperfine structure in electron paramagnetic resonance spectra, chemical shift in Mössbauer spectra and in nuclear magnetic resonance spectra arise. This is a peculiarity not only of 1s

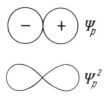

Fig. 14. Representation of p orbital as ψ_p or ψ_p^2

electrons, but also of 2s, 3s and in general ns electrons. Thus electron shells of atoms interact with their nucleus by means of their s electrons.

Radial dependence of s orbitals is shown in Figure 11: (1) as ψ wave function (equal here to R_{nl} multiplied by the normalizing constant of the angular part Y_{lm_l} of s orbitals, see Table 5) depending on radius r, (2) as ψ^2 – electron density and (3) as $4\pi r^2 \cdot \psi^2$ – electron density radical distribution (see also Chapter 1.3). Note that the boundary surface covering about 95% of charge cloud volume (Figs. 8, 10) has the radius of about 2 Å for the 1s orbital of the hydrogen atom, while the maximum of electron density (in the $\psi^2 \cdot 4\pi r^2$ function curve) corresponds to $r_{max} = 0.528$ Å, i.e., to the radius of the first circular Bohr orbit.

As distinct from Bohr orbits, where an electron motion restricted only by these orbits with forbidden space between them, the electron density in atomic orbital description is distributed in an atomic volume, is not equal to zero at the nucleus (for s electrons) and continues far beyond r_{max}.

p Orbitals (see Table 5 and Fig. 10 for angular part and Table 4 and Fig. 12 for radial part), i.e., the orbitals with $l = 1$, exist beginning with the principal quantum number $n = 2$ (see Table 3). Since for $l = 1$ there are three m_l values ($m_l = 1, 0, -1$), three p orbitals occur: p_1, p_0, p_{-1}. More often instead of p_1, p_0, p_{-1}, three linear combinations are used: $p_1 + p_{-1} = p_x$, $p_1 - p_{-1} = p_y$ and $p_0 = p_z$.

The p-orbitals are pictured by means of the angular part Y_{lm_l} of wave function ψ_{nlm_l} as two adjoined spheres with different signs or by means of $Y_{lm_l}^2$ (or $\psi_{nlm_l}^2$) as boundary surfaces of dumb-bell shape and, since it represents the square of wave functions, without signs (Fig. 14). However, although it is often the $\psi_{nlm_l}^2$ functions which are pictured (since the real meaning of electron density has a ψ^2 function), at the same time the signs are indicated, thus assuming that these signs are related to the wave functions.

d Orbitals (see Table 5 and Fig. 10 for angular part and Table 4 and Fig. 13 for radial part), i.e., the orbitals with $l = 2$; exist beginning with the principal quantum number value $n = 3$ (see Table 3). Since for $l = 2$ there are five possible values for $m_l = 2, 1, 0, -1, -2$ five d orbitals occur: $d_2, d_1, d_0, d_{-1}, d_{-2}$. Usually instead of these orbitals representing (see Table 5) complex functions and imaginary forms (which are depicted in [79]) their linear combinations are used: $d_2 + d_{-2} = d_{x^2-y^2}$, $d_2 - d_{-2} = d_{xy}$, $d_1 - d_{-1} = d_{xz}$, $d_1 + d_{-1} = d_{yz}$, $d_0 = d_{3z^2-r}$ (or simply d_{z^2}).

Note that four d orbitals have the same form (Fig. 10) with different orientation relative to coordinate axes, but $d_{z^2}(d_{3z^2-r^2})$ orbital only has a different shape. It is connected with the fact that it is impossible to construct five independent d orbitals of the same form. Indeed, there are six positions of d-orbitals with respect to coordinate axes: $d_{xy}, d_{yz}, d_{xz}, d_{x^2y^2}, d_{z^2-x^2}, d_{z^2-y^2}$. Of these last three four-lobe orbitals, two only are independent. Hence linear combination of the last two orbitals is taken, which provides $d_{3z^2-r^2}=1/\sqrt{2}(d_{z^2-x^2}+d_{z^2-y^2})$; (here $1/\sqrt{2}$ is the normalizing coefficient).

The $d_{3z^2-r^2}$ designation is obtained as follows: $(z^2-x^2)+(z^2-y^2)=2z^2-(x^2+y^2)$. From $r^2=x^2+y^2+z^2$ (see Chap. 1.3) we have $(x^2+y^2)=r^2-z^2$, hence $2z^2-(x^2+y^2)=3z^2-r^2$.

f Orbitals (with $l=3$) can be formed with $n=4$ and 5 (in ground state) and exist in seven types with $m_l=3, 2, 1, p, -1, -2, -3$, which are written usually as the following linear combinations [59, 94]:

f_{xyz}, $f_{x(x^2-y^2)}$, $(f_{y(x^2-z^2)}$, $f_{z(x^2-y^2)}$, $f_{z^3-3/5zr^2}$, $f_{y^3-3/5yr^2}$, $f_{x^3-3/5xr^2}$.

1.5 Orbital Radii (Solutions of the Schrödinger Equation)

1.5.1 Concerning the Calculations of Electronic Structure of Many-Electron Atoms

All quantum-mechanical concepts discussed so far have been considered with relation to the hydrogen atom.

Now passing from the hydrogen atom to many-electron atoms, we must take into account that: (1) nuclear charges are $+Ze$ (not $+e$ as in the hydrogen atom) where Z is the atomic number; then the Coulomb interaction of an electron and the nucleus is $V=-Ze^2/r$; (2) there are not one but Z electrons interacting with the nucleus and with each other; the interaction between an electron and all other electrons depends on their instantaneous configuration.

In calculations of many-electron atom structure it is assumed that each electron behavior can be described by an one-electron wave function characterized by the quantum numbers n, l, m_l and depending on the coordinate of this electron only. It is called the one-electron approximation, or atomic orbital method.

A field acting on a given electron is assumed to be: (1) a central (i.e., its charge is concentrated in a center) and a spherical one (i.e., averaged over the ϑ and φ angles) and thus depending on the radius r only, (2) determined by the potential of the interaction between an electron and the nucleus ($V=-Ze^2/r$) and by the additional potential $V_{add}(r)$ representing an interaction with other electrons.

The calculations of many-electron atom structures [33, 51, 54] were made on the basis of these assumptions.

The analytical hydrogen-like wave functions method is the simplest and was widely used earlier in quantum-chemical calculations for approximate estimations of atomic structures.

The central field acting on the electron is assumed to be determined by the effective nuclear charge Z^*e with $Z^*=Z-s$, where s is the screening constant.

Here the effects of electrons, which are inner ones with respect to a given electron, are reduced to the screening of the nuclear charge and are accounted for by the screening constant.

Then the principal quantum number n is replaced by the effective n^*.

Thus the radial part of a hydrogen-like wave function of any electron can be written as the so-called Slater atomic orbital:

$$R_{nl}(r)=N_{nl}\cdot r^{n^*-1}\cdot e^{-Z^*r/n^*}.$$

Compare this with the R_{nl} function for the hydrogen atom ($n^* = n$ and $Z^* = Z = 1$):

$$R_{nl}(r) = N_{nl} \cdot r^{n-1} \cdot e^{-r/n},$$

i.e., it is the radial part of the hydrogen atom orbitals given in Table 4.

There are two ways of obtaining the numerical expressions for the Slater orbitals of various atoms.

1. Slater empirical rules for determination of the effective r^* and $Z^* = Z - s$ values [51].

   a) Instead of $n = 1, 2, 3, 4, 5, 6 \ldots$ one takes $n^* = 1, 2, 3, 3.7, 4, 4.2 \ldots$, respectively.

   b) The screening constant s is found as follows: for electrons with the same $n: s = 0.35$ (only for 1s electrons $s = 0.30$)
   for each electron with $(n-1)$: $s = 0.85$,
   for each electron with $(n-2)$, $(n-3)$ etc.: $s = 1.00$,
   for each d and f electron: $s = 1.00$.

For example, the screening constant for the $3p$ electron of the phosphorus atom is the sum of the contributions from 14 other electrons ($Z = 15$ for P and 15 electrons are $1s^2 2s^2 2p^6 3s^2 3p^3$): $s = 2 \cdot 0.35 + 2 \cdot 0.35 + 6 \cdot 0.85 + 2 \cdot 0.85 + 2 \cdot 1.00 = 10.20$ (the calculation begins from the outer shells). Then the effective charge $Z^* = 15 - 10.20 = 4.80$.

Inserting the Z^* and n^* values thus found in the radial part of the Slater atomic orbital, one obtains its numerical expression which can be used for the construction of the radial electron density distribution plot (see Figs. 11–13).

2. Computations of the semi-empirical coefficients and exponent in hydrogen-like functions (such as, for example, $R_{1s} = \sqrt{\alpha^3/\pi} \cdot e^{-\alpha r}$) give more accurate results [19].

Self-consistent field methods provide most accurate atomic structure calculations [27].

a) Hartree self-consistent field method without exchange.

Instead of the Schrödinger equation for the hydrogen atom (see Chap. 1.3):

$$-\frac{\hbar^2}{2m}\nabla^2 \psi + V\psi = E\psi,$$

the Hartree equation for each electron (since it is the one-electron approximation) is written:

$$-\frac{\hbar^2}{2m}\nabla^2 \psi + [V + V_H]\psi = E\psi,$$

where to the potential $V = -Ze^2/r$ the potential V_H is added, which describes an interaction of one electron with all other electrons of the atom.

In order to be able to write the $[V + V_H]$ value, we need to know the wave functions ψ_j for all the electrons of the atom which must be determined in turn. Thus the ψ_j wave functions are first taken as, for example, hydrogen-like functions (zeroth approximation). Using them one computes V_H. Using the obtained V_H value in the Hartree equation, the wave functions of first approximation for each electron are found. These functions are used for the calculation of the new V_H value, which is utilized to determine the wave function of second approximation.

These cycles are repeated until further iterations yield no change of the ψ.

b) Hartree-Fock self-consistent field method with exchange.

Since electron position is not independent of the probability of finding other electrons in the same point (which is connected with Pauli exclusion principle, see Chap. 1.7), it is necessary also to take into account the possibility of electron exchange between orbitals.

For this Fock proposed the introduction of the so-called exchange term in the Coulomb interaction operator. Thus the Hartree-Fock principle can be written schematically as

$$+\frac{\hbar^2}{2m}\nabla^2\psi + [V + V_H + V_F]\psi - E\psi.$$

A simplified form of this procedure, known as the Hartree-Fock-Slater method, considers the exchange term in a form which saves about 80% computer time.

Relativistic extensions. The Hartree-Fock method does not take into account relativistic terms (corresponding to the phenomena which are attributed in nonrelativistic quantum mechanics to spin and spin-orbital interaction, see Chaps. 1.6 and 2.2). In heavy atoms these relativistic terms are more important than exchange interactions. Hence the atomic structure calculations can be made with the same one-electron approximation and within the framework of the same self-consistent field as in the Hartree method but by means of the relativistic Dirac equation (i.e., the Schrödinger equation complying with the theory of relativity). By calculation of this equation (with Slater's simplified form of Fock's exchange terms) the self-consistent Dirac-Slater wave functions were computed for all atoms [33].

The data for the atomic wave functions so far obtained can be classified as follows:

1. The hydrogen-like wave functions: (a) with Slater's empirical coefficients: easy obtainable using Slater's rules (see above); (b) with semi-empirical coefficients: computed for atoms with $1s^2 2s^k 2p^n$ configurations [20].

2. Self-consistent field method (nonrelativistic): (A) without exchange: computed for all atoms, all positive and some negative ions [47]; (B) with exchange (Hartree-Fock method): (a) in the form of numerical solutions: obtained for atoms and for singly and doubly charged positive ions from $Z=2$(He) to $Z=18$(Ar) [19]; (b) in the form of analytical functions (tables of exponent coefficients): obtained for atoms from He to $4d$ atoms; for atoms with $Z=2-36$ [46], for atoms with $3p$ shells and, for atoms with $3d$, $4s$, $4p$, $4d$ shells; (c) with Slater's simplified exchange: the Herman-Skillman tables of numerical wave function values for all atoms from $Z=2$ to $Z=103$ [28].

3. Relativistic methods: for all atoms and many ions the calculations made by Lieberman et al. [33] with Slater's simplified form of exchange interaction provide the most accurate data so far. (They compare also the accuracy of the different self-consistent field methods.)

All these methods of atomic structure calculations give not only qualitatively similar results but also volumes numerically very close to each other (with the exception of the 1a empirical rules method.) The accuracy of the atomic wave function calculations is determined by comparison of the calculated energy values corresponding to these functions with the experimental energy values obtained from the atomic X-ray and optical spectra.

1.5.2 Orbital Radii and the Wave Functions of Atoms

The tables of numerical values of atomic wave functions [19, 28] contain the $R_{nl}(r)$ radial function values (in atomic units, where the atomic unit of length is the first Bohr circular orbit radius equal to 0.529 Å) for different distances r from the nucleus. These data can be plotted in form of electron-density radial distribution $R_{nl}^2(r) \cdot r^2$ diagram (see Figs. 15–17 and also Figs. 11–13).

Orbital sizes are described usually by means of the radii called orbital radii and corresponding to main maxima of the electron-density distribution (Fig. 15).

The values of the orbital radii for outer orbitals of free atoms and ions [58] are readily obtainable from the plots such as Figure 15 or directly from the tables of atomic wave functions.

It must be noted that the orbital radii do not restrict the sizes of free atoms and ions. The electron density distribution curve extends beyond the maximum, i.e., beyond the orbital radius. For outer orbitals these curves stretch for many Å. Therefore, free atoms have no sharp boundaries and their sizes are conveniently determined by a radii of boundary surfaces covering about 95% of electron density.

Let us compare the sizes of the free atoms and corresponding ions. Sizes of cations, i.e., of atoms with removed valence electrons, are abruptly diminished because of the disppearance of the outer orbitals. However for other (inner) orbitals fully occupied by electrons, there are only very slight differences. For example, the changes in 2s and 2p curves for the Na^+ ion in comparison with the Na atom are so slight that they cannot be shown on the scale of Figure 15 (However the total electron density diminishes and that plays a role in wave function calculations of ions in crystals). Sizes of anions hardly change (increase very little) in comparison with atom sizes, since in this case the same p shell fills (but the electron density increases).

In the case of isoelectron atoms, for instance, Na^+ and F^- with the same electron configurations (see Chap. 1.7) $1s^2 2s^2 2p^6$, the size of F^- anion having lesser nuclear charge ($Z=9$) is larger than the size of Na^+ cation having greater nuclear charge ($Z=11$).

Relative sizes of atoms in the periodic system of elements are governed by the two tendencies: (1) they increase abruptly with the building up of consecutive electron shells, (2) they decrease with increase of the nuclear charge.

Thus in the same row of the periodic system, the alkali atoms are the largest, and the halogen atoms the smallest (because of the increase of the nuclear charge). In the same period of the periodic system (for example, Li–Na–K–Rb–Cs–Fr) the atomic sizes increase from lighter to heavier elements (because of the building up

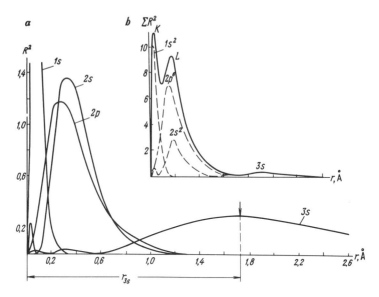

Fig. 15a and b. Radial distribution of electron density for the orbitals **a** and of the total electron density **b** for Na atom

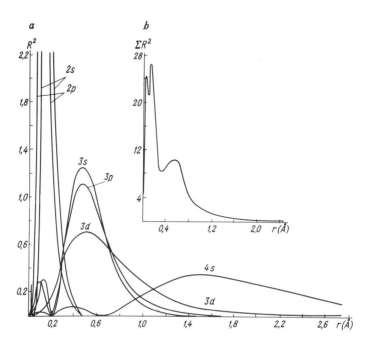

Fig. 16a and b. Radial distribution of electron density for the orbitals **a** and of the total electron density **b** for Ti atom

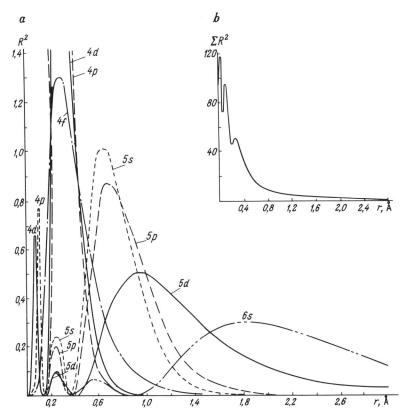

Fig. 17a and b. Radial distribution of electron density for the orbitals **a** and of the total electron density **b** for Gd atom

of the electron shells from 2s to 7s). However the size, for example, of the U atom (Z=92) with 92 electrons in the six shells is not much larger than that of lighter atoms because of the extremely close sequence of the inner shells.

The orbital radii and orbital energy values for the same orbital of different atoms range between the values for the ground and excited states of the hydrogen atom (the largest radii and the lowest energies) and those for the actinide atoms.

1.6 Electron Spin

Electron spin, i.e., the spin quantum numbers equal to 1/2 for one electron (and multiple of 1/2 for several electrons), and the magnetic spin quantum number $m_s = \pm 1/2$, were primary considered together with the quantum numbers n, l, m_l in the framework of the old quantum theory (see Chap. 1.2). At that time (see Chap. 1.3) the quantum numbers n, l, m_l were obtained and defined as a result of the solution of the Schrödinger equation.

However, the spin is not considered in Schrödinger quantum mechanics (see Chaps. 1.3–1.5), since the spin is the relativistic property of an electron (i.e., related to effects considered by the relativity theory, while the Schrödinger equation does not take relativity into account.

However, in addition to Schrödinger (nonrelativistic) quantum mechanics, Dirac relativistic quantum mechanics exist, which combine quantum theory with relativity.

The Dirac wave equation, instead of the ψ_{nlm_l} wave function of the Schrödinger equation, comprises a more abstract function depending on four and not three variables. Thus the spin quantum number arises as a result of the Dirac equation solution, just as the quantum numbers n, l, m_l arise from the Schrödinger equation solution.

Electron states are described in the Dirac formalism by an abstract value called "ket-vector" and designated as $|\rangle$. The ket-vector $|nlm_l\rangle$ has the same physical meaning as the ψ_{nlm_l} wave function, but is connected with a different mathematic apparatus.

The ψ wave function is a complex one and is a conjugate of the ψ^* function; for example, an integral $\int \psi H \psi^*$ (where H is an energy operator) including ψ and ψ^* is often used in quantum mechanics.

Similarly there exists a bra-vector $\langle|$ conjugated to the ket-vector. Therefore the expression $\langle|H|\rangle$ corresponds to the integral $\int \psi H \psi^*$. Thus the names bra- and ket-vectors are readily understandable as originating from "bracket". In the Dirac theory, the spin remains an abstract value derived from the mathematical formalism. This represents the only rigorous definition of spin, having no adequate physical picture and no analogs in classical mechanics.

The picture of an electron spinning around "its own axis" (see Chap. 1.2) and moving simultaneously in an orbit (as the earth rotates on its own axis while moving in orbit around the sun) is a crude approximation to the actual physical phenomenon. However, it is the only available physical model with its spin properties (the possibility for the spin angular momentum to be aligned parallel or antiparallel to the magnetic or electric field direction with the spin quantum numbers $+1/2$ and $-1/2$) and since it does not contradict experimental observations, this model is used in interpretations of many atomic properties.

It is now important to consider how spin characteristics can be used in nonrelativistic quantum mechanics, the most important method of solid state theories.

The ψ_{nlm_l} wave function is the coordinate function, i.e., $\psi_{nlm_l}(x, y, z)$ or $\psi_{nlm_l}(r, \vartheta, \varphi)$, and completely determines the state of an electron moving in the space around the nucleus. In order to take the spin into account, it must be supplemented with the spin function χ_{m_s}, i.e., the "spin coordinate" function, is not connected with the electron motion and position, but is associated with the "proper" (or "inner") angular momentum of the electron. The square of this spin function gives the distribution probability of $+1/2$ and $-1/2$ values with respect to a fixed direction. In the absence of a fixed direction, the states with $m_s = +1/2$ and $-1/2$ are equivalent and the system is spin-degenerated.

The wave function taking the spin into account is written as $\psi_{nlm_l m_s} = \psi_{nlm_l} \cdot \chi_{m_s}$ or in expanded form $\psi_{nlm_l m_s} = R_{nl}(r) \cdot Y_{lm_l}(\vartheta, \varphi) \cdot \chi_{m_s}(\sigma)$, where σ is the spin coordinate.

The two kinds of designation of an electron state were discussed earlier (see p. 32): (1) by means of m_l, for example, for $l=1$ there are p_1, p_0, p_{-1} orbitals (i.e., the z-axis projection of l) and (2) by means of the orbital momentum components in the Cartesian coordinate: p_x, p_y, p_z (obtained as the linear combination of p_1, p_0, p_{-1}).

Similarly, besides the electron spin state designations by means of m_s (equal to $+1/2$ and $-1/2$ for a single electron and representing the z axis projection of the spin quantum number s), it is possible to use spin components in Cartesian coordinates s_x, s_y, s_z (connected with the former by the linear combinations $s_+ = s_x + is_y$ and $s_- = s_x - is_y$).

The Schrödinger equation solutions are obtained without the spin, but then the spin is taken into account in the form of the spin-orbital interaction considered as a perturbation. Hence to the Hamiltonian (see p. 21) $H = -\hbar^2/2m - V(r)$ one adds the term describing the spin-orbital interaction $H_{so} = \xi \cdot l \cdot s$, where ξ is the spin-orbital coupling constant.

Then the Hamiltonian is written as

$$H = -\frac{\hbar^2}{2m} - V(r) + \xi ls.$$

Thus, summarizing this consideration of electron spin, we conclude that it is necessary to combine: (1) the model of spin as a rotating electron (being aware, however, of the conventionality of this model: it is a metaphor rather than the real picture), (2) the concept of the relativistic nature of spin and its connection with the Dirac formalism, (3) the presentation of spin in the Schrödinger treatment by means of the spin function complementary to the coordinate function and by the insertion in the Hamiltonian of the spin-orbital interaction.

The understanding of electron spin (as of other principal concepts of quantum mechanics) will become deeper and its different aspects will be more concrete when considering optical absorption spectra, magnetic properties, electron paramagnetic (spin) resonance, etc.

1.7 Electronic Configurations and the Periodic System of Elements

Points of Departure. Considerations of the electron shell structure of the elements in the periodic system is based on the same approximations as calculations of many-electron atomic structures (see Chap. 1.5). According to these approximations, it is possible to "distinguish" every electron in the many-electron atom, to describe the electron by means of a proper wave function called the atomic orbital and obtained as the result of the Schrödinger equation by the Hartree-Fock or any other method.

Atoms are composed of nuclei and electrons; the nuclei differ in various atoms (consisting of different numbers of protons and neutrons), the electrons are similar, with identical values of mass, charge and spin. At the same time every electron has its own individual state.

In order to describe the behavior of an electron in the atom, the four quantum number values (n, l, m_l, m_s) need to be indicated, as they determine the electron state unambiguously. According to the Pauli exclusion principle there are no two electrons in an atomic system with an identical set of all these four quantum numbers.

The number of electrons in the atomic shell is readily obtainable from the knowledge of the possible values and combinations of n, l, m_l, m_s and the Pauli principle. In addition to the Table 3, the possibility of two spin states must be taken into account. The shells are determined by the principal quantum number $n = 1, 2, 3, 4, 5, 6\ldots$ (K, L, M, N, O, $P\ldots$), the subshells by the orbital quantum number $l = 0, 1, 2, 3, 4\ldots$ (s, p, d, f, $g\ldots$), the number of electrons in the shells and subshells is determined by the magnetic orbital and magnetic spin quantum numbers m_l and m_s. The distribution of electrons in the shells and subshells is referred to as the electronic configuration.

The electronic configurations of atoms are indicated by the atomic orbitals ($1s$, $2s$, $2p$, $3d$ etc.) with the number of electrons in each orbital as a superscript: $1s^2$ for two $1s$ electrons, $3d^5$ for five $3d$ electrons and so on. Thus, for example, the electronic configuration for Na is written in the form $1s^2 2s^2 2p^6 3s^1$.

The electronic configurations are described also by means of a quantum box diagram representation:

The s-orbitals are indicated by a single box [↑↓] corresponding to the nondegenerate orbital state with two possible kinds of spin: ↑ means $m_s = +1/2$ and ↓ $m_s = -1/2$ (in one s-orbital no more than two electrons can be placed).

The p-orbitals: three boxes [↑↓|↑↓|↑↓] corresponding to $m = 1, 0, -1$ (or to p_x, p_y, p_z) with one or two electrons in each box; the maximum number of electrons is six.

The d-orbitals: five boxes [↑↓|↑↓|↑↓|↑↓|↑↓] ($m_l = 2, 1, 0, -1, -2$, or d_{xy}, d_{yz}, d_{xz}, $d_{x^2-y^2}$, d_{z^2}); ten electrons.

The f-orbitals: seven boxes ($m_l = 3, 2, 1, 0, -1, -2, -3$ or f_{xyz} etc.); 14 electrons.

To determine the order of electron population of atomic orbitals, we need to know their energy for every given atom. To some extent the quantum number n value (and to a lesser degree l value) indicates the relative sequence of the orbital energies. Calculations and experimental data give the following order of electron occupation of the orbitals: $1s$, $2s$, $2p$, $3s$, $3p$, $4s$, $3d$, $4p$, $5s$, $4d$, $5p$, $6s$, $4f$, $5d$, $6p$, $7s$, $5f$, $6d$.

The orbital energies are calculated: (a) for the hydrogen atom by means of the simplest expression: $E = -e^2/2r_a n^2$ (see Chap. 1.2); (b) for many-electron atoms by means of the Hartree-Fock and other methods (see Chap. 1.5) giving besides the electron density radial distributions (Chap. 1.5) the orbital energy for each atom.

The experimental data for the orbital energies are obtainable: (a) from X-ray spectra: for inner shells; transitions between the inner electron levels correspond to the X-ray region of spectrum; (b) from optical spectra: for outer (valence) electrons and their excited states.

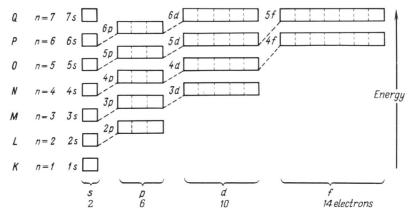

Fig. 18. Relative filling order of atomic orbitals

Fig. 19

Fig. 20

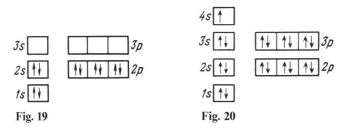

Fig. 19. Electronic configuration $1s^2 2s^2 2p^6$ (Ne and isoelectronic ions)

Fig. 20. Electronic configuration of K atom ($Z = 19$)

Fig. 21

Fig. 22

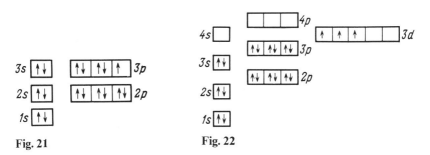

Fig. 21. Electronic configuration of Cl atom ($Z = 17$)

Fig. 22. Electronic configuration of Cr^3 ion

The electron distribution in the orbitals also obeys Hund's rule: electrons occupy as many orbitals as possible with parallel (unpaired) spins; only after all the boxes of each orbital type are occupied by a single (unpaired) electron does the occupation continue by two electrons with antiparallel (paired) spins. The relative orbital energies of atoms depending on the n and l quantum numbers are shown in Figure 18.

Electronic Configurations and the Periodic System of Elements

	K	L		M			N	
	1s	2s	2p	3s	3p	3d	4s	4p
1 H	1							
2 He	2							
3 Li	2	1						
4 Be	2	2						
5 B	2	2	1					
6 C	2	2	2					
7 N	2	2	3					
8 O	2	2	4					
9 F	2	2	5					
10 Ne	2	2	6					
11 Na	2	2	6	1				
12 Mg	2	2	6	2				
13 Al	2	2	6	2	1			
14 Si	2	2	6	2	2			
15 P	2	2	6	2	3			
16 S	2	2	6	2	4			
17 Cl	2	2	6	2	5			
18 A	2	2	6	2	6			
19 K	2	2	6	2	6		1	
20 Ca	2	2	6	2	6		2	
21 Sc	2	2	6	2	6	1	2	
22 Ti	2	2	6	2	6	2	2	
23 V	2	2	6	2	6	3	2	
24 Cr	2	2	6	2	6	5	1	
25 Mn	2	2	6	2	6	5	2	
26 Fe	2	2	6	2	6	6	2	
27 Co	2	2	6	2	6	7	2	
28 Ni	2	2	6	2	6	8	2	
29 Cu	2	2	6	2	6	10	1	
30 Zn	2	2	6	2	6	10	2	
31 Ga	2	2	6	2	6	10	2	1
32 Ge	2	2	6	2	6	10	2	2
33 As	2	2	6	2	6	10	2	3
34 Se	2	2	6	2	6	10	2	4
35 Br	2	2	6	2	6	10	2	5
36 Kr	2	2	6	2	6	10	2	6

	N				O				P
	4s	4p	4d	4f	5s	5p	5d	5f	6s
37 Rb	2	6	—	—	1	—	—	—	—
38 Sr	2	6	—	—	2	—	—	—	—
39 Y	2	6	1	—	2	—	—	—	—
40 Zr	2	6	2	—	2	—	—	—	—
41 Nb	2	6	4	—	1	—	—	—	—
42 Mo	2	6	5	—	1	—	—	—	—
43 Tc	2	6	6	—	1	—	—	—	—
44 Ru	2	6	7	—	1	—	—	—	—
45 Rh	2	6	8	—	1	—	—	—	—
46 Pd	2	6	10	—	—	—	—	—	—
47 Ag	2	6	10	—	1	—	—	—	—
48 Cd	2	6	10	—	2	—	—	—	—
49 In	2	6	10	—	2	1	—	—	—
50 Sn	2	6	10	—	2	2	—	—	—
51 Sb	2	6	10	—	2	3	—	—	—
52 Te	2	6	10	—	2	4	—	—	—
53 I	2	6	10	—	2	5	—	—	—
54 Xe	2	6	10	—	2	6	—	—	—
55 Cs	2	6	10	—	2	6	—	—	1
56 Ba	2	6	10	—	2	6	—	—	2
57 La	2	6	10	—	2	6	1	—	2
58 Ce	2	6	10	2	2	6	—	—	2
59 Pr	2	6	10	3	2	6	—	—	2
60 Nd	2	6	10	4	2	6	—	—	2
61 Pm	2	6	10	5	2	6	—	—	2
62 Sm	2	6	10	6	2	6	—	—	2
63 Eu	2	6	10	7	2	6	—	—	2
64 Gd	2	6	10	7	2	6	1	—	2
65 Tb	2	6	10	8	2	6	1	—	2
66 Dy	2	6	10	10	2	6	—	—	2
67 Ho	2	6	10	11	2	6	—	—	2
68 Er	2	6	10	12	2	6	—	—	2
69 Tm	2	6	10	13	2	6	—	—	2
70 Yb	2	6	10	14	2	6	—	—	2
71 Lu	2	6	10	14	2	6	1	—	2
72 Hf	2	6	10	14	2	6	2	—	2

	O			P			Q
	5p	5d	5f	6s	6p	6d	7s
73 Ta	6	3	—	2			
74 W	6	4	—	2			
75 Re	6	5	—	2			
76 Os	6	6	—	2			
77 Ir	6	9	—	0			
78 Pt	6	9	—	1			
79 Au	6	10	—	1			
80 Hg	6	10	—	2			
81 Tl	6	10	—	2	1		
82 Pb	6	10	—	2	2		
83 Bi	6	10	—	2	3		
84 Po	6	10	—	2	4		
85 At	6	10	—	2	4		
86 Rn	6	10	—	2	6		
87 Fr	6	10	—	2	6	—	1
88 Ra	6	10	—	2	6	—	2
89 Ac	6	10	—	2	6	1	2
90 Th	6	10	—	2	6	2	2
91 Pa	6	10	3	2	6	1	2
92 U	6	10	5	2	6	1	2
93 Np	6	10	5	2	6	1	2
94 Pu	6	10	6	2	6	—	2
95 Am	6	10	7	2	6	—	2
96 Cm	6	10	7	2	6	1	2
97 Bk	6	10	8	2	6	1	2
98 Cf	6	10	9	2	6	1	2
99 Es	6	10	10	2	6	1	2
100 Fm	6	10	11	2	6	1	2

The Periodic Table of Elements Construction. Using these rules and considerations, the electronic configuration for any atom can be found. It is sufficient to indicate the atomic number Z equal to the number of electrons and to fill then the orbital boxes one after another, beginning from the lowest and obeying the Pauli principle and Hund's rule.

The order of occupation (Figs. 19–22) corresponds to the positions of atoms in the periodic system. Up to $3p^6$ configuration the sequence of occupation of the boxes goes parallel with increasing quantum numbers n and l, but additionally the $4s$, $5s$, $6s$, $7s$ orbitals are filled before the $3d$, $4d$, $5d$, $6d$ orbitals respectively. Beginning with $5d$ configuration, the d and f orbital energies become so close together that the simple order of occupation does not apply.

Electronic configurations of the elements of the periodic system are given in Table 6.

1.8 Term Symbols and Atomic States

1.8.1 Description of Atomic States

The Vector Model for Interactions in Atoms. The data considered in preceding sections were related immediately to the description of an electron behavior in atom, but not to an atom as a whole.

Electronic configurations given in Table 6 describe the atomic structures in one-electron approximation, assuming mutual independence of electrons.

Descriptions of the states of an atom as a whole require consideration of interelectronic repulsion in addition to the interaction between electrons and the nucleus.

The method for the diagrammatical representation of these interactions is given by the vector model of an atom (Figs. 23, 24).

There are two cases of interaction in atoms.

1. For atoms with $Z<30$ the electron–electron interaction is described by means of the quantum numbers L and S, which determine the resultant angular and spin momenta, respectively, and of the quantum number I, which determines the total resultant momentum for the atom. This case is referred as LS (or Russell-Saunders) coupling. First the resultant angular L and resultant spin S momenta of the atom are determined (as the vector sums of the orbital and spin momenta of the electrons) then the total resultant momentum of the atom I is obtained:

$$L = l_1 + l_2 + \cdots + = \Sigma l,$$
$$S = s_1 + s_2 + \cdots + = \Sigma s,$$
$$I = L + S$$

(here L, S, I are the vector sums).

2. For heavier atoms the jj-scheme of interaction is used: first the resultant momenta j of each electron are determined, and then the total resultant momentum of the atom I is obtained as the vector sum of the individual j.

Description of Atomic States

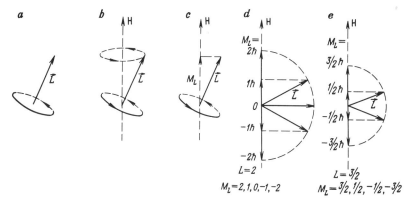

Fig. 23a–e. Vector model of atom: orbital momentum of atom and its quantization. **a** Presentation of orbital momentum of atom by vector L (by the same manner: l, j, I). **b** Precession of orbital momentum L about an applied magnetic field H. **c** Projection of orbital momentum L on the magnetic field H direction. **d, e** Possible (quantized) projections M_L of orbital momentum L on the magnetic field H direction for $L=2$ and $L=3/2$ (in the same manner projections of l, j, S, I can be obtained)

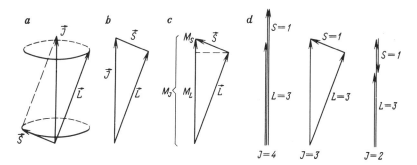

Fig. 24a–d. Vector model of atom; vector summation of orbital L and spin S momenta of atom. **a** Precession of L and S about common direction I. **b** Vector addition of $L+S=I$ (or $l_1+l_2=L$, or $s_1+s_2=S$, or $l+s=j$, or $j_1+j_r=I$). **c** Vector addition of projection of momenta $M_L+M_S=M_I$ (or $m_l^1+m_l^2=M_L$, or $m_s^1+m_s^2=M_S$, or $m_l+m_s=m_j$, or $m_j^1+m_j^2=M_I$). **d** Example of vector addition: $L+S=I$ for $L=3$ and $S=1$; takes the values from $(L+S)$ to $(L-S)$; the terms $^3F_4, \ ^3F_3, \ ^3F_2$

The values of the resultant angular momentum L are designated by the same letters as the values of the orbital quantum numbers l of an electron, but by capitals instead of small letters for l:

$l = s \quad p \quad d \quad f \quad g \quad h \quad i \quad \ldots$ (for an electron),
$ 0 \quad 1 \quad 2 \quad 3 \quad 4 \quad 5 \quad 6 \quad \ldots$
$L = S \quad P \quad D \quad F \quad G \quad H \quad I \quad \ldots$ (for an atom).

It is clear that unlike the spin quantum number of an electron which always equals $s=1/2$, the resultant spin quantum numbers of an atom can be integers and half-integers: $S=0, 1/2, 1, 3/2, 2, 5/2 \ldots$

The magnetic quantum numbers M_L, M_S and M_I determine projection of the resultant angular, spin, and total momenta of an atom on the magnetic (or electric) field direction respectively.

Whereas the n, l, m_l, m_s quantum numbers determine an electron state, those for an atom are L, S, I, M_I.

Let us write the quantum numbers describing the electron and atomic states and their relations.

Electron quantum numbers:
- n: principal; $n = 1, 2, 3, 4, 5, 6, \ldots$;
- l: orbital; $l = s, p, d, f, \ldots$ $(0, 1, 2, 3 \ldots)$;
- m_l: magnetic orbital; $m_l = l, l-1, \ldots -l$;
- s: spin; $s = 1/2$;
- m_s: magnetic spin; $m_s = \pm 1/2$;
- j: resultant; $j = l \pm s$;
- m_j: magnetic resultant; $m_j = j, j-1, \ldots -j$.

Atom quantum numbers:
- L: orbital $(L = l_1 + l_2 + \ldots + = \Sigma l)$; $L = S, P, D, F \ldots$;
- M_L: magnetic orbital: $M_L = L, L-1, \ldots -L$;
- S: spin $(S = s_1 + s_2 + \ldots + = \Sigma s)$; $S = 0, 1/2, 1, 3/2 \ldots$;
- M_S: magnetic spin; $M_S = S, S-1, \ldots -S$;
- I: total; $I = (L+S)$ till $(L-S)$;
- M_I: magnetic total; $M_I = I, I-1, \ldots, -I$.

The quantum numbers l, s, j of the electron and L, S, I of the atom determine the corresponding angular, spin, and total momenta following the general relation $x^*\hbar = \sqrt{x(x+1)} \cdot \hbar$, where x^* designates a momentum, x the corresponding quantum number, $\hbar = h/2\pi$. That is:

$$x^*\hbar = \sqrt{x(x+1)} \cdot \hbar$$

if $x = l$: $l^*\hbar = \sqrt{l(l+1)} \cdot \hbar$ orbital ⎫
s: $s^*\hbar = \sqrt{s(s+1)} \cdot \hbar$ spin ⎬ momentum of an electron
j: $j^*\hbar = \sqrt{j(j+1)} \cdot \hbar$ resultant ⎭

L: $L^*\hbar = \sqrt{L(L+1)} \cdot \hbar$ orbital ⎫
S: $S^*\hbar = \sqrt{S(S+1)} \cdot \hbar$ spin ⎬ momentum of an atom
I: $I^*\hbar = \sqrt{I(I+1)} \cdot \hbar$ resultant ⎭

The value $\hbar$ has the dimensions of an angular momentum, hence all the momenta, (orbital, spin, resultant) are obtained by multiplication of $\hbar$ by the corresponding quantum number.

Projections of all the electron and atom momenta on the magnetic (or electric) field direction (see Fig. 23) are obtainable on the general scheme by means of the corresponding magnetic quantum numbers: $m_l\hbar$, $m_s\hbar$, $m_j\hbar$, $M_L\hbar$, $M_S\hbar$, $M_I\hbar$. The

Term Derivation from Electron Configurations 49

number of the possible values of a magnetic quantum number is equal to $(2x+1)$ where $x = l, s, j, L, S, I$. For instance, for L number of the M_L values is $(2L+1)$.

The total momenta of an atom I (vector sum $L+S$) and of an electron $(l+s)$ and the resultant momenta $L(\Sigma l)$ and $S(\Sigma s)$ are obtainable according to the general scheme of the vector summation shown in Figure 24. Projections of the resultant momenta are obtainable as the sums of the corresponding projections of momenta: for example $M_I = M_L + M_S$.

The vector model shown in Figure 24 is the general method of representing the interactions of the atom: it is used in the following description of interelectron repulsion leading to LS terms, of the spin-orbital coupling leading to multiplets characterized by the I values, of the interactions with an external magnetic field characterized by the M_I values. The same model is employed in considerations of atom magnetic properties.

1.8.2 Term Symbols

Terms are the designations of atomic states taking into account interelectron interactions and described by the L and S values. Term symbols are written as

$$^{2S+1}L$$

where $L = S, P, D, F, G, H, I$; S is the resultant spin quantum number (or simply spin) of the atom, and $(2S+1)$ is the spin multimplicity.

For example, the ^4F term of the Cr^{3+} ion corresponds to the state with $L = F = 3$; $2S+1 = 4$ and therefore $S = 3/2$.

For designation of an multiplet level we add the I value to the term symbol:

$$^{2S+1}L_I$$

where the total quantum number of an atom I takes the values from $(L+S)$ to $(L-S)$. Thus for the ^3F term $(L+S) = 4$ and $(L-S) = 2$. Hence for the ^3F term there are three multiplet levels with $I = 4, 3, 2$, that is $^3F_4, ^3F_3, ^3F_2$. (The spin multimplicity $2S+1$ in the upper left-hand corner indicates the number of these multiplet levels).

Similarly, the $^6H_{5/2}$ term (for Sm^{3+} ion, for example) corresponds to $L = H = 5$, $S = 5/2, I = 5/2$; there are six possible multiplet levels $^6H_{15/2}, ^6H_{13/2}, ^6H_{11/2}, ^6H_{6/2}, ^6H_{7/2}, ^6H_{5/2}$.

1.8.3 Term Derivation from Electron Configurations

Ground state terms are readily obtainable by means of the electron distribution in the quantum boxes according to Hund's rule (as many parallel spins as possible, see Chap. 1.7).

Let us consider the typical examples of the term derivation from some electron configurations.

The $3d^3$ electron configuration (Cr^{3+} ion and others). For d electrons $l=d=2$; $m_l=2, 1, 0, -1, -2$, i.e., there are $(2l+1)=5 m_l$ values. Place the three d electrons into five quantum boxes:

$m_l = \quad 2 \quad 1 \quad 0 \quad -1 \quad -2$

| ↑ | ↑ | ↑ | | | $\quad M_L = \Sigma m_l = 2+1+0 = 3$.

Since $M_L \geq L$ we obtain $L=3=F$. From $M_s = \Sigma m_s = 1/2+1/2+1/2 = 3/2$ (three electrons with $m_s = 1/2$) we obtain $S=3/2$ and the spin multiplicity $2S+1=4$.

Thus the ground state term for the $3d^3$ electron configuration is ^4F.

The J values range from $(L+S) = 3+3/2 = 9/2$ to $(L-S) = 3-3/2 = 3/2$, that is $J = 9/2, 7/2, 5/2, 3/2$. Hence the multiplet levels for the $3d^3$ configuration are $^4F_{9/2}$, $^4F_{7/2}$, $^4F_{5/2}$, $^4F_{3/2}$.

The np^2 electron configuration:

$m_l = \quad 1 \quad 0 \quad -1$

| ↑ | ↑ | | $\quad M_L = \Sigma m_l = 1+0 = 1;\quad L=1=P;$
$\quad\quad\quad\quad\quad M_S = \Sigma m_s = 1;\quad S=1;\quad 2S+1 = 3$.

The term $^{2S+1}L = {}^3P$.

$J = (L+S) = 2;\quad J = (L-S) = 0;\quad J = 2, 1, 0$.

The multiplet levels are $^3P_2$, $^3P_1$, $^3P_0$.

Note the three important particular cases.

1. For all the complete subshells ns^2, np^6, nd^{10}, nf^{14} the ground term is $^1S_0$ (because the sum of the positive m_l values is equal to the sum of the negative m_l values: $M_L = \Sigma m_l = 0$; $L=0=S$, and because in each of the quantum boxes there are two paired electrons; $M_S = \Sigma m_s = 0$; $S=0$; $2S+1=1$; $J = L \pm S = 0$). Thus in determining atom and ion terms one takes into account only electrons from incomplete subshells.

2. For the half-filled subshells ns^1, np^3, nd^5, nf^7 there is in each quantum box one electron, hence $\Sigma m_l = M_L = 0$ and the spin multiplicity $2S+1$ is determined by the number of electrons, i.e., one obtains respectively the ground terms ^2S, ^4S, ^6S, ^8S.

3. For the one-electron configurations ns^1, np^1, nd^1, nf^1 the ground term is obtained most simply: here $L=l$, $S=s=1/2$; $2S+1=2$ and the ground terms are ^2S, ^2P, ^2D, ^2F respectively.

Note that the terms for the p^1 and p^5, d^1 and d^9, f^1 and f^{13} configurations are identical (because Σm_l and Σm_s are identical). For example,

$m_l = \quad 2 \quad 1 \quad 0 \quad -1 \quad -2$

d^1 | ↑ | | | | | $\quad\quad \Sigma m_l = 2;\quad\quad \Sigma m_s = 1/2$

d^9 | ↑↓ | ↑↓ | ↑↓ | ↑↓ | ↑ | $\quad \Sigma m_l = -2;\quad \Sigma m_s = 1/2$.

Term Derivation from Electron Configurations 51

Table 7. Terms and multiplets of the ground states of ns^q and np^q configurations

s^1	^2S	$^2S_{1/2}$	H, Li, Na, K, Rb, Cs; Cu, Ag, Au
s^2	^1S	$^1S_0$	Be, Mg, Ca, Sr, Ba; Zn, Cd, Hg
p^1, p^5	^2P	$^2P_{1/2}$ $^2P_{3/2}$	B, Al, Ga, In, Tl (p^1) F, Cl, Br, I (p^5)
p^2, p^4	$^3P\,^1D\,^1S$	$^3P_0$ $^3P_2$	C, Si, Ge, Sn, Pb (p^2) O, S, Se, Te (p^4)
p^3	$^4S\,^2D\,^2P$	$^4S_{3/2}$	N, P, As, Sb, Bi
p^6	^1S	$^1S_0$	O^{2-}, F^-, Ne, S^{2-}, Cl^-, Ar etc.

It must be mentioned also that the majority of ions in crystals in their usual valencies (with exception of many transition metal ions with d and f subshells) have the electron configurations with complete subshells and consequently, have the single (for the ground configuration) term $^1S_0$ (see Table 7).

Excited state terms. The knowledge not only of ground term but also of all the excited terms derived from a given electron configuration (and from excited electron configurations) is necessary in spectroscopy of atom and solids.

One considers first the total number of the states corresponding to an electron configuration.

Let us recall that electron configurations represent only the nl-systematics of electrons indicating the electron shell (determined by the principal quantum number n) and subshell (determined by the orbital quantum number l). The four quantum numbers n, l, m_l, m_s determine unambiguously the electron state.

Several states having different m_l and m_s values correspond to each nl configuration. For example, in the case of the $3d^1$ configuration (Ti^{3+}, V^{4+} and other ions; see Chap. 1.7) the single valence d electron can occur in ten possible states which differ from each other only by m_l and m_s values. One then determines for each state with a given nlm_lm_s the $M_L = \Sigma m_l$ and $M_S = \Sigma m_s$ values, the possible L and S values are obtained and the corresponding terms are written [21–23 and others]. Ground term and ground multiplet level with lowest energy and an ordering of the terms and levels are determined according to Hund's rule.

The ground term is characterized (1) by maximum multiplicity $(2S+1)$, i.e., the highest spin S of the atom, (2) for a given multiplicity by the highest L value. For example, the ground term for the d^2 configuration is ^3F, further $^3P, ^1G, ^1D, ^1S$ follow.

Ground multiplet levels for the less than half-filled subshells (i.e., less than p^3, d^5, f^7) are the levels with minimum J value, for the more than half-filled subshells whose with maximum J value. For example, for the d^2 the ground level is $^3F_2$ (further $^3F_3, ^3F_4$), for the d^8 the ground level is $^3F_4$ (further $^3F_3, ^3F_2$).

Let us consider only the simplest cases: the nd^1 and np^1 configurations.

Calculate the number of the states for these configurations. For nd^1: $M_L = m_l = 2$; $M_S = m_s = \pm 1/2$; hence $L = 2 = D$; $S = 1/2$; $(2S+1) = 2$ and the term is ^2D. The J values: $J = L + S = 2 + 1/2 = 5/2$ and $J = L - S = 2 - 1/2 = 3/2$, hence there are two

Table 8. Terms of ions with d^n configurations

$d^1 d^9$	^2D	$Ti^{3+}, V^{4+} (d^1); Cu^{2+} (d^9)$
$d^2 d^8$	$^3F\,^3P$ $^1G\,^1D\,^1S$	$V^{3+} (d^2); Ni^{2+} (d^8)$
$d^3 d^7$	$^4F\,^4P$ $^2H\,^2G\,^2F\,^2D\,^2D\,^2P$	$V^{2+}, Cr^{3+}, Mn^{4+} (d^3)$ $Co^{2+} (d^7)$
$d^4 d^6$	^5D $^3H\,^3G\,^3F\,^3F\,^3D\,^3P\,^3P$ $^1I\,^1G\,^1G\,^1F\,^1D\,^1D\,^1S\,^1S$	$Mn^{3+} (d^4)\ Fe^{2+} (d^6)$
d^5	^6S $^4G\,^4F\,^4D\,^4P$ $^2I\,^2H\,^2G\,^2G\,^2F\,^2F\,^2D\,^2D\,^2D\,^2P\,^2S$	Mn^{2+}, Fe^{3+}

multiplets $^2D_{5/2}$ and $^2D_{3/2}$. For the $^2D_{5/2}\,I=5/2$, then $M_I=5/2, 3/2, 1/2, -1/2, -3/2, -5/2$, total $(2I+1)=6$ states. For the $^2D_{3/2}$ multiplet: $I=3/2$, then $M=3/2, 1/2, -1/2, -3/2$, total $(2I+1)=4$ states. Thus, there are ten states corresponding to this term: $(2L+1)(2S+1)=5\cdot 2=10$.

For np^1 there are six states. $M_L = =1, M_S=m_s=1/2; L=1:S=1/2$. The term is ^2P, the multiplets are $^2P_{3/2}$ (4 states with different M_I values) and $^2P_{1/2}$ (2 states with different M_I values).

The terms and the ground multiplet level for the ns^q, np^q and d^n configurations are given in Tables 7 and 8.

1.8.4 Free Atom Energy Levels and Hamiltonian

All the interactions corresponding to derivation of the terms and multiplets are plotted in energy level diagrams (Fig. 25) and described by a Hamiltonian operator.

The consideration is the framework of the perturbation theory: one obtains in consecutive order approximate solutions going from stronger interactions to weaker ones. Each following interaction is treated as a perturbation suerposed on the preceding stronger interaction. Thus different atomic interactions are arranged in the order of transitions from higher to lower energy ones. In the same order a splitting of the levels corresponding to the energy states of an atom arising from the different quantized interactions are observed, and in the same order the free atom Hamiltonian terms are written:

$$H_F = \underbrace{H_0 + H_{ee}}_{10^5\ \text{cm}^{-1}} + \underbrace{H_{LS}}_{10^2 \div 10^3\ \text{cm}^{-1}}.$$

(The approximate magnitude of the interactions is shown under the corresponding Hamiltonian terms.)

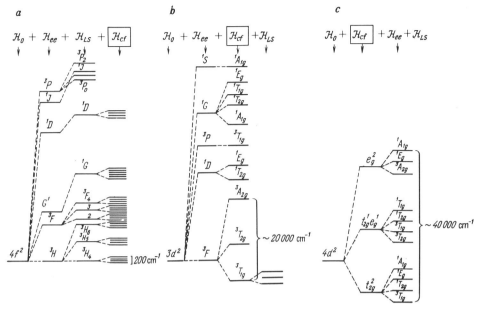

Fig. 25a–c. Free ion energy level diagram showing the levels arising from $3d^2$(V^{3+} ion) electronic configuration. Terms ^{2S+1}L arise from Coulomb interaction of electrons (electrostatic interelectronic repulsion); multiplet levels $^{2S+1}L_J$ result from spin-orbit (magnetic dipole-dipole) interaction; levels in magnetic field M_J arise from the splitting by the external magnetic field

H_0 represents the most substantial part of the interactions and is expressed as:

$$H_0 = -\Sigma \frac{\hbar^2 \nabla^2}{2m} - \Sigma \frac{Ze^2}{r_i}$$

where the first term describes the electron kinetic motion, the second the potential energy determined by the Coulomb (electrostatic) interaction between ith electron and the nucleus; r_i – the distance between the electron and the nucleus; $H_{ee} = \Sigma \frac{e^2}{r_{ij}}$ is the Coulomb interaction between ith and jth electrons; r_{ij} is the distance between these electrons; H_{LS} is the spin-orbit interaction (see Chap. 2.2).

In the following chapters, behavior of ions in crystals (not of the free atoms and ions) will be considered. Then a Hamiltonian of an ion in a crystal can be written as

$$H = H_F + H_{CF}$$

where H_{CF} represents the interaction of an ion with the crystal field (see Chap. 2.2).

Free Atom and Ion Term Energy Values as obtained from spectroscopic observations are compiled by Moore [38]. The data not only on atoms but also on the ionized states of free atoms are included in this survey. For example, the numerical term values are given not only for the lead atom Pb I, but also for Pb II, Pb III, Pb IV, Pb V (this spectroscopic labeling corresponds to Pb^0, Pb^+, Pb^{2+}, Pb^{3+}, Pb^{4+}).

The energy level values are given not only for a ground state electron configuration but also for excited configurations. For example, for V^{3+} there are all the term and level values derived not only from the $3d^2$ configuration but also all the terms and levels derived from the excited $3d4s$, $3d4p$ and others configurations.

Comprehensive treatment of atomic spectroscopy is given in many textbooks and monographs [21, 22, 23, 26, 29, 30, 41, 43, 44, 48].

1.8.5 Atomic Spectroscopy and Spectrochemical Analysis of Minerals, Rocks, and Ores; Spectroscopy – Cosmochemistry – Astrophysics

The atomic structure and free atom spectra so far considered are only the introduction to theories and spectra of solids, but in two cases they present definitive conclusions: (1) in spectrochemical analysis in the course of which the material excited by an electric arc or spark exists in the form of atoms and ions, and (2) in the spectra of sun, stars and interstellar matter. In these cases only, the spectra are those of free atoms and ions and can be interpreted immediately according to the atomic spectra theory.

Let us consider the most general principles of atomic spectroscopy as a logical consequence of the preceding discussions.

Divide first of all the transitions (1) between outer (valence) electron levels and (2) those between inner electron levels. In both, excitation of atoms is necessary in order to obtain a spectrogram.

1. In the case of outer electrons spectral transitions occur between the ground term and excited terms of the ground electron configuration and terms of excited electron configurations.

The transitions with emission of radiation energy are used in emission spectrochemical analysis [34, 36 and others], the transitions with absorption of the energy in atomic absorption spectrometry [17].

In the case of inner electrons spectral transitions occur between the ground state and the atomic state where one of the inner level electrons is removed. These transitions between inner electron levels are used in X-ray spectrochemical analysis.

Existence of many terms for outer electrons explains the large number of lines in emission atomic spectra. Already for the hydrogen atom with a single electron, the number of lines is quite large, while for the rare earths and many other electron atoms, linear spectra consist of thousands of lines. The spectrogram of a mineral or rock consists of superposition of regular sets of lines of elements composing these minerals and rocks.

In atomic absorption spectra there are by now only few lines since the transitions occur from one (ground) level to the nearest excited ones.

In a particular region of spectrum of X-ray spectra there are one or two lines since the differences between inner electron ($1s$, $2s$, $2p$, ...) energies is very large and therefore, the transitions for example, from 2p level to 1s level (with one electron removed from latter) and from 3p to 1s level fall in different regions of the X-ray spectrum.

Spectrochemical analysis is of a great importance for geology [34]. The discovery of several elements (Tl, In, Ge and others) in the course of mineral studies, trace elements investigations in geochemistry, mineralogy and petrology are connected with spectrochemical analysis. However, the majority of spectrochemical determinations are carried out in ore prospecting and especially in metallometry. Recently additional possibilities have arisen with atomic absorption spectrometry [17].

However, spectrochemical analysis is not only the method for studying the composition of earth's minerals. It is also the main method for qualitative and quantitative determination of the chemical composition of the universe and is widely used in astrophysics [16, 37]. The first great discovery in astrophysics was the determination, from spectrochemical analysis data, of the identity of the chemical composition of the whole of universe matter [16].

New possibilities for astrophysics and cosmic spectroscopy appear with earth satellites, geophysical rockets and interplanetary vessels [36]. Earth atmosphere is in fact transparent without essential absorption only for radiation in the region of 3000–8000 Å. Spectra obtained with the aid of spectrometers carried beyond the earth's atmosphere by satellites and rockets revealed a great number of lines in far UV and soft X-ray regions.

All the spectral line elements of the periodic system in different valencies are known and calculated. There is unique correspondence between the periodic system of elements and a system of spectral lines. However, the attempt to assign these new lines to any of elements in any ionization state failed. Similarly, about twenty lines (for instance, the 5303 Å line known from 1870) observed in sun crown spectrum resisted interpretation.

The atomic spectroscopy theory application allowed assignment of these lines to the highly ionized states of elements, for example, to the hydrogen-like (i.e., with a single electron) ions such as C VI, N VII, O VIII, Ne X, helium-like C V, N VI, O VII, highly ionized Fe XVII, Fe XVIII and Fe XXV; the 5303 Å lines was, in particular, assigned to Fe XIV [36]. From these spectra the sun crown temperature was determined as about one million degrees.

2. Crystal Field Theory

The behavior of ions in crystals is a complex phenomenon that cannot be covered by any single theory; various aspects of this phenomenon are described by several theories which supplement each other: the crystal field theory is considered in this chapter, and the molecular orbital theory, valence bond theory and band theory will be considered in the following chapters.

Ions in crystals acquire special individual properties in each compound depending on the type, number and arrangement of the surrounding ions. Thus coordination polyhedra (e.g., octahedra, tetrahedra...) which are used in crystal structure models can also be considered as spectroscopic entities. In polyhedra one distinguishes a central metal ion and, directly bonded with it, negative ions located at the vertices of the polyhedron, referred to as ligand ions or simply ligands.

In the crystal field theory, as distinct from other theories, the ligands are treated as point charges whose role is to create the electrostatic potential that is called the crystalline field. This potential gives rise to the splitting of the central ion energy levels (the mechanism of this splitting will be discussed in the next section, Chap. 2.1). This represents one type of the Stark effect noted in atomic spectroscopy, and thus is also known as Stark splitting.

Since only energy levels of ions with incomplete subshells are subject to splitting, with d and f electrons, crystal field theory is thus the theory of transition metal ion behavior in crystals and molecules.

The main characteristic of the crystal field giving rise to the splitting of d and f electron levels is its symmetry. This corresponds to symmetry in the arrangement of ligand ions and is identical to the point group of symmetry at the central ion site. The type of ligand ions and the distances between the central and ligand ions determine the strength (the intensity) of the crystal field.

The symmetry corresponding to a regular octahedron or tetrahedron occurs in very few structural types. Mostly the polyhedra are distorted and have lower symmetry. Thus for every case of d^n and f^n electron configurations, the characteristics of splitting of the energy levels in crystal fields with different point group symmetry must be determined.

This is carried out by means of group theory. Wave functions must be transformed into symmetry types, i.e., they must represented in group theory language. In this way, the number, symmetry type, and relative energies for energy levels can be derived.

In the same way as fundamental knowledge of the structure of free atoms was obtained from their emission spectra, so information about the atoms in crystals can be obtained from their optical, X-ray, Mössbauer and radio-frequency spectra. Thus, crystal field theory is the application of group theory to electron transition spectroscopy.

Ligand ions are assumed to be the point charges which do not exchange electrons with the central metal ion; this case corresponds to the pure ionic without any covalent contribution. However, there is no obstacle that the experimental parameters (as obtained from optical and other spectra) describing the crystal field effects reveal a dependence on chemical bonding; even in such covalent compounds as sulfide minerals these parameters give quite satisfactory agreement with theoretical predictions.

Taking into account the overlapping of atomic orbitals between central ion and ligands, the crystalline field is referred to as a ligand field; thus the ligand field theory is considered as a molecular orbital theory for cases of ions with unfilled d and f subshells.

The main aim of this chapter is: (1) to provide an outline for applications of the group theory apparatus to transformations of atomic orbitals occurring in crystals and describing electronic behavior in the atom, and of terms describing the total atom's state; all theories of atoms in solids involve in one form or another both the orbital and term symmetry transformations, (2) to present final schemes for the optical absorption spectra interpretation.

The beginning of the crystal field theory (1929) is connected with Bethe's paper [51], which presents the solutions, based on group theory, for the problem of term splitting in crystal fields of different symmetries. Later Van Vleck (1932) used these solutions for interpretations and calculations of magnetic suscpeibility. Only in the 1950s, however, did crystal field theory applications to optical absorption spectra and electron paramagnetic resonance spectra greatly extend their significance for transition metal chemistry, spectroscopy of solids and also for mineralogy and geochemistry.

There are now a number of reviews and monographs which deal with various aspects of the crystal ligand field theory and its applications in chemistry [49, 55–62, 68, 69, 72–74] and in mineralogy [52].

2.1 The Actions of the Crystal Field upon the Atomic Orbitals and Terms

2.1.1 Symmetry of Atomic Orbitals in the Crystal Field; the Concepts of Characters and Irreducible Representations

Two principal models underlie the crystal field theory: one is connected with behavior of atomic d orbitals (see Figs. 26–28), and the second related to the behavior of terms (see Fig. 29). Five four-lobe d orbitals of a free atom with spherical symmetry are by themselves quite equivalent and even their number is determined by the number of the possible states of the electron with the orbital quantum number $l=2=d$, that is $m_l = +2, +1, 0, -2, -2$ (see Table 3).

Nonequivalency of the d orbitals arises in crystals or molecules where ions with d electrons are placed in surroundings of ligands ions.

In the case of octahedral coordination (Fig. 26), i.e., a coordination polyhedra with six oxygen ions (or H_2O, F^-, Cl and other ligands) located on the corners of

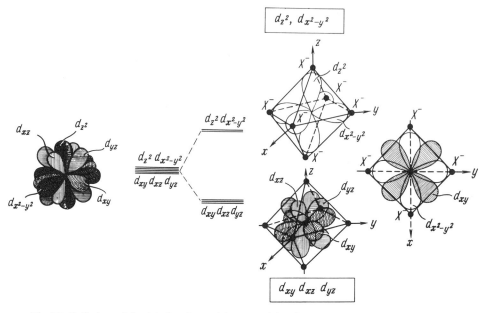

Fig. 26. Splitting of *d* orbitals of transition metal ion in octahedral cristalline field created by ligand ions X⁻

an octahedron, the five *d* orbitals are divided into two groups: (1) three four-lobe orbitals d_{xy}, d_{yz}, d_{xz} pointing between the Cartesian axes (which coincide here with the directions metal – ligands): d_{xy} between x and y axes (in the xy plane), d_{yz} between y and z axes (in the yz plane), and d_{xz} between x and z axes (in the xz plane); (2) two orbitals of the same type but pointing towards the ligands and disposed along the Cartesian coordinate: $d_{x^2-y^2}$ along x and y axes, d_{z^2} along z axis.

The $d_{x^2-y^2}$, d_{z^2} orbitals are subjected to the greater repulsion by the ligands than the d_{xy}, d_{yz}, d_{xz} orbitals. Then the energy level diagram (Fig. 26) shows: (1) the splitting of the *d* orbitals into two groups, (2) the relative energy of these orbitals: the $d_{x^2-y^2}$, d_{z^2} levels are higher in energy than the d_{xy}, f_{yz}, d_{xz} orbitals.

The *d* orbitals in a free atom are referred to as the five-fold degenerated orbitals (i.e., there is a single energy level corresponding to five *d* orbitals with identical energies).

In octahedral environment this degeneracy is partly removed and two energy levels arise: the lower triply degenerated level (i.e., the single level for the d_{xy}, d_{xz}, d_{yz} orbitals) and the higher doubly degenerated one (i.e., the single level for the $d_{x^2y^2}$, d_{z^2} orbitals).

These two sets of orbitals are labeled by different designations which can be compared as follows:

$(d_{x^2-y^2}, d_{z^2})$: e_g or d_γ or γ_3 or Γ_3;

(d_{xy}, d_{yz}, d_{xz}): t_{2g} or d_ε or γ_5 or Γ_5.

Here the symbols e_g and t_{2g} are introduced by Mulliken for symmetry types for a single electron, as γ_3 and γ_5 by Bethe, the Γ_3 and Γ_5 being Bethe's general designations for the same types of symmetry for an electron and for an atom (see also p. 68).

In tetrahedral coordination (Fig. 27), that is with four ligands represented as four point charges at the vertices of a tetrahedron: (1) the d orbitals split into the same two sets: $(d_{x^2y^2}, d_{z^2})$ and $d_{xy}, d_{xz}, d_{yz})$ labeled as in octahedron with the exception that instead of e_g and t_{2g} one writes e and t_2 (i.e., the subscript indicating the behavior of an orbital with respect to the center of symmetry is absent because there is no center of symmetry in the tetrahedron); (2) relative energies of these sets of orbitals are reversed as compared with the octahedron, since here the $d_{x^2y^2}, d_{z^2}$ orbitals become energetically less favorable than the d_{xy}, d_{xz}, d_{yz} orbitals (Fig. 27); (3) energy differences between the e and t_2 levels in the tetrahedron is lower than those between the e_g and t_{2g} levels in the octahedron.

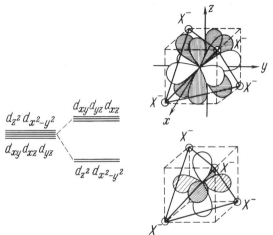

Fig. 27. Splitting of d orbital of transition metal ion in tetrahedral crystalline field created by ligand ions X^-

In cubic coordination, that is with eight ligands at the vertices of a cube (see Fig. 27 and add four more ions at the four vertices of the cube), energy level positions are the same as in the case of the tetrahedron and reversed as compared with the octahedron, but the distance between the levels is twice as much as in the case of the tetrahedron because of the doubling of the ligand ions number acting on the four-lobe d orbitals.

The energy difference between the d_{xy}, d_{xz}, d_{yz} orbitals and the $d_{x^2y^2}, d_{z^2}$ orbitals is designated as Δ or $10\,Dq$ (Fig. 28); zero value of the energy is taken as a mean weight value of the energy of the five d orbitals.

The energies of the orbitals in the octahedral field are:

for e_g: $+3/5\Delta$ or $+6\,Dq$,
for t_{2g}: $-2/5\Delta$ or $+4\,Dq$,

in the tetrahedral field:

for t_2: $+3/5\Delta$ or $+6\,Dq$,
for e: $-2/5\Delta$ or $+4\,Dq$,
but $Dq_{\text{tetr}} = 4/9\,Dq_{\text{oct}}$,

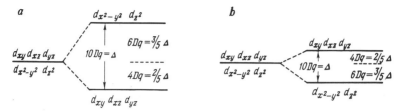

Fig. 28a and b. Comparison of d orbital splitting in a octahedral and b tetrahedral fields

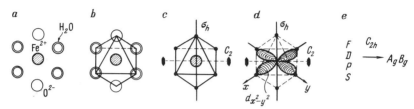

Fig. 29a–e. Origin of the crystal field model. a A portion of crystal structure of vivianite: Fe^{2+} in six-fold coordination of $4H_2O$ and $2O^{2-}$, i.e., $Fe^2(H_2O)_4O_2$. b The same in the form of coordination polyhedron (distorted octahedron). c Ligand ions (two oxygen and four H_2O molecules) are substituted for point charges creating crystalline (electric) field at the Fe^{2+} site with C_{2h} symmetry. d Central transition metal ion is represented by atomic orbitals: $d_{x^2-y^2}$ (shown in the figure), d_{z^2}, d_{xy}, d_{xz}, d_{yz} describing electron states in the C_{2h} field. e Terms S, P, D, F... describing states of free ion transform into irreducible representations A_g, B_g describing states of the whole ion (but not individual electrons) in crystalline field with the C_{2h} point group

in the cubical field the position of the levels is the same as in the tetrahedral field, but

$$Dq_{cub} = 8/9\, Dq_{oct} = 2\, Dq_{tetr}.$$

The strength of the crystal field Dq varies in different complexes and is an important spectroscopic characteristic of the crystal determined from its optical absorption spectrum.

Distortions of the regular octahedra, tetrahedra and cubes lead to a further splitting of the d orbitals' energy levels.

For example, compression of the octahedron (see Fig. 26) along the z axis leads to lesser energy of the $d_{x^2-y^2}$ orbital than the d_{z^2} orbital and to lesser energy of the d_{xz}, d_{yz} orbitals than the d_{xy} orbital. Elongation of an octahedron along the z axis leads to similar splittings but with reversed position of the levels. Splittings of atomic states with a different number of d electrons in the crystal field of different symmetries representing the heart of the crystal field theory are discussed later in Chapter 2.3.

For descriptions of the atomic states (by means of terms) the crystal field model is arranged in the form to be considered by group theory (Fig. 29), i.e., the problem of the behavior of atoms in crystals is reduced to the problem of the general theory of symmetry group.

Symmetry and applications of the theory of symmetry group to the description of atomic states in crystals. The symmetry of the arrangement of crystal faces is described by 32 point groups; the point group is a set of symmetry elements intersecting at a common point. The symmetry of crystal structure, i.e., of an arrangement of atoms in the crystal structure is described by 230 space groups, which involve in addition to normal symmetry axes and planes screw axes and glide planes. However, these space groups do not describe the symmetry of those atoms which are represented in structural crystallography as points or balls, but determine the arrangements of atoms relatively to each another.

Spectroscopic and chemical considerations of the behavior of atoms in crystal structure, that involve the description of symmetry of the very atom and in particular the symmetry of its d orbitals, are selecting those symmetry elements which intersect in the point corresponding to the site of the atom and compose thus the point group for this atom site.

Descriptions of electron and atom states in crystals, systematics of their energy levels, determination of the allowed transitions between the levels in optical absorption and reflection spectra, luminescence and EPR etc. are very closely connected with symmetry.

In the crystal field theory, symmetry is related only to atomic orbitals and atomic states (since in this approximation atoms still keep their individuality) while in the molecular orbital theory, the symmetry operations are related to the molecule as a whole, to the entire AB, AB_2, AB_3, AB_4' complexes, to the formation of these molecules and complexes and to the linear combination of the atomic orbitals (LCAO, see Chap. 3.2).

The symmetry operations in crystal field theory are related to the descriptions (1) of the one-electron s, p, d, f orbital states and (2) of the whole-atom states represented by the terms.

The one-electron s, p, d, f orbitals in this approximation retain in crystals the same forms but acquire the orientation (p_x, p_y, p_z and $d_{xy}, d_{xz}, d_{yz}, d_{x^2-y^2}, d_{z^2}$) and various energies in dependence on their position with respect to the ligands; they transform into the one-electron states designated by the symmetry types: $a_1, a_2, b_1, b_2, e, t_1, t_2$ ... Multielectron atom states described by the terms S, P, D, F ... are split in crystals and transform into the states designated by the symmetry types: $A_1, A_2, B_1, B_2, E, T_1, T_2$...

Similarly one considers not only electron states of atoms in crystals but also vibrational and rotational states of atoms described by the same symmetry types and determining infrared, Raman and microwave spectra of crystals.

Thus, the following designations can be compared.

Electron states:

s, p, d, f: one-electron orbitals of free atoms; spherical symmetry;

σ, π, Δ: one-electron molecular orbitals of linear molecules; axial symmetry;

$a_1, a_2, b_1, b_2, e, t_1, t_2$: one-electron orbitals of atoms in crystals in crystal field theory approximation or one-electron molecular orbitals (a single unpaired electron in free radicals etc); different point groups of symmetry.

Atom and molecule states:

S, P, D, F: free atom terms; spherical symmetry (representations of infinite grous of rotation; S: one-dimensional; P three-dimensional; D: five-dimensional etc.);

$\Sigma, \Pi, \Delta, \Phi \ldots$: molecular orbitals of linear molecules; axial symmetry (representations of inifinite rotation group; Σ: one-dimensional; others: two-dimensional);

$A_1, A_2, B_1, B_2, E, T_1, T_2$: atomic states in crystal field theory approximation or molecular orbitals (see Chap. 3.2) in different point groups of symmetry (representations of finite groups; A, B: one-dimensional; E: two-dimensional; T: three-dimensional).

All the applications of symmetry groups (to description of crystal forms, of crystal structure, of atomic electron states, vibrational and rotational states and so on) are the particular cases of the general group theory [53, 54, 58, 63, 64, 66, 67, 69, 71]; some necessary principles of group theory are considered below.

Coordination Number and Local Symmetry. The most common coordination for transition metal ions is the octahedral, less common coordinations are the tetrahedral and cubic (often for trace quantities of these ions detected by EPR). However, octahedra, tetrahedra, and cubes in most structures are designations only for coordination, but do not correspond to the regular forms of these polyhedra and their actual symmetry. More or less significant distortions lead to lowering of their symmetry and consequently to lowering of the crystal field symmetry (i.e., of the local symmetry). Identity of the crystal field symmetry and of point group symmetry in a given ion site is rigorously preserved. Thus, in cubic garnets and spinels only one three-fold axis C_3 passes through Fe, Mn, and Cr ions in octahedral coordination, and thus local symmetry is lowered to trigonal (and no lower than trigonal) and the octahedra are distorted along the C_3 axis.

Only in some simple structures (NaCl–MgO, CaF_2, CsCl, ZnS) are there sites with cubic symmetry and only in these cases does full octahedral, tetrahedral or cubic symmetry exist.

In tetragonal, trigonal and hexagonal crystals the local symmetry cannot be higher than tetragonal and trigonal respectively (the C_6 axis does not occur among the point groups of local symmetry of inorganic crystals) and in the crystals of lower symmetry the local symmetry is not higher than ortho-rhombic or monoclinic. The different point group designations are given in Table 9.

In models of rock-forming silicate structures (olivines, pyroxenes, amphiboles, micas etc.) the octahedra and tetrahedra used are in fact always distorted to rhombic, monoclinic and triclinic local symmetries.

In close packing structures the distortions perpendicular to the close packing plane lead to octahedra compressed along the C_3 axis (because only cations can only move anions apart forming the close packing).

The octahedra have for example in calcite the C_{3v} symmetry, in beryl D_{3d}, in corundum C_3, in amphiboles and micas C_{2h} and C_2, in vivianite (see Fig. 29) C_{2h} and C_2.

Table 9. Comparison of the some symmetry point group designations

Schönflis	Hermann-Mauguin	Geometrical crystallography	Detailed designations used in group theory
Cubic symmetry			
O_h	$m3m$	$4L_6^3 3L_4 6L_2 9PC$	$8C_3 3C_2 6C_4 6C_2' i 8iC_3 3iC_2 6iC_4 6iC_2$
T_d	$\bar{4}3m$	$4L_3 L_4^2 6P$	$8C_3 3C_2 6S_4 6\sigma_d$
Trigonal symmetry			
D_{3d}	$\bar{3}m$	$L_6^3 3L_2 3PC$	$2C_3 3C_2 i 2S_6 3\sigma_d$
$[D_{3h}$	$\bar{6}m2$	$L_3 3L_2 4P$	$2C_3 3C_2 \sigma_h 2S_3 3\sigma_v]$
D_3	32	$L_3 3L_2$	$2C3C_2$
C_{3v}	$3m$	$L_3 P$	$2m 3\sigma_v$
$[C_{3h}$	$\bar{6} = 3/m$	$L_3 P$	$C_3 C_3^2 \sigma_h S_3 S_3^2]$
C_3	3	L_3	$C_3 C_3^2$
$C_{3i} = S_6$	$\bar{3}$	$L_6^3 C$	$C_3 C_3^2 i S_6^5 S_6$
Tetragonal symmetry			
D_{4h}	$4/mmm$	$L_4 4L_2 5PC$	$2C_4 C_2 2C_2' C_2'' i 2S_4 \sigma_h 2\sigma_v 2\sigma_d$
D_4	422	$L_4 4L_2$	$2C_4 C_4^2 = C_2 2C_2' 2C_2''$
C_{4v}	$4mm$	$L_4 4P$	$2C_4 C_2 2\sigma_v 2\sigma_d$
C_{4h}	$4/m$	$L_4 PC$	$C_4 C_2 C_4^3 i S_4^3 \sigma_h S_4$
C_4	4	L_4	$C_4 C_2 C_4^3$
S_4	$\bar{4}$	L_4^2	$C_4 C_2 S_4^3$
$D_{2d} = V_d$	$\bar{4}2m$	$L_4^2 2L_2 2P$	$2S_4 C_2 2C_2 2\sigma_d$
Rhombic symmetry			
$D_{2h} = V_h = D_{2v}$	mmm	$3L_2 3PC$	$C_2(z) C_2(y) C_2(x) i\sigma(xy) \sigma(xz) \sigma(yz)$
$D_2 = V$	222	$3L_2$	$C_2(z) C_2(y) C_2(x)$
C_{2v}	$mm2$	$L2P$	$C_2 \sigma_v(zx) \sigma_v(zy)$
Monoclinic symmetry			
C_{2h}	$2/m$	$L_2 P = V$	$C_2 \sigma_h i$
C_2	2	L_2	C_2
$C_s = C_h$	m	P	σ_h
Triclinic symmetry			
C_τ	$\bar{1}$	C	i
C_1	1	L'	C_1

The Action of the Symmetry Operations of a Point Group upon the Atomic Orbitals: the Concept of the Characters and Irreducible Representatives ("the Types of Symmetry"). Consider the action of the symmetry operations by the example of the two point groups, C_{2h} (with the nondegenerate types of symmetry) and C_{4v} (with the degenerate types of symmetry).

1. Given are (a) the C_{2h} group (see Fig. 30) containing two-fold axis C_2, the symmetry plane σ_h perpendicular to C_2, the symmetry center, the identity operation I (its meaning is clarified later by the example of the C_{4v} group) and (b) the atomic orbitals d_{xy}, d_{xz}, d_{yz}, $d_{x^2-y^2}$, d_{z^2}.

The full procedure requires the treatment of the actions of each symmetry operation in the point group for each orbital. Figure 30 illustrates this only for d_{xy} and $d_{x^2-y^2}$, since for the other d orbitals, as well as for p and s orbitals, the operations are similar.

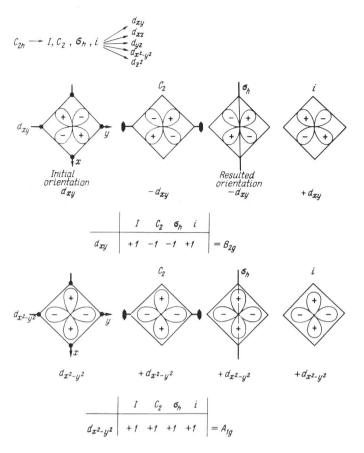

Fig. 30. Action of symmetry operation C_2, σ_h, i of the point group C_{2h} upon atomic orbitals $d_{x^2-y^2}$ and d_{xy} (Determination of B_{2g} and A_{1g} symmetry types). Symmetry types (=irreducible representations) of the C_{2h} point group: with respect to C_2: symmetric (+1) A; antisymmetric (−1) B; with respect to σ_h: symmetric (+1) $A_1 B_1$; antisymmetric (−1) A_2, B_2; with respect to i: symmetric (+1) $A_g B_g$; antisymmetric (−1) $A_u B_u$

The initial orientation of the orbitals depicted in Figure 30 differs from that obtained by the actions of the C_2 symmetry operations only in sign, i.e., the atomic orbitals retaining their notations change signs: the C_2 operation converts d_{xy} into $-1 d_{xy}$ and d_{x-y} into $+1 d_{x^2-y^2}$. Hence, the symmetry operation here may be termed singly: either the atomic orbital is symmetric with respect to this particular operation (factor +1), or antisymmetric (factor −1). The behavior of a given orbital with respect to a given symmetry operation is written in the form of the character tables for a given point group (for C_{2h} see Table 10).

The characters are the numbers (for C_{2h} +1 or −1) describing the relation between the initial and resultant atomic orbitals under the action of a given symmetry operation. They indicate the effect of given symmetry operation upon the atomic orbital.

Symmetry of Atomic Orbitals, Characters and Irreducible Representations

Table 10. Characters of the C_{2h} point group

C_{2h}	I	C_2	σ_h	i	
A_g	+1	+1	+1	+1	$d_{x^2-y^2}d_{xz}d_{yz}d_{z^2}s$
A_u	+1	+1	−1	−1	p_z
B_u	+1	−1	+1	−1	$p_x p_y$
B_g	+1	−1	−1	+1	d_{xy}

Table 11. Characters of the C_{4v} point group

C_{4v}	I	$2C_4$	C_2	$2\sigma_v$	$2\sigma_d$	
A_1	+1	+1	+1	+1	+1	$d_{z^2} p_z$
A_2	+1	+1	+1	−1	−1	—
B_1	+1	−1	+1	+1	−1	$d_{x^2-y^2}$
B_2	+1	−1	+1	−1	+1	d_{xy}
E	+1	0	−2	0	0	d_{xz}, d_{yz}, p_x, p_y

The set of characters (horizontal rows in Table 10) describing the effects of all symmetry operations of the point group upon a given orbital determines the type of symmetry of the orbital, or irreducible representation (this term, taken from the mathematical theory of groups, is more common). For instance, d_{xy} characters in C_{2h} (see Fig. 30 and Table 10) are +1, −1, −1, +1, corresponding to the effects of the I, C_2, σ_h, i symmetry operations, and hence d_{xy} type of symmetry in C_{2h} (d_{xy} irreducible representation in C_{2h}) is labeled as B_g in Table 10 (see the notation of the irreducible representations below).

Thus, the irreducible representations appear to be a new means of labeling (A_g, A_u, B_g, B_u in Table 10) of the electron and atom states in the crystalline field used here instead of the free atom electron orbitals (s, p_x, p_y, p_z, d_{xy}...) and terms (S, P, D, F...). They represent the symmetry behavior either of the orbital or of the terms in the crystalline field of a given symmetry. States of electrons and atoms in the crystalline environment of a certain symmetry are determined assigning them to the corresponding irreducible representations.

The number of symmetry types in the given point group depends on the possible number of character combinations. Of the three symmetry elements in the C_{2h} group (C_2, σ_h, i) any two are independent, the third one being their resultant. Hence, we can obtain four types of symmetry:

C_2	σ_h	
+	+	A_{1g}
+	−	A_{2u}
−	+	B_{1u}
−	−	B_{2g}

2. Consider the same treatment for the group C_{4v}. The elements of the symmetry of this group are listed in Table 11. The actions of the three symmetry elements upon the d_{xz} orbital are depicted in Figure 31. For some orbitals, such as d_{z^2}, $d_{x^2-y^2}$, d_{xy}, p_z (see Table 11) the results of the symmetry operations are

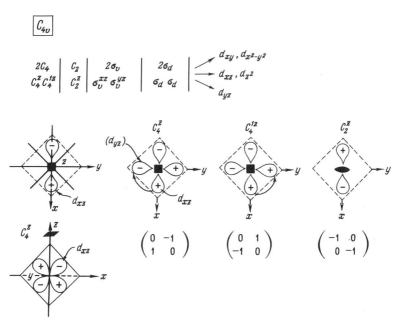

Fig. 31. Action of symmetry operation C_4^z, C_4^{1z}, C_2^z of the point group C_{4v} upon atomic orbital d_{xy} (determination of the symmetry type E)

described similarly to the preceding group, by the single-termed determination: symmetric ($+1$) or anti-symmetric (-1). However, other orbitals (d_{xy}, d_{yz}, p_x, p_y) fail to be defined singly. Thus, the rotation of the d_{xz} orbital about the C_4^z at $90°$ brings it (Fig. 31) into a position which can neither be termed symmetric nor antisymmetric; the operation C_4^z here mixes the d_{xz} and d_{yz} orbitals. The characters for the orbitals d_{xz} with the d lobes in xz plane can then be obtained by the following procedure:

a) Express the symmetry operation through the linear conversion of the coordinates. The clockwise rotation about C_4^z converts x into y and y into x:
a new sense x (denoted as x') arises from former y

$$x' = 0 \cdot x + 1 \cdot y$$

a new sense y (denoted as y') arises from former x

$$y' = -1 \cdot x + 0 \cdot y.$$

b) Written in the matrix form:

$$\begin{matrix} 0 & 1 \\ -1 & 0 \end{matrix}$$

c) The trace of this matrix (the sum of the elements on the principal diagonal top left to bottom right) defines the character of the d_{xz} orbital in the C_{4v} group under the C_4^z rotation, in this case being zero.

Symmetry of Atomic Orbitals, Characters and Irreducible Representations 67

Table 12. Transformation matrices and characters of the d_{xz} orbital

Symmetry operation	I	C_4^z	C_4^{1z}	C_2^z	$\sigma_v^{x\pi}$	$\sigma_v^{y\pi}$	σ_d	σ_d
Transformation matrices	$\begin{pmatrix}1 & 0\\0 & 1\end{pmatrix}$	$\begin{pmatrix}0 & -1\\1 & 0\end{pmatrix}$	$\begin{pmatrix}0 & 1\\-0 & 1\end{pmatrix}$	$\begin{pmatrix}-1 & 0\\0 & -1\end{pmatrix}$	$\begin{pmatrix}-1 & 0\\-1 & 0\end{pmatrix}$	$\begin{pmatrix}1 & 0\\0 & -1\end{pmatrix}$	$\begin{pmatrix}-1 & 0\\0 & 1\end{pmatrix}$	$\begin{pmatrix}1 & 0\\0 & -1\end{pmatrix}$
Characters	2	0	0	-2	0	0	0	0

The same treatment is applied for the rotation of the d_{xz} orbital about the C_2^z (at 180°):

$$\begin{array}{ll} x \to -x & x' = -1 \cdot x + 0 \cdot y \\ y \to -y & y' = 0 \cdot y + -1 \cdot y \end{array}$$

$$\begin{array}{cc} -1 & 0 \\ 0 & -1 \end{array}$$

and the character is -2.

The identity symmetry operation I leaves the d_{xz} orbital unchanged, i.e., converts it into itself:

$$\begin{array}{ll} x \to y & x' = 1 \cdot x + 0 \cdot y \\ y \to y & y' = 0 \cdot x + 1 \cdot y \end{array}$$

$$\begin{array}{cc} 1 & 0 \\ 0 & 1 \end{array}$$

and the character is 2.

The set of these transformation matrices (see Table 12 for the d_{xz} orbital) constitutes the set of the characters ($=$ of the traces of the matrices) and determines the irreducible representation of the orbital in the given group (for d_{xz} in C_{4v} see the last row in Table 11; the representation is labeled there as E, the characters being $+2, 0, -2, 0, 0$).

Having summarized the data obtained we come to understand the following: (1) why three symmetry axes, C_4^z, $C_4^{'z}$ and C_2^z, coinciding with the axes z are distinguished (transformation matrices for C_4^z and $C_4^{'z}$ are different, C_4^z and $C_4^{'z}$ correspond to the clockwise and anticlockwise rotation respectively, their characters remaining identically zero, and thus are united into $2C_4$, the character for C_2^z being -2); (2) why σ_v^{xz} and σ_v^{yz} (and $2\sigma_d$) differ; (3) what the identity operation involves: it determines the number of the interconnected coordinates: unchanged coordinates correspond to character 1, two mixed coordinates – character 2, three coordinates – character 3; this number also indicates the degree of degeneracy: if the orbital transforms into itself, it is a one-dimensional representation, nondegenerate state; two mixed orbitals correspond to two-dimensional representation, doubly degenerate state; three orbitals produce three-dimensional representation, triply degenerate state.

The presence of the degenerate types of symmetry distinguishes C_{4v} group as well as all the groups with the symmetry axes higher than the two-fold C_2 axis (all the groups of the trigonal, tetragonal and cubic systems) from the groups of the lower symmetry systems having a single C_2 axis and comprising only nondegenerate types.

The doubly degenerate E symmetry type mixes two states d_{xz} and d_{yz} in the C_{4v} group ($d_{x^2-y^2}$ and d_{z^2} in the octahedral O_h group). The triply degenerate T symmetry type mixes three states existing only in the cubic system, where d_{xy}, d_{xz}, d_{yz} orbitals are mixed.

For example, two states connected by the C_4 or C_3 symmetriy axis (d_{xz} and d_{yz} in the C_{4v} group) are doubly degenerated.

With the help of the character tables available in many published works [54, 63, 66, 69, 75] it is possible to determine to which particular type of symmetry the states belong, as well as to determine the selection rules for the transition between these states.

The notations of the irreducible representations (=types of symmetry) are as follows:

1. With respect to the degree of degeneracy: (*A*) nondegenerate one-dimensional type: the rotation about the principal axis does not change the sign of an orbital (character $+1$); (*B*) nondegenerate one-dimensional type; the rotation about the principal axis changes the sign of a orbital (character -1); (*E*) doubly degenerate type of symmetry (two-dimensional representation); (*T*) triply degenerate type of symmetry (three-dimensional representation).

2. With respect to the rotation about the C_2 axis perpendicular to the principal C_n axis and to the rotation in the σ_v plane containing the principal C_n axis:

A_1, B_1, T_1: symmetric with respect to $C_2 \perp C_n$ or σ_v;
A_2, B_2, T_2: antisymmetric with respect to $C_2 \perp C_n$;

with respect to the reflection in the σ_h plane perpendicular to the principal axis

A', B': symmetric with respect to σ_h (character $+1$);
A'', B'': antisymmetric with respect to σ_h (character -1);

with respect to the center of symmetry:

$A_{1g} A_{2g} B_{1g} B_{2g} E_g T_{1g} T_{2g}$: even;
$A_{1u} A_{2u} B_{1u} B_{2u} E_u T_{1u} T_{2u}$: uneven;

(g from the German "gerade" meaning even, and u for "ungerade", uneven); if the group has no symmetry center, subscripts g and u are lacking; s and d orbitals are even ($+$ and $-$ lobes of the d orbitals are centrosymmetric): p-orbitals are odd (two-lobe p orbitals have one half of the orbital positive, the other negative, and so are noncentrosymmetric). Therefore in the centrosymmetric crystal field, the representations of s and d orbitals and the terms derived from them have subscript g, and those of p orbitals and their derived terms have subscript u.

A_1 is a totally symmetric representations, as it is symmetric with respect to all the symmetry operations (all the characters being $+1$).

The small letters denote the representation of the single-electron orbitals ($a_1 a_2 b_1 b_2 ...$), the same capital letters are used for the representations of the terms ($A_1, A_2, B_1, B_2 ...$).

There are two alternative types of labeling:

$A_1 \; A_2 \; E \; T_1 \; T_2$ – Mulliken,
$\Gamma_1 \; \Gamma_2 \; \Gamma_3 \; \Gamma_4 \; \Gamma_5$ – Bethe,
1   1   2   3   3 – degree of degeneracy.

2.1.2 Correlation Tables for the Symmetry Types in Various Point Groups

The transformations of the one-electron s, p, d, f orbitals are easy derived by means of the simple pictorial model as made for d orbitals in Figures 30 and 31. For terms (describing states of the whole atoms) there is no such model to derive symmetry types (cf. Fig. 29d and e).

Table 13. Comparison of splitting of s, p, d, f orbitals and S, P, D, F states of atoms in the octahedral crystal field

Electron states					Atomic states			
	l	2^{l+1}	O_h	Degeneracy	L	2^{L+1}	O_h	Degeneracy
s	0	2	a_{1g}	1	S	0	1	A_{1g}	1
p	1	3	t_{1u}	3	P	1	3	T_{1u}	3
d	2	5	$e_g + t_{2g}$	2+3	D	2	5	$E_g + T_{2g}$	2+3
f	3	7	$a_{2g} + t_{1g} + t_{2g}$	1+3+3	F	3	7	$A_{2g} + T_{1g} + T_{2g}$	1+3+3

Table 14. Symmetry types (irreducible representations) for S, P, D, F term transformations in different symmetry point groups

		O_h	$D_{3d} D_3 C_{3v} C_3 S_6$	$D_{4h} D_4 C_{4v} C_{4h} C_4 S_4$	$D_{2v} D_2 C_{2v}$	$C_{2h} C_2 C_s$	C_i
S		A_{1g}	$A_{1g} A_1 A_1\ A\ A_g$	$A_{1g} A_1 A_1\ A_g\ A\ A$	$A_g\ A\ A_1$	$A_g\ A\ A'$	A_g
P	T_{1g}		$A_{2g} A_2 A_2\ A\ A_g$ $E_g\ E\ E\ E\ E_g$	$A_{2g} A_2 A_1\ A_g\ A\ A$ $E_g\ E\ E\ E_g\ E\ E$	$B_{3g} B_3 B_1$ $B_{2g} B_2 B_2$ $B_{1g} B_1 A_1$	$B_g\ B\ A'$ $A_g\ A\ A''$ $A_g\ A\ A''$	A_g A_g A_g
D	T_{2g}		$A_{1g} A\ A_1\ A\ A_g$ $E_g\ E\ E\ E\ E_g$	$B_{2g} B_2 B_2\ B_g\ B\ B$ $E_g\ E\ E\ E_g\ E\ E$	$B_{1g} B_1 A_2$ $B_{2g} B_2 B_1$ $B_{3g} B_3 B_2$	$A_g\ A\ A'$ $B_g\ B\ A''$ $B_g\ B\ A''$	A_g A_g A_g
	E_g		$E_g\ E\ E\ E\ E_g$	$A_{1g} A_1 A_1\ A_g\ A\ A$ $B_{1g} B_1 B_1\ B_g\ B\ B$	$A_g\ A\ A_1$ $A_g\ A\ A_1$	$A_g\ A\ A'$ $A_g\ A\ A'$	A_g A_g
F	A_{2g}		$A_{2g} A_2 A\ A\ A_g$	$A_{2g} A_2 A_2\ A_g\ A\ A$	$B_{1g} B_1 A_2$	$A_g\ A\ S'$	A_g
	T_{2g}		$A_{1g} A_1 A_1\ A\ A_g$ $E_g\ E\ E\ E\ E_g$	$A_{1g} A_1 A_1\ A_g\ A\ A$ $E_g\ E\ E\ E_g\ E\ E$	$A_g\ A\ A_1$ $B_{2g} B_2 B_1$ $B_{3g} B_3 B_2$	$A_g\ A\ A'$ $B_g\ B\ A''$ $B_g\ B\ A''$	A_g A_g A_g
	T_{1g}		$A_{2g} A_2 A_2\ A\ A_g$ $E_g\ E\ E\ E\ E_g$	$A_{2g} A_2 A_2\ A_g\ A\ A$ $E_g\ E\ E\ E_g\ E\ E$	$B_{1g} B_1 A_2$ $B_{2g} B_2 B_1$ $B_{3g} B_3 B_3$	$A_g\ A\ A'$ $B_g\ B\ A''$ $B_g\ B\ A''$	A_g A_g A_g

1. In T_d group (tetrahedral) there are the same symmetry types as in O_h except of omission of "g" sign.
2. Term G in O_h transforms into symmetry types A_{1g}, E_g, T_{1g}, T_{2g}; term H into E_g, $2T_{1g}$, T_{2g}; term I into A_{1g}, A_{2g}, E_g, T_{1g}, $2T_{2g}$; their further splittings are identical with those of A_{1g}, A_{2g}, E_g, T_{1g}, T_{2g}.
3. Spin multiplicity of free atoms retains in symmetry types derived from corresponding terms; for example, ^6S in O_h transform into $^6A_{1g}$ while ^3F into $^3A_{2g}$, $^3T_{1g}$, $^3T_{2g}$ etc.

Table 15. Symmetry type correlations for cubic, trigonal, and tetragonal symmetries

Cubic	Trigonal	Tetragonal
A	A	A
E	E	$A+B$
T_1	$A+E$	$A+E$
T_2	$A+E$	$A+E$

However the $S, P, D, F\ldots$ terms represent only the designations by letters of the whole atom orbital quantum number, L being 0, 1, 2, 3... as the s, p, d, f orbitals represent the designations by letters, the electron orbital quantum number l being 0, 1, 2, 3.... The degree of degeneracy for one d electron ($l=2=d$) being 5 (i.e., $2l+1=5$) corresponds to the five m_l values (i.e., 2, 1, -1, -2 or $d_{xy}, d_{xz}, d_{yz}, d_{x^2-y^2}, d_{z^2}$). Similarly atomic states described by the D term ($L=2=D$) have five-fold degeneracy ($2L+1=5$), corresponding to the five M_L values (i.e., 2, 1, 0, -1, -2). Thus the D term splittings are completely identical to those for the one-electron d orbitals. The same identity exists for the splitting shcemes for p orbitals and P atomic states, for s orbitals and S terms, for f orbitals and F terms (see Table 13).

The term splittings in the crystal fields of most important point groups are given in Table 14 (this table can be used also for one-electron orbitals: it is sufficient to replace the small letters by capitals).

The $S, P, D, F\ldots$ term transformations (Table 14) does not depend on the electron configuration from which the terms were derived (i.e., from f^n, d^n, p^n or s^n configurations). For the terms derived from p^n configurations we need only to change the subscript g (where it is i.e., in centrosymmetrical point groups) for u since s, d, f orbitals are even and p is odd.

The number of symmetry types and their degeneracy are the same for all point groups of a given system (i.e., of cubic, trigonal, tetragonal, rhombic, monoclinic triclinic symmetries); in various groups the symmetry types differ only by the designations indicating existence of the center of symmetry (the g or u subscripts are dropped in noncentrosymmetrical groups), relation to symmetry planes parallel or perpendicular to the principal C_n axis, relation to the C_2 axis perpendicular to the principal C_n axis.

The symmetry type correlations for cubic, trigonal and tetragonal symmetries are given in Table 15; for $D_{3d}, S_6, D_{4h}, C_{4h}$ groups the terms transform as A_g and E_g, in D_3, C_{3v}, D_4, C_{4v} as $A_1, A_2, (B_1 B_2)$, in C_3, C_4, S_4 as $A(B)$.

2.1.3 Selection Rules Related to Symmetry Types

From the correlation table of the symmetry types (or from the character tables) it is possible to determine between which of the atomic states, represented by the symmetry types, transitions giving rise to optical absorption spectra are allowed and in which directions they are polarized. This can be derived by means of the selection rules obtained as follows.

Selection Rules Related to Symmetry Types

Table 16. Rules for the direct product of symmetry types evaluation (after [59], p. 139)

$A \times A = A$	$B \times A = B$	$B \times A = E$	$T \times A = T$	$g \times g = g$	$' \times ' = '$	$1 \times 1 = 1$
$\times B = B$	$\times B = A$	$\times B = E$	$\times B = T$	$u \times u = u$	$' \times '' = ''$	$1 \times 2 = 2$
$\times E = E$	$\times E = E$	$\times E =$ ᵃ	$\times E = T_1 + T_2$	$u \times g = u$	$'' \times ' = ''$	$2 \times 1 = 2$
$\times T = T$	$\times T = T$	$\times T = T_1 + T_2$	$\times T =$ ᵇ	$u \times u = g$	$'' \times '' = '$	$2 \times 2 = 1$ ᶜ

ᵃ O_h, T_d, C_{3v}: $E_1 \times E_1 = E_2 \times E_2 = A_1 + A_2 + E_2$; $E_1 \times E_2 = E_2 \times E_1 = B_1 + B_2 + E_1 C_{4v}$, D_4 etc.
  $E \times E = A_1 + A_2 + B_1 + B_2$.
ᵇ $T_1 \times T_2 = T_2 \times T_2 = A_1 + E + T_1 + T_2$; $T_1 \times T_2 = T_2 \times T_1 = A_2 + E + T_1 + T_2$.
ᶜ D_{2h}, D_2: $1 \times 2 = 3$; $2 \times 3 = 1$; $1 \times 3 = 2$.

Optical transitions between two states: ground state with the energy E_1 and the wave function ψ_1, and an excited state (with E_2^* and ψ_2^*) are described by means of the two conditions:

1. Transition frequency v being equal to energy difference of these states

$$h\nu = E_2'' - E_1.$$

2. Intensity of the transition is determined as transition moment by the wave functions of the ground and excited states ψ_1 and ψ_2^* and by the dipole moment operator P:

$$I = \int \psi_1 P \psi_2^* d\tau.$$

The integral $\int \psi_1 P \psi_2^* d\tau$ determines final absorption intensity for the given transition. If the integral is zero, there is no transition between given states, i.e., this transition is forbidden; if the integral is nonzero, the transition is allowed.

Therefore, consider first if the transition between the two states is allowed, i.e., $\int \psi_1 P \psi_2^* d\tau \neq 0$ or forbidden, i.e., $\int \psi_1 P \psi_2^* d\tau = 0$.

Translate these expressions into group theory language:

1. Transform the ψ_1 and ψ_2^* states (they can be one-electron $d_{xy}, d_{x^2-y^2}$... wave functions or many-electron atom states S, P, D, F...) to irreducible representations (=symmetry types) for a given point group (see Table 14) $A_1, A_2, B_1, B_2, E, T_1, T_2$.

2. Expand the transition moment vector P into components along the x, y, z coordinates ($P \rightarrow P_x + P_y + P_z$) and transform also these components as p_x, p_y, p_z orbitals (see, for example Tables 10 and 11) into irreducible representations of the same point group[2].

3. Obtain the product of the irreducible representations (so-called direct product) of $\psi_1 \psi_2^*$ and P, which in its turn is also one of the irreducible representations of the same point group; the evaluation of the direct product can be made by two ways: (a) by means of multiplication of characters ($+1$ and -1, $+2, 0$ and -2, see Tables 10 and 11) for each of the symmetry operations for $\psi_1 \psi_2^*$ and P or (b) by means of multiplication of the very irreducible representations with the aid of simple rules for the evaluation of direct products (see Table 16).

---

[2] This is for electric dipole transitions (spectra of the ions of the iron group are mostly electric dipole transitions). Quadrupole transition components $Q_{xx}, Q_{yy}, Q_{zz}, Q_{xy}, Q_{yz}, Q_{xz}$ transform like the coordinates $x^2, y^2, z^2, xy, yz, xz$ ($d_{x^2-y^2}, d_{z^2}$ orbitals). Magnetic dipole transition components $\mu_x \mu_y \mu_z$ transform like yz, xz, xy coordinates ([29], p. 110).

Table 17. Character able for C_{2v} point group

C_{2v}	I	$C_2(z)$	$\sigma_v(xz)$	$\sigma_v(yz)$	
A_1	+1	+1	+1	+1	$d_{z^2}, d_{x^2-y^2}, p_z$
A_2	+1	+1	−1	−1	d_{xy}
B_1	+1	−1	+1	−1	d_{xz}, p_x
B_2	+1	−1	−1	+1	d_{yz}, p_y

4. If $\psi_1 \cdot P \cdot \psi_2^* \neq A_1$ the integral $\int \psi_1 \cdot P \psi_2^* d\tau = 0$ and the corresponding transition is forbidden; if $\psi_1 \cdot P \cdot \psi_2^* = A_1$ the integral $\int \psi_1 \cdot P \cdot \psi_2^* d\tau \neq 0$ and the transition is allowed. It means that only if the direct product of the three irreducible representations, to which $\psi_1 \psi_2^*$ and P transform, is the irreducible representation A_1, than the integral determining intensity of a transition is nonzero and the transition is allowed.

The irreducible representation A_1 (or A_g, A_1', A in other groups) is a special one: it is the only totally symmetric representation, i.e., it is symmetric with respect to all the symmetry operations, and all its characters being +1. However, if this direct product $\psi_1 \cdot P \cdot \psi_2^*$ does not belong to A_1 then (see for example Tables 10–12) one of the characters will always be −1 (i.e., there will be a symmetry operation connecting a state with the similar one but of opposite sign) and the integrand will be zero.

Consider two examples of the derivation of selection rules in the C_{2v} point group. There are two variants of the problem:

a) Given are the characters of this point group (Table 17); the character table could be derived by the same procedure as was done for the C_{2h} group (see Fig. 30); character tables for all the point groups are available also from a number of works [54, 63, 69, 75].

b) Given are the irreducible representations of this group (see Table 14).

1. Determine the direct products for all the pairs of the irreducible representations: $A_1 \times A_2$, $A_1 \times B_1$, $A_1 \times B_2$, $A_2 \times A_1$, and so on (it corresponds to $\psi_1 \cdot \psi_2^*$ from the integral $\int \psi_1 \cdot P \cdot \psi_2^* d\tau$).

Obtain, for example, the direct product for $B_1 \times A_2$. If the initial data are represented as the character table, then for each of the symmetry operations I, $C_2(z)$, $\sigma_v(xz)$, $\sigma_v(yz)$ multiply the corresponding characters of B_1 and characters of A_2:

	I	$C_2(z)$	$\sigma_v(xz)$	$\sigma_v(yz)$
$B_1 \times A_2$	$(+1)\times(+1)=+1$	$(-1)\times(+1)=-1$	$(+1)\times(-1)=-1$	$(-1)\times(-1)=+1$

The characters obtained +1, −1, −1− +1 correspond (see Table 17) to the B_2 irreducible representation, hence $B_1 \times A_2 = B_2$. Write down the result in the Table 18. In the same way obtain the rest of the direct products.

As for the initial data in form of irreducible representations (the case b) this result can be readily obtained with the aid of the rules for the evaluation of direct products (Table 16).

Selection Rules Related to Symmetry Types

Table 18. Direct products for point group C_{2v}

C_{2v}	A_1	A_2	B_1	B_2	
A_1	A_1	A_2	B_1	B_2	$P_x \to B_1$
A_2	A_2	A_1	B_2	B_1	$\times P_y \to B_2$
B_1	B_1	B_2	A_1	A_2	$P_z \to A_1$
B_2	B_2	B_1	A_2	A_1	

Table 19. Selection rules for point group C_{2v}

C_{2v}	A_1	A_2	B_1	B_2
A_1	z	—	x	y
A_2	—	z	y	x
B_1	y	y	z	—
B_2	y	x	—	z

2. Find from the character table (see Table 18) that the transition moment components transform in the C_{2v} point group to the following irreducible representations:

$$P_x - B_1,$$
$$P_y - B_2,$$
$$P_z - A_1.$$

3. Multiply each of the direct products of the irreducible representations of ψ_1 and ψ_2^* (i.e., the results given in Table 18) by B_1 (to which P_x transforms), by B_2 (P_y) and $A_1(P_z)$.

Those of the products of the three irreducible representations ($\psi_1 \cdot \psi_2^* \cdot P$) which prove to be A_1, will correspond to allowed transitions that are polarized along the definite direction: along the x axis if A_1 results from multiplication by $P_x(B_1)$, along the y axis if A_1 is obtained from multiplication by $P_y(B_2)$ and along the z axis if A_1 is obtained from multiplication by $P_z(A_1)$. However this multiplication by $P_x(B_1), P_y(B_2), P_z(A_1)$ is not necessary if we note that A_1 can be obtained only by multiplication of an irreducible representation by itself. Then from Table 18 we find the products ($\psi_1 \cdot \psi_2^*$) equal to B_1: when multiplied by $P_x(B_1)$ they become equal to A_1. Therefore the transitions $B_1 \to A_1$, $B_2 \to A_2$, $A_1 \to B_1$, $A_2 \to B_2$ are allowed along the x axis direction. The products equal to B_2, when multiplied by $P_y(B_2)$ give A_1, and correspond to the transitions $A_1 \to B_2$, $A_2 \to B_1$, $B_1 \to A_2$, $B_2 \to A_1$ allowed along the y direction and the products equal to $A_1(P_z - A_1) = A_1$ correspond to the transitions allowed along the z axis direction ($A_1 \to A_1$, $A_2 \to A_2$, $B_1 \to B_1$, $B_2 \to B_2$).

Table 19 gives the selection rules for the C_{2v} group. This table indicates the directions along which the transitions are allowed between the states determined by given irreducible representations.

Table 20. Direct products of point group C_{4v}

C_{4v}	A_1	A_2	B_1	B_2	E	
A_1	A_1	A_2	B_1	B_2	E	$\times \begin{array}{l} P_x, P_y \to E\,(\perp z) \\ P_z \to A_1\,(\parallel z) \end{array}$
A_2	A_2	A_1	B_2	B_1	E	
B_1	B_1	B_2	A_1	A_2	E	
B_2	B_2	B_1	A_2	A_1	E	
E	E	E	E	E	E	

Table 21. Selection rules for point group C_{4v}

C_{4v}	A_1	A_2	B_1	B_2	E
A_1	$\parallel$	—	—	—	$\perp$
A_2	—	$\parallel$	—	—	$\perp$
B_1	—	—	$\parallel$	—	$\perp$
B_2	—	—	—	$\parallel$	$\perp$
E	$\perp$	$\perp$	$\perp$	$\perp$	$\parallel$

The C_{4v} Group. The character table for this group was considered earlier (see Table 11). The results of multiplication of the $\psi_1 \cdot \psi_2^*$ states (Table 20) as well as results of multiplcation by P_x, P_y, P_z of products thus obtained (Table 21) need only one comment: the multiplication $E \times E$ (with the aid of the character table) gives the characters +4, 0, +4, 0, 0 corresponding to a reducible representation which is

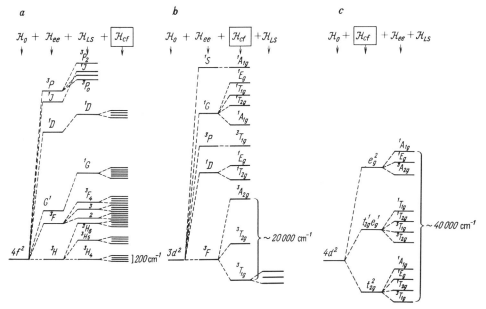

Fig. 32a–c. Three cases of interaction with crystalline field. **a** Weak crystalline field: $H_{ee} > H_{LS} > H_{CF}$ (multiplets split). **b** Medium crystalline field: $H_{ee} > H_{CF} > H_{LS}$ (terms split). **c** Strong crystalline field: $H_{CF} > H_{ee} > H_{LS}$ (atomic orbitals split)

expanded according to the rules in Table 16 into the irreducible representations: $E \times E = A_1 + A_2 + B_1 + B_2$ ([101] and others). From Table 20 we take only the results equal to E which, when multiplied by P_x, $P_y(E)$ give A_1, i.e., the transitions allowed along the x, y axes (i.e., perpendicular to z) and the results equal to A_1 which, when multiplied by $P_z(A_1)$ give A_1, i.e., the transitions allowed parallel to the z axis.

2.2 Three Types of Ion Behavior in Crystal Fields: Weak, Medium, and Strong Crystalline Fields

Crystal field theory deals with the ions with the infilled d and f electron shells. However, the behavior of ions with the $3d^n$ electrons (iron group) is rather distinct from that of the ions with $4d^n$ and $5d^n$ electrons (palladium and platinum groups) and of the $4f^n$ and $5f^n$ ions (lanthanides and actinides). This difference is due to the difference in the interactions of $3d, 4d, 5d, 4f, 5f$ electrons with the crystal field.

All types of interaction in ions are described by the Hamiltonian (Fig. 32; see also Chap. 1.8), in which to the free ion terms the crystal field interaction is added:

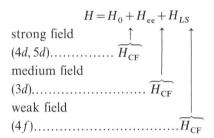

where H_0 represents the kinetic motion of an electron and the Coulomb (Ze^2/r) interaction of each electron with the central nucleus field; this interaction leads to atomic orbital types ns, np, nd, nf and to electron configurations (see also Chap. 1.8); H_{ee} is the interelection (Coulomb, electrostatic) interaction $e_1 e_2/r_{1-2}$; this interaction leads to term formation, see Chap. 1.8; H_{LS} is the spin-orbit interaction, i.e., the interaction between the orbital (L is the whole atom orbital quantum number) momenta; it corresponds to the multiplet levels formation; H_{CF} is the interaction with the crystalline field.

Weak crystalline field (rare earths and actinides, i.e., $4f$ and $5f$ ions):

$$H_{ee} > H_{LS} > H_{CF}.$$

The crystalline field effect is here weaker than other interactions (than the interactions in a free ion); the crystal field splitting is of the order of 50–400 cm^{-1}.

First, one considers term and multiplet derivation from electron configurations and then the effects of the crystal field on the multiplet levels. For example, for the

$4f^2$ configuration (Pr^{3+} ion): (1) find from Table 14 the terms ^3H, ^3F, ^1G, ^1D, ^1I, ^3P, ^1S; (2) obtain for each of the terms the possible multiplet levels, for example: $^3H \to {}^3H_6, {}^3H_5, {}^3H_4$ where the total quantum number I is 6, 5, 4, respectively (see Chap. 1.8); (3) determine for each of the multiplet levels from the total quantum number I the number and degeneracy of the levels in crystal fields of various symmetries.

Medium crystalline field (iron group, i.e., $3d$ ions):

$$H_{ee} > H_{CF} > H_{LS}.$$

The crystalline field effect is less than the Coulomb interelectron interaction, but is greater than the spin-orbit interaction, the crystal field splittings are of the order of $10000-20000 \text{ cm}^{-1}$. Because of this the consideration does not as far as to multiplet levels for the iron group ions: crystalline field acts on the free ion terms, giving new terms corresponding to irreducible representations of symmetry point groups (Table 14).

For example, consider $3d^2$ configuration (V^{3+} ion, see Fig. 32): (1) find from Table 8 the terms ^3F, ^1D, ^3P, ^1G, ^1S; (2) obtain from the Table 14 for each of the terms the new terms, i.e., the irreducible representations (=symmetry types) in which the free ion terms split in the crystal field of octahedral symmetry (as in Fig. 32) or of any other symmetry; (3) consider the splitting of these new terms by the weak spin-orbit interaction (see later, Fig. 37).

Strong crystalline field (Pd and Pt groups, i.e., $4d$ and $5d$ ions, and also rarely observed low-spin states of the iron group ions. see below, Fig. 36):

$$H_{CF} > H_{ee} > H_{LS}.$$

The crystalline field is of the same order as the inter-electronic interaction (H_{ee}) and so the crystal field effect is considered before or with the interelectronic interaction, i.e., one considers its action immediately on electron configurations.

For example. the $4d^2$ configuration (see Fig. 32) transforms in the crystalline field of cubic symmetry to two irreducible representations: t_{2g} and e_g (see the same Table 14 but with d replacing D and t_{2g}, e_g replacing T_{2g}, E_g). For the two d electrons in $4d^2$ there are three states corresponding to three possible distributions of these two electrons: t_{2g}^2 (both d electrons are in t_{2g} orbital), $t_{2g}^1 e_g^1$ (one electron in t_{2g} and other in e_g orbitals) and e_g^2 (both are in e_g orbital).

The ion states in strong crystalline fields are derived by means of the direct products of one-electron states. Thus t_{2g}^2 corresponds to two one-electron states, each characterized by the T_{2g} symmetry type. The direct product $T_{2g} \times T_{2g}$ is readily obtained with the aid of rules for evaluation of direct products of irreducible representations (Table 16) being $A_{1g} + E_g + T_{2g} + T_{1g}$, thus corresponding to the $d^2 \to t_{2g}^2$ states in the strong crystalline field (see Fig. 32).

Similarly (from Table 16) for $t_{2g}^1 e_g^1$ obtain $T_{2g} \times E_g = T_{1g} + T_{2g}$, for $e_g^2 E_g \times E_g = A_{1g} + A_{2g} + E_g$. The set of states derived from the d^2 configuration in cases of the medium and strong field ($3d^2$ and $4d^2$) is the same (see Fig. 32) but the level positions are different. The strong field case is rarely considered in a straight form.

2.3 Iron Group: Term Splitting by Crystal Field

2.3.1 Electron Configurations, Terms, Cubic Field Splitting (Qualitative Schemes)

States of atoms in crystals are derived in crystal field approach from two points of departure: a number of electrons in the atom (ion) and a symmetry of the ion site in the crystal. The number of electrons in the atom (i.e., atom position in the periodic system) distributed in electron shells in strict order determines an electron configuration unambiguously (see Table 6). From an electron configuration, possible terms for a ground and excited states are derived (see Table 8 for the $3d^n$ ions). From terms (not from multiplets, see Fig. 32) schemes of their splitting in the crystalline fields of any symmetry are obtained (see Table 14).

All these steps are brought together for the iron group ions in Figure 33.

In the general form the electron configuration for the iron group ions (see also Table 6) can be written as $[1s^2 2s^2 2p^6 3s^2 3p^6] 3d^n 4s^{1-2}$. Consider two examples:

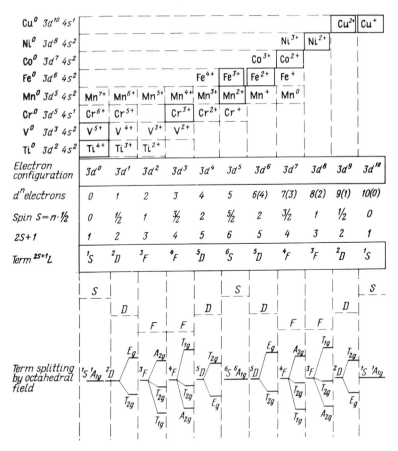

Fig. 33. Valence states of iron group ions, their electronic configuration, terms and term splitting in the cubic (octahedral) crystalline field

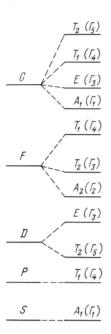

Fig. 34. Splitting of S, P, D, F, I orbital states in the cubic field (see Table 14)

Ti: $[1s^2 2s^2 2p^6 3s^2 3p^6] 3d^2 4s^2$. Trivalent titanium Ti^{3+} has the $3d^1$ configurations, i.e., the single d electron besides the filled electron shells: the orbital quantum number of the whole atom is equal to the orbital quantum number l of this single d electron: $L=l=d=2=D$; the spin quantum number S is equal to the spin of the single electron s, i.e., $S=s=1/2$. The multiplicity $2S+1=2$. Then the term ^{2S+1}L of the Ti^{3+} ground state is ^2D. In the octahedral crystalline field the ^2D term splits into two states: ^2E_g and $^2T_{2g}$ (see Table 14 and Fig. 33).

V: $[1s^2 2s^2 2p^6 3s^2 3p^6] 3d^3 4s^2$. The V^{3+} ion has the $3d^2$ configuration, i.e., there are two d electrons besides the filled electron shells: $L=l_1+l_2=d+d=3$ (according to the rules of vector summation, see Chap. 1.8), i.e., $L=3=F$; $S=s_1+s_2=1/2+1/2=1$; the multiplicity is $2S+1=3$. Then the term is ^3F. In the octahedral field the term splits into three states $^3A_{2g}, ^3T_{2g}, ^3T_{1g}$ (see Table 14).

All other ions of the iron group have in the ground state one of the three orbital states (see Fig. 33): D, F and S (terms ^2D and ^5D, ^3F and ^4F, ^6S and ^1A). Such a limited number of possible ground states is explained by the fact that all the electron configurations of iron group ions can be reduced to one or two d electrons besides the filled d^0 and d^{10} or the half-filled d^5 subshells (Fig. 34).

S states: d^0 (Ti^{4+}, V^{5+}) and d^{10} (Cu^+, Zn^2) represent completely filled subshells; there are no unpaired electrons, hence $L=0$; $S=0$; $2S+1=1$ and the term is ^1S; it is the only possible term for completely filled subshells (see Table 8);

d^5: all five possible states for the d electrons ($2L+1=5$) are occupied; this corresponds to a spherically symmetric charge distribution and to the resultant moment with the orbital quantum number $L=0$, but there are here five unpaired electrons giving the total spin of atoms $S=5\cdot 1/2=5/2$ and the multiplicity $2S+1=6$. The term ^6S is that of a ground state, and there are additionally a number of excited state terms for the d^5 configuration (see Table 8).

The S states are orbitally nondegenerate states ($2L+1=1$), hence in the crystalline fields of any symmetry they transform to A_1 states retaining the free ion multiplicity: $^6S \to {}^6A_1$ and $^1S \to {}^1A_1$; for centrosymmetric crystal fields the subscript g indicating the parity of the state is added: $^6S \to {}^6A_{1g}$.

D states: The d^1, $d^4 (=d^{5-1})$, $d^6 (=d^{5+1})$, $d^9 (=d^{10-1})$ electron configurations give the D states (Fig. 33). These configurations can be considered as those having a single electron above spherically symmetric distribution of the d^0, d^5 configurations or a single hole in the spherically symmetric d^5, d^{10} configurations. Hence all these four configurations have the same D orbital state (as Ti^{3+} with $3d^1$). In cubic crystalline field it splits into two levels, E_g and T_{2g}. However, each of these four states is discerned either by multiplicity (^2D for d^1 and d^9, ^5D for d^4 and d^6) or by level ordering derived from the D states in their splitting by the octahedral field: upper ^2E_g and lower $^2T_{2g}$ for d^1 (or upper ^5E_g and lower $^5T_{2g}$ for d^6) but inverse ordering for d^4 and d^9: upper $^5T_{2g}$ and lower ^5E_g for d^4 (or upper $^2T_{2g}$ and lower ^2E_g for d^9).

F states: The orbital F states are derived from configurations with two d electrons (or holes) above d^0, d^{10} and d^5 subshells: d^2, $d^3 (=d^{5-2})$, $d^7 (=d^{5+2})$, $d^8 (=d^{10-2})$. These four configurations giving the F orbital moment are discerned by multiplicity (^3F for d^2 and d^8, ^4F for d^3 and d^7) and by level ordering in the octahedral field: $^3T_{1g}$ (the lower), $^3T_{1g}$ and $^3A_{2g}$ for d^2 (or $^4T_{1g}$, $^4T_{2g}$, $^4A_{2g}$ for d^7) but an inverse ordering for d^3 and d^8.

It is obvious that ions with the same electron configuration (for example, Mn^{2+} and Fe^{3+} with $3d^5$ or Co^{2+} and Ni^{3+} with $3d^7$, see Fig. 33) have the same terms and the same splitting schemes in the crystal fields; the differences in their absorption spectra are connected with the differences in crystal field parameters (see below).

In cubic fields (octahedral, tetrahedral and cubal) for a given ion (for a given electron configuration) the same levels arise, but the level ordering for tetrahedra and cubes is the inverse of that for octahedra (moreover, for tetrahedra the subscript g is dropped). For example, for Fe^{2+} ($3d^6$, ^5D) there are two levels in the octahedral field ^5E_g and $^5T_{2g}$ with lower $^5T_{2g}$, in the tetrahedral field ^5E is lower than $^5T_2$ (in the cubal field ^5E_g is lower than $^5T_{2g}$).

It should be noted also (Fig. 33) that the level order for Fe^{2+} ($3d^6$, ^5D) in the octahedral field (^5E_g, $^5T_{2g}$) is the same as for Mn^{3+} ($3d^4$, ^5D) in the tetrahedral field (^5E_g, $^5T_{2g}$) and the same as for Ti^{3+} ($3d^1$, ^2D) in the octahedral field, but with different multiplicity, (^2E_g, $^2T_{2g}$), and for Cu^{2+} ($3d^9$, ^2D) in the tetrahedral field (^2E_g, $^2T_{2g}$).

In a general form (see Fig. 35): d^n_{Oct} = direct scheme for d^{10-n}_{tetr} = inverse d^{10-n}_{oct} (the same multiplicity) = direct d^{5+n}_{oct} = inverse d^{5+n}_{tetr} (different multiplicity) = direct d^{5-n}_{tetr} = inverse d^{5-n}_{Oct} (different multiplicity). The notable particularity of the $3d^5$ (^6S) configuration is that both for octahedral and tetrahedral coordinations and for any of their distortions the single $^6A_{1g}$ (or $^6A_1$, ^6A) level exists always.

A total degree of degeneracy of the crystal field levels is composed from: (a) orbital degeneracy ($2L+1$) determined by term (for $L=S=0$ the degeneracy is $2L+1=1$, for $L=P=1$ the degeneracy is 3, for $L=D=2$ the degeneracy is 5, for $L=F=3$ the degeneracy is 7) or by term in the crystal field (A and B designate the

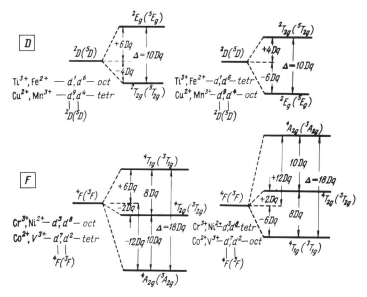

Fig. 35. Splitting of ground terms S, D, F of the ions with $3d^n$ configurations in the octahedral and tetrahedral field (the sign g, even, must be omitted for tetrahedral fields)

degeneracy being 1, $E-2$, $T-3$); (b) spin degeneracy $(2S+1)$ determined by term multiplicity (^3F designates three-fold spin degeneracy; ^6S six-fold spin degeneracy) or by the same multiplicity of the crystal field state ($^3A_{2g}$, $^3T_{2g}$, $^3T_{1g}$ derived from ^3F have three-fold spin degeneracy, $^6A_{1g}$ derived from ^6S has six-fold spin degeneracy).

The degree of orbital degeneracy determines a number of levels into which a term can split in cubic field (for example, ^2D or ^3D into five levels, ^3F or ^4F into seven levels) or a number of levels into which the cubic field term can split in lower symmetry fields (for example, the octahedral term $^2T_{2g}$ split into three lower symmetry terms; in C_{2v} symmetry field $^2T_{2g}$ split into $^2A_2$, $^2B_1$, $^2B_2$ see Table 14).

The degree of spin degeneracy determines a number of spin sublevels into which the orbital level can split in the crystal and magnetic fields. Spin multiplicity determines a number of levels to which electron paramagnetic resonance spectra and magnetic properties of crystals are related. In optical absorption spectra, spin multiplicity is connected with additional selection rules: transitions between states with identical multiplicity (for example, $^5T_{2g} \rightarrow {}^5E_g$) are "spin-allowed", while transitions between states with different multiplicity (for example, $^4A_{2g} \rightarrow {}^2T_{2g}$) are "spin-forbidden" (see also Chap. 6.2).

For example, the $^6A_{1g}$ level is the orbital singlet and spin sextet, the total degeneracy being $1 \times 6 = 6$; the $^4T_{2g}$ level is the orbital triplet and spin quartet, the total degeneracy being $3 \times 4 = 12$; ^5E_g, is the orbital doublet and spin quintet, the total degeneracy being $2 \times 5 = 10$. From the ^4F term three spin quartets are derived: the orbital singlet ^4A_g ($4 \times 1 = 4$), the two orbital triplets $^4T_{2g}$ and $^4T_{1g}$ ($4 \times 3 = 12$ and $4 \cdot 3 = 12$), the total degeneracy for the ^4F term being $4 \times 7 = 28$.

The number of transitions between levels derived from a term in the crystal field (these levels are those with the same multiplicity because they are formed from the same term and thus are spin-allowed and have a strong intensity) determines the number of corresponding bands in an optical absorption spectrum:

D state ions give one band:
  Ti^{3+}, Fe^{2+} : $T_{2g} \to E_g$ in octahedra, $E \to T_2$ in tetrahedra,
  Mn^{3+}, Cu^{2+} : $E_g \to T_{2g}$ in octahedra, $T_2 \to E$ in tetrahedra.

F state ions give two bands:
  V^{3+}, Co^{2+} : $T_{1g} \to T_{2g}$ and $T_{1g} \to A_{2g}$ in octahedra,
  $A_2 \to T_2$ and $A_2 \to T_1$ in tetrahedra,
  Cr^{3+}, Ni^{2+} : $A_{2g} \to T_{2g}$ and $A_{2g} \to T_{1g}$ in octahedra,
  $T_1 \to T_2$ and $T_1 \to A_2$ in tetrahedra.

S state ions give no spin-allowed bands:
  Fe^{3+}, Mn^{2+} : $^6A_{1g}$ is the single level formed from ^6S.

2.3.2 Crystal Field Parameters; Tanabe-Sugano Diagrams

After determination of the qualitative schemes for the term splitting in the crystalline fields, we go on to a quantitative estimation of separations between the crystal field levels. All the electronic interaction stages in an ion shown by means of energy level splitting schemes (see Figs. 28–32) and by the introduction of additional terms in the Hamiltonian can be represented quantitatively by means of the few parameters:

$$H = H_0 + H_{ee} + H_{CF} + H_{LS}$$
$$\quad\quad\quad\; \downarrow \quad\;\; \downarrow \quad\;\; \downarrow$$
$$\quad\quad\quad B,C \quad Dq \quad \xi$$

where B, C are interelectronic repulson (H_{ee}) parameters governing term separation due to Coulomb repulsion of electrons; Dq is the cubic crystalline field parameter determining term splitting by cubic crystalline field (H_{CF}); ξ is spin-orbit coupling parameter, taking into account further splittings of ion levels in crystals due to an interaction between orbital and spin momenta of an ion (H_{LS}).

Interelectron interaction parameters are taken in optical the interpretation of absorption spectra in the form of Racah B and C parameters (The Racah A parameter corresponds to the parallel shift of all terms of a given electron configuration, hence in optical transition considerations it is not taken into account). In free ion spectra calculations the Slater-Condon-Shortley F_0, F_2, F_4 parameters are used that are related to the Racah parameters by the simple expressions: $B = F_2 - 5F_4$; $C = 35F_4$; $A = F_0 - 49F_4$.

Free ion term energies expressed by the Racah parameters are represented by simple equations; using experimental data on free ion optical spectra, the numerical values for these parameters can be obtained from these equations. For example, there are for the d^2 configuration (V^{3+} ion):

a) The terms: ^3F, ^3P, ^1G, ^1D, ^1S (see Table 14 and Fig. 32).

b) These term energies obtained from the optical spectrum; they can be found in Moore's compilation [38]:

$^3F_2$ 0 $\qquad$ $^3P_0$ 13121
$^3F_3$ 318 $\qquad$ $^3P_1$ 13283
$^3F_4$ 730 $\qquad$ $^3P_2$ 13435
$^1D_2$ 10960 $\qquad$ $^1G_4$ 18389 (in cm^{-1}).

c) Express these terms energies E by the Racah parameters [49]:

$$E(^3F) = A - 8B$$
$$E(^3P) = A + 7B$$
$$E(^1D) = A - 3B + 2C$$
$$E(^1G) = A + 4B + 2C.$$

Hence obtain:

$$^3P - ^3F = 15B = 13000 \text{ cm}^{-1}$$
$$^1D - ^3F = 5B + 2C = 10600 \text{ cm}^{-1}.$$

Thus the B and C parameters for the V^{3+} free ion are: $B = 866$ cm^{-1}, $C = 3175$ cm^{-1}. Note that the energy separations for the terms of the same multiplitity (say, $^3F - ^3P$ or $^1D - ^1G$) are determined only by the B parameter but those for the terms of different multiplicity (say, $^1D - ^3F$) by the B and C parameters. In particular the $^nP - ^nF$ energy sepration is equal to $15B$.

The C/B relation is usually close to 4 and thus in many cases interelectron interaction is taken into account by means of the B parameter only. For crystals the free ions B and C values are often taken (as in the Tanabe-Sugano diagrams considered below). The values measured from optical absorption spectra of crystals B and C are always less than those for free ions. The change of the B and C values in crystals in comparison with free ions and their dependence on the transition metal ion and on the ligand ion are understandable from a comparison of the physical meanings of these parameters with the meaning of the interactions attendant on formation of transition metal–ligand complexes.

While in free ions the B and C parameters are a measure of interelectron repulsion, in crystals they become also a measure of bond covalency, thus representing an empirical deviation from the point charges model.

Both for a free ion and for an ion in a crystal, interelectron interactions depend on the ion size (on the $3d$ orbital size): the greater the ion size, the more distant the electrons from each other, the less repulsion between them, and vice versa.

Hence B and C decrease: (a) with decreasing of oxidation state (the M^{2+} size is larger than that of M^{3+} and the interactions in the M^{2+} are weaker than in M^{3+}; (b) from the first transition series to the second and the third series (when sizes of ions with the same number of d electrons increase); (c) from the first to the last ions within each of the series (ion sizes increase with an increasing number of the d electrons).

For an ion in crystal the overlapping of transition metal and ligand orbitals entails two effects, both leading to the decrease of B and C: (a) the partial transfer of the d electron into ligand orbitals means increasing the extent of the d orbitals, increasing their average radius, and decreasing the effective charge; (b) a "penetration" of ligand orbitals into d subshells enlarges the shielding effect, decreases the effective nuclear charge, and leads to the radial widening of the d orbitals.

In other words, the more covalent the bonding, the less B is in crystals in comparison with a free ion.

For a given transition metal ion the bond coavalency depends on the ligand ions which are arranged in the order of B decreasing in so-called nephelauxetic series (from the Greek: electron "cloud-expanding" series):

B free ion: $F > O^{2-} > Cl^- > Br^- > S^{2-} > I^- > Se^{2-}$.

This method of ordering has, however, a number of exceptions and can only be considered as the empirically established tendency.

Decreased B values in crystals are usually about 0.7–0.8 of the free ion B value but be come as low as 0.3 B. The reduction of B values indicates the contraction of the term separations, from which crystal field levels arise.

The cubic crystalline field parameter Dq (crystal field strength) is the quantitative measure for the split term separations.

The splitting of the D terms (^2D, ^5D) derived from configurations with the single unpaired d electron above the d^0, d^5, d^{10} shells, i.e., from the d^1, d^4, d^6, d^9 configurations, is determined by the crystal field strength Dq only (while transitions to levels derived from other terms of the same configurations are described by the three parameters: Dq, B and C).

Splitting of the F terms (^3F, ^4F) and also of excited G, H... terms derived from d^2, d^3, d^7, d^8 configurations are expressed by means of the Dq, B and C values.

The Dq value (as B and C values) is determined from optical absorption spectra and represents one of the most important spectroscopic characteristics for an ion in a given crystal.

The crystal field strength is determined by the effective change on the ligands Q, by the average radius r of the d orbital and by the metal–ligand distance R:

$$Dq = \text{const} \frac{Q \langle r^4 \rangle}{R^5}.$$

This expression entails the following dependence for Dq:

1. Dependence on the kind of transition metal ion. Dq is greater for M^{3+} than for M^{2+} (for M^{3+} $Dq = 1400$–2500 cm^{-1}, for M^{2+} $Dq = 750$–1250 cm^{-1}); for ions

of the same valency and in the same coordination Dq varies within comparatively close ranges.

For the first transition series Dq is $\sim 30\%$ less than for the second transition series and Dq for the latter $\sim 30\%$ less than Dq for the third transition series:

$$Dq(3d) < Dq(4d) < Dq(5d).$$

Dq for the d^5 ions is less than for d^4 and d^6 ions. The spectrochemical series exists according to the order of increasing Dq:

$$Mn^{2+} < Ni^{2+} < Co^{2+} < Fe^{2+} < V^{2+} \sim Cu^{2+} < Fe^{3+}$$
$$< Cr^{3+} < V^{3+} < Co^{3+} < Ti^{3+} < Mn^{3+}.$$

2. Dependence on the ligands arranged also in the spectrochemical series in the order of increasing Dq:

$$I^- < Br^- < Cl^- < F^- \leqq O^{2-} \leqq H_2O.$$

3. Dependence on the metal–ligand distances R: with decreasing R Dq increases but not always proportional to the five powers of R.

4. Dependence on the coordination: $Dq_{oct} : Dq_{cub} : Dq_{tetr} = 1 : 8/9 : 4/9$.

Determination of Dq, B, C from the Optical Absorption Spectra. After assigning absorption bands to the transitions between levels formed due to the splitting of a ground state term by the cubic crystalline field we can determine the parameters Dq, B, C from the following simple relations [68].

a) For the d^1, d^4, d^6, d^9 configurations (ground states ^2D, ^5D) only, $\Delta = 10\,Dq$ is determined from the allowed transition:

$d^1(^2D)$: $^2T_2 \rightarrow {}^2E = \Delta$,
$d^4(^5D)$: $^5E \rightarrow {}^5T_2 = \Delta$,
$d^6(^5D)$: $^5T_2 \rightarrow {}^5E = \Delta$,
$d^9(^2D)$: $^2E \rightarrow {}^2T_2 = \Delta$.

b) For the d^2, d^3, d^7, d^8 configurations (ground states ^3F, ^4F), one determines $\Delta = 10\,Dq$ and B:

$d^2(^3F)$: $^3T_1 \rightarrow {}^3T_2 = -7.5B + 0.5\Delta + (b^+)$,
 $^3T_1 \rightarrow {}^3A_2 = -7.5B + 1.5\Delta + (b^+)$,
 $^3T_1 \rightarrow {}^3T_1(^3P) = 2(b^+)$,
 $^3T_2(^3F) - {}^3A_2(^3F) = \Delta$;

$d^7(^4F)$: $^4T_1 \rightarrow {}^4T_2 = -7.5B + 0.5\Delta - (b^+)$,
 $^4T_1 \rightarrow {}^4A_2 = -7.5B + 1.5\Delta - (b^+)$,
 $^4T_1 \rightarrow {}^4T_1(^4P) = 2(b^+)$,
 $^4T_2(^4F) - {}^4A_2(^4F) = \Delta$;

$d^3(^4F)$: $^4A_2 \rightarrow {}^4T_2 = \Delta$,
 $^4A_2 \rightarrow {}^4T_1 = 7.5B + 1.5\Delta - (b^-)$,
 $^4A_2 \rightarrow {}^4T_1(^4P) = 7.5B + 1.5\Delta + (b^-)$,
 $^4T_1(^4F) - {}^4T_1(^4P) = 2(b^-)$;

$d^8(^3F)$: $^3A_2 \to {}^3T_2 = \Delta$,
$\qquad {}^3A_2 \to {}^3T_1 = 7.5B + 1.5\Delta - (b^-)$,
$\qquad {}^3A_2 \to {}^3T_1(^3P) = 7.5B + 1.5\Delta + (b^-)$,
$\qquad {}^3T_1(^3F) - {}^3T_1(^3P) = 2(b^-)$,

where

$(b^+) = 1/2[(9B+\Delta)^2 + 144B^2]^{1/2}$,
$(b^-) = 1/2[(9B-\Delta)^2 + 144B^2]^{1/2}$.

c) For the d^5 configuration the ^6S ground state transforms into the single level $^6A_1$ (i.e., does not split by crystal fields); the values B, C_1 and $\Delta = 10\,Dq$ are determined from transitions between the ground level ^6A, and excited levels derived from other terms:

$d^5(^6S)$: $^6A_1 \to {}^4T_1(^4G) = -\Delta + 10B + 6C - 26B^2/\Delta$,
$\qquad {}^6A_1 \to {}^4T_2(^4G) = -\Delta + 18B + 6C - 26B^2/\Delta$.

Complete schemes of levels in the cubic field for all terms derived from d^n configurations are presented in the form of the Tanabe-Sugano diagrams ([74] see later descriptions of transition metal ion spectra, Chap. 6.3). These diagrams are plots of the energies of the crystal field levels (units of E/B) as a function of the crystal field parameter $10\,Dq$ (units of Dq/B) for the given B and C/B values. The ground term coincides with the horizontal coordinate in these diagrams, while other free ion term positions contracted due to the change of interelectron interaction between the terms in crystal (described by changes in B, C parameters in crystals in comparison with a free ion) are shown at the vertical coordinate. From these ground and excited free ion terms the crystal field levels are derived.

The restrictions of these diagrams are connected with: (a) restrictions of the very model of crystalline field, (b) the fact that the diagrams are calculated only for one B and C values, (c) the fact that free ion values for B and C are taken which differ significantly from B and C in crystals, (d) the fact that splitting by cubic crystalline field only is shown (and without spin-orbital splitting, see Chap. 2.4) while the crystalline fields of lower symmetries are the most common.

All these restrictions make the diagrams semiquantitative.

However these are the only diagrams which give a general plot for the behavior of all levels in the crystalline field, the relative positions of the levels and their energy variations as functions of the crystal field strength.

Levels in the cubic field represent a starting point for consideration of further splittings by the crystal fields of lower symmetries.

The levels, independent of crystal field strength, are plotted in these diagrams (see Chap. 6.3) parallel to the horizontal coordinate (for instance the ^4E, $^4A_2$ levels in the d^5 configuration); transitions to these levels give rise to the sharp absorption bands whose positions vary little in different crystals.

The levels which are unique of a given type (for instance, the $^3T_2$ and $^3A_2$ levels in the d^2 configuration) show linear dependence on Dq.

The levels having analogs formed from different terms (for instance, two ^1E levels arising from the terms ^1D and ^1G in d^2 configuration) interact with each other, never intersect and are shown by divergent curves.

Transitions from ground state to excited states can be spin-allowed and spin-forbidden. Transitions are spin-allowed between states of the same multiplicity, i.e., usually between the levels arising due to splitting of the same ground term; transitions between the states of different multiplicity are spin-forbidden. The allowed transitions give rise to intense absorption bands, the forbidden ones correspond to weak bands often not observed at all.

Since the scheme of levels in the tetrahedral field is inverted as compared with that of the octahedral field (Fig. 35), the Tanabe-Sugano diagrams can be used also for ions in tetrahedra coordination, i.e., for the d^n configuration in tetrahedral coordination one uses the diagram for d^{10-n} configuration in octahedral configuration. For example, the diagram for $Fe^{2+}(d^6)$ in the octahedron can be used for Mn^{4+} ($d^4 = d^{10-6}$) in the tetrahedron. However, in this case the ranges of Dq value for tetrahedral coordination will be in another region because $Dq_{tetr} = 4/9\, Dq_{oct}$.

High-Spin and Low-Spin States. Distribution of d electrons between t_2 and e_g states arising in the cubic crystal field (Fig. 36) can be different in their dependence on the crystal field strength.

In the cases of one to three electrons (d^1, d^2, d^3 configurations) it is only lower orbitals t_{2g} (d_{xy}, d_{yz}, d_{xz}) which are occupied in octahedral coordination.

In the cases of four to seven electrons, there are the two possibilities for the t_{2g} and e_g states occupation, depending on the electron spin arrangement: parallel (unpaired electrons occupying different t_{2g} orbitals: d_{xy}, d_{yz} or d_{xz}) and antiparallel (paired electrons, occupying the same orbital but having opposite spin directions). For example, for d^4 two variants are possible (Fig. 36): (a) three electrons are in the lower level t_{2g} (one by one in d_{xy}, d_{yz}, d_{xz} orbitals) while the fourth is in the higher e_g level (d_{z^2} or $d_{x^2-y^2}$); all the electrons are unpaired and the electron configuration is

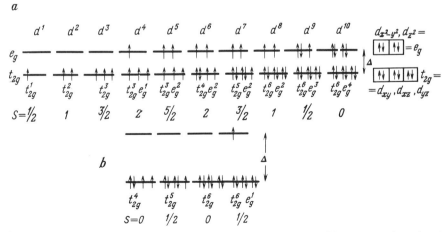

Fig. 36. a High-spin and **b** low-spin electronic configurations of iron group ions (ground states)

$t_{2g}^3 e_g$; (b) all the electrons are in t_{2g} level with the fourth electron paired with one of the other electrons; the electron configuration is t_{2g}^4.

In the cases of eight to ten electrons, their arrangement in the ground state, as in the d^1, d^2, d^3, is always unique: the t_{2g} orbitals are occupied completely by six electrons while others replace the e_g orbitals.

Realization of one of two possible $d^4 - d^7$ configurations is determined by competition of the two opposite tendencies: (a) interelectron repulsion (H_{ee} in the Hamiltonian operator) favor arising of unpaired electrons (with parallel spins); when in the t_{2g} level there are three unpaired electrons, this interaction forces the fourth and following electrons to rise in the higher e_g level in order to prevent the arising of a spin-paired state; this corresponds to the empirical Hund's rule (see Chap. 1.8): the ground state is that with the maximum spin, i.e., with the maximum number of electrons; (b) on the other hand, the crystal field strength $\Delta = 10\,Dq$ (H_{CF} in Hamiltonian operator) determining the separation of the t_{2g} and e_g levels opposes the transfer of the electron in the higher e_g state, thus favoring maximum electron occupancy (by means of electron-pairing) of the lower state.

In the case of the medium crystalline field (see Chap. 2.3) with $H_{ee} > H_{CF}$ the high-spin states arise, i.e., the states with maximum spin, maximum number of unpaired electrons (in accordance with Hund's rule).

In the case of a strong crystalline field with $H_{ee} < H_{CF}$ the low-spin states arise with the spins S equal to 0 or 1 due to pairing of electrons (Fig. 36).

Thus, the low-spin states: (1) correspond to the case of strong crystalline field, (2) are possible in the $d^4 - d^7$ configurations only (while in the $d^1 - d^3$ and $d^8 - d^{10}$ occupancies of t_{2g} and e_g are unique), (3) occur mainly in ions of the second and third transition metal groups; in the iron group they occur only in some sulfide minerals and in compounds with such ligands as CN and others which do not exist in minerals.

It is the high-spin states that usually occur in minerals. The low-spin state of Fe^{2+} (d^6) was found for example, in pyrite by means of Mössbauer spectroscopy and the magnetical properties.

2.3.3 Splitting by Spin-Orbit Interaction, Jahn-Teller Effect, and Lowering of Symmetry

The splitting by the cubic crystal field described by the parameters Dq, B, C and correlated by the Tanabe-Sugano diagrams, although it represents the main part of the ion interactions in crystal as one to two orders greater than other interactions, however never occurs without some other superimposed interactions.

Spin-orbit interaction of the $3d^n$ ion states split by the crystal field is the same coupling (see Chap. 1.8) between orbital momentum (L) and spin momentum (S) which is determined by spin-orbit coupling parameter λ (i.e., $\lambda\,LS$), and which leads to the quantum number I in the case of free ions (term $^{2S+1}L \rightarrow$ multiplet $^{2S+1}L_I$), but it is considered here after the crystal field action (see Fig. 32) and is applied to the states denoted by the symmetry type in the crystal field (T, E, A represent the designations of the orbitally triply, doubly and nondegenerate states). The spin multiplicity of the states, for instance ^3T, ^3E, ^3A..., signifies, as in free ions, the spin degeneracy equal to $(2S+1)$ where S is the ion spin.

Only the behavior of orbital states in crystals has been thus far considered: spin states are inherited by crystal field levels from the free ion terms.

The interaction of orbital and spin momenta of ion leads to (a) further small splitting of the cubic field levels that fives rise to widening of the absorption bands or – at low temperatures – to fine structure in absorption spectra; if the local symmetry of the ion position is not cubic, the spin-orbit coupling acts together with lowering of symmetry, leading to level splitting; (b) weakly allowed transitions between the states of different multiplicity which are otherwise forbidden: instead of the selection rule $\Delta S = 0$ (i.e., by spin quantum number) the possible transitions are governed by the selection rule $\Delta I = \pm 1.0$ (i.e., by the quantum number $I = L \pm S$); (c) non-coincidence of the magnetic susceptibility values with the pure spin values; (d) initial splitting of spin sublevels in electron paramagnetic resonance spectroscopy which determines the g factor values.

The spin-orbit coupling value can be expressed by one of the two parameters: ξ_{3d} is one-electron parameter of spin-orbit coupling describing the strength of the interaction between spin and orbital momenta of a single electron of a given electron configuration; it represents the property of the given electron configuration (i.e., property of the ion); ξ_{3d} is positive and the same for all the terms arising from a given configuration; it is connected with effective charge of the nucleus Z_{eff} and average radius of the $3d^n$ orbital r:

$$\xi_{3d} = (Z_{eff} \cdot e^2 / 2m^2c^2)/r^{-3} ;$$

λ is a parameter of spin-orbit coupling related to the term of the ion: it is $2S$ times less than the one-electron parameter ξ_{3d}:

$$\lambda = \pm \xi/2S .$$

For example, for $Cr^{3+}(3d^3)$ $\xi_{3d} = 270$ cm^{-1} (see Table 22); for the term ^4F (or $^4T_2$, $^4T_1$) arising from this configuration $\lambda = 1/3 \cdot \xi_{3d} = 90$ cm^{-1} (because $2S+1=4$; $S=3/2$; $\lambda = \xi_{3d}/2S$); for the term ^2G (or ^2E, $^2T_1$) $\lambda = \xi_{3d} = 270$ cm^{-1}.

Thus parameter λ is the ion property; the strength of spin-orbit coupling of a term is $\lambda \cdot LS$.

Table 22. One-electron parameter ξ_{3d} of spin-orbit coupling of iron group ions

Ion	ξ_{3d}, CM^{-1}	Ion	ξ_{3d}, CM^{-1}	Ion	ξ_{3d}, CM^{-1}
Ti^{2+}	120	Cr^{2+}	230	Co^{2+}	515
Ti^{3+}	155	Cr^{3+}	275	Co^{3+}	580
V^{2+}	170	Mn^{2+}	300	Ni^{2+}	630
V^{3+}	210	Mn^{3+}	255	Ni^{3+}	715
V^{4+}	(250)	Fe^{2+}	400	Cu^{2+}	830
		Fe^{3+}	460		

In order to obtain λ divide the ξ values by $2S$ (for S value see Fig. 33).

Splitting by Spin-Orbit Interaction, Jahn-Teller Effect, and Lowering of Symmetry

In contrast to ξ_{3d} which is always a positive parameter, λ can be positive and negative ($\lambda = \pm \xi_{3d}/2S$). It is positive for the electron shells less than half-filled ($d^1 - d^4, p^1, p^2$) and it is negative for the shells more than half-filled ($d^6 - d^9, p^4, p^5$); for d^5 and p^3 λ is zero. (It is related to the g factor value which can be more and less than 2). The change of the sign of λ is related to the fact that the same spin value for d^n and d^{10-n} configurations is determined for $n=5$ according to the number of electrons while for $n>5$ according to the number of holes being considered here as "positive electrons". The λ values in crystals are close to the corresponding λ values of the free ion decreasing slightly in crystals (as Racah parameters B and C): for example, λ of free ion Fe^{2+} is 100 cm^{-1}, while λ of Fe^{2+} in spinell is 98 cm^{-1}, in ZnS 92 cm^{-1}, in CdTe 64 cm^{-1}.

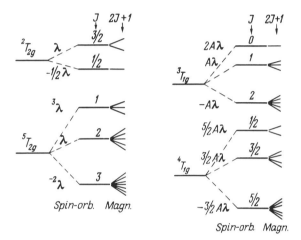

Fig. 37. Splitting of T terms by spin-orbit interaction (and by external magnetic field)

The spin-orbit coupling in the crystal field splits in first approximation only the T terms, in second approximation the E terms and does not split the A terms. The number of levels in which the T terms are split (Fig. 37) can be determined by the same method as the number of multiplets of free ions (see Chap. 1.8) if one considers the t_{2g} orbitals (d_{xy}, d_{xz}, d_{yz}) as equivalent with the p orbitals (the same triply degenerated state: p_x, p_y, p_z) and the T terms as equivalent to the P terms. Then, for example for the ^3T term:

$L = T = P = 1; \quad 2S+1 = 3; \quad S = I;$

I vary from $L+S$ till $L-S$,

i.e. from $1+1 = 2$ till $1-1 = 0$,

i.e., the ^3T term is split into three levels with $I = 2, 1, 0$. For the ^4T and ^5T terms this analogy with the P terms becomes only formal because there are no P terms with these multiplicities.

The spin-orbit level energy is determined from the simple expression:

$\varepsilon = -1/2 \cdot \xi_{3d} \cdot 1/2 [I(I+1) - L(L+1) - S(S+1)].$

The Jahn-Teller effect (or configuration instability, or intrinsic asymmetry) represents the property of the very electronic configurations for which the degenerate state in crystals even in the case of cubic local symmetry are unstable. The general idea of this effect can be considered with the example of the ion $Cr^{3+}(3d^9)$ configuration in the octahedral crystal field, i.e., the configuration $t_{2g}^6 e_g^3$. In the e_g^3 group, represented by two equivalent orbitals $d_{x^2-y^2}$ and d_{z^2}, there are three electrons; these orbitals pointed towards the ligand ions ($d_{x^2-y^2}$ along the x and y axis, d_{z^2} along the z axes) interact with the ligands unequally: if two electrons occur in the $d_{x^2-y^2}$ orbital and one electron in the d_{z^2} orbital, the ligand negative charges are more screened from electrostatic attraction of the Cu^{2+} ion along the x and y axes than along the z axis, hence the ligands occur nearer to the Cu^{2+} ion along the z axis than the ligands along the x and y axes. In consequence of this, the e_g^3 orbital is split into two levels which have become nonequivalent. More detailed consideration [50, 73] shows that the mechanism of the Jahn-Teller effect is related to the interaction of electron and vibrational states.

Consider here only some consequences of the Jahn-Teller effect and its peculiarities in different configurations and coordinations.

From the three orbital states in the cubic field T, E, A, the Jahn-Teller effect does not occur naturally in the case of the nondegenerate A state and is more pronounced in E states than in T: the e_g orbitals are directed towards ligands and are more respondent to nonequivalency of interactions with the ligands than the t_{2g} orbitals, directed between ligands (Fig. 26). In the tetrahedron, this difference is less pronounced.

E states, due to the Jahn-Teller effect exhibit a tetragonal distortion, T states a tetragonal and trigonal.

One discerns a static and a dynamic Jahn-Teller effect. The static effect is related to the configuration instability of the ground state and causes splitting of this ground state, thus removing its degeneracy. It means that a given ion must be coordinated by the distorted octahedron or distorted tetrahedron. However, if this complex does not have cubic local symmetry, one can only discuss if this noncubic symmetry is the consequence of the static Jahn-Teller effect or simply a result of the packing of ions in the crystal structure with noncubic symmetry.

In the iron series, the ground state E most liable to the static Jahn-Teller effect occurs in octahedral coordination in the cases of the ions Cu^{2+} ($3d^9$, ^2E) and Mn^{3+} ($3d^4$, ^5E). These ions are in fact always in noncubic coordinations.

Direct experimental evidence for the static Jahn-Teller effect could be the observation of noncubic spectrum (optical, EPR or Mössbauer) from an admixture ion in the site of cubic local symmetry.

The dynamic Jahn-Teller effect occurs as a configuration instability of excited states. The splitting of the excited state leads to the appearance in the optical absorption spectrum instead of one absorption band, due to transition from the ground state to the excited one, of two absorption bands (or one very braod band) due to transitions in two split excited levels.

The excited E state occurs in octahedral coordination in the cases of ions $Ti^{3+}(3d^1, {}^2T_{2g} \to {}^2E_g)$ and $Fe^{2+}(3d^6, {}^5T_{2g} \to {}^5E_g)$. If these ions are in the crystal structures in sites with noncubic symmetry, the splitting of the excited E levels can be attributed to that noncubic symmetry of the crystal structure or to the combined action of the lowering of symmetry and the Jahn-Teller effect. However, the appearance of the split bands in the optical absorption spectra of some cubic crystals is related unequivocally to this effect of intrinsic asymmetry. For example, in the optical absorption spectrum of the Fe^{2+} ion in the octahedral position in MgO (with NaCl crystal structure) one observed instead of one absorption band $^5T_{2g} \rightarrow {}^5E_g$ (with the average frequency 10,300 cm^{-1}), two bands split by 1,800 cm^{-1}.

Lowering of symmetry. Thus far the behavior of ion levels in crystal fields has been considered for cases of cubic symmetry. However, in most crystal structures of minerals and inorganic compounds, octahedron and tetrahedron indicate only the coordination polyhedra, but not the local symmetry which is in most cases noncubic.

However, the splittings due to the distortions of the octahedral and tetrahedral coordination polyhedra are of one–two orders less than the splitting by the cubic fields and are usually similar in the magnitude to spin-orbit splitting. That is why the distortions due to the lowering of symmetry can be treated as perturbations, as further steps following splitting by the cubic field.

In optical absorption spectra the lowering of symmetry leads to the appearance of two bands or one broad band. The polarized spectra are governed by the selection rules: some transitions are allowed and other are forbidden in one orientation and vice versa.

By gowing gradually from one orientation to another there are no changes in the positions of the absorption bands, but one observes the gradual decrease in intensity of the allowed transition until in the perpendicular orientation it becomes forbidden and either vanishes or its intensity takes the minimum value. At the same time the absorption band caused by other transitions can be growing in intensity when this transition becomes allowed and it reaches the maximum of its intensity with the minimum of intensity of the first band.

If the splitting is small, then in the intermediate orientations these two bands can overlap and one can observe the apparent displacement of the absorption band, which is in fact the superimposition of the two bands changing in relative intensity but not in their position.

Directions of maximum and minimum intensity of the absorption bands are that of the crystal field axes connected with the elements of local symmetry which may not coincide with the crystal axes and with the optical indicatrix axes. However, the local crystal field axes are usually the same in optical, electron paramagnetic resonance and Mössbauer spectra.

Number and symmetry type of the levels in thecrystal fields of different symmetry are given for the most important point groups in Table 11.

The allowed transitions between these levels are determined by the selection rules represented in Tables 23–27.

For all the point groups of the same system there is an equal number of levels distinguished only by small peculiarities of symmetry: for example, the nondegenerate level can transform into A_1, A_{1g}, A'_1, A_1, B, B_{1g}, B_1). This explains the different selection rules for these point groups.

Table 23. Selection rules for triclinic point groups

C_i	A_g	A_u
A_g	−	+
A_u	+	−

C_1	A
A	+

Table 24. Selection rules for monoclinic point groups

C_2	A	B
A	∥	⊥
B	⊥	∥

C_s	A'	A''
A'	⊥	∥
A''	∥	⊥

C_{2h}	A_g	A_u	B_g	B_u
A_g	−	∥	−	⊥
A_u	∥	−	⊥	−
B_g	−	⊥	−	∥
B_u	⊥	−	∥	−

Table 25. Selections rules for rhombic point groups

C_{2v}	A_1	A_2	B_1	B_2
A_1	z	—	x	y
A_2	—	z	y	x
B_1	x	y	z	—
B_2	y	x	—	z

D_2	A	B_1	B_2	B_3
A	—	z	y	x
B_1	z	—	—	—
B_2	y	—	—	—
B_3	x	—	—	—

Table 26. Selection rules for tetragonal point groups

C_4	A	B	E
A	∥	—	⊥
B	—	∥	⊥
E	⊥	⊥	—

D_4	A_1	A_2	B_1	B_2	E
A_1	—	∥	—	—	⊥
A_2	∥	—	—	—	⊥
B_1	—	—	—	∥	⊥
B_2	—	—	∥	—	⊥
E	⊥	⊥	⊥	⊥	∥

S_4	A	B	E
A	—	∥	⊥
B	∥	—	⊥
E	⊥	⊥	∥, ⊥

D_{2d}	A_1	A_2	B_1	B_2	E
A_1	—	—	—	∥	⊥
A_2	—	—	∥	—	⊥
B_1	—	∥	—	—	⊥
B_2	∥	—	—	—	⊥
E	⊥	⊥	⊥	⊥	∥

C_{4h}	A_g	B_g	E_g	A_u	B_u	E_u
A_g	—	—	—	—	∥	⊥
B_g	—	—	—	—	∥	⊥
E_g	—	—	—	⊥	⊥	∥
A_u	—	—	⊥	—	—	—
B_u	∥	∥	⊥	—	—	—
E_u	⊥	⊥	∥	—	—	—

It is in the orthorhombic system where the orbital degeneracy is already completely removed (i.e., only the levels A, B remain) and further changes in local symmetry are reflected only in the selection rules.

The existence or absence of the center of symmetry is especially important: the absence of the center of symmetry removes the otherwise forbidden, which leads to a greatly increased intensity of absorption.

Table 27. Selection rules for trigonal point groups

D_{3d}	A_{1g}	A_{1u}	A_{2g}	A_{2u}	E_g	E_u
A_{1g}	—	—	—	∥	—	⊥
A_{1u}	—	—	∥	—	⊥	—
A_{2g}	—	∥	—	—	—	⊥
A_{2u}	∥	—	—	—	⊥	—
E_g	—	⊥	—	⊥	—	⊥
E_u	⊥	—	⊥	—	⊥	—

D_3	A_1	A_2	E
A_1	—	∥	⊥
A_2	∥	—	⊥
E	⊥	⊥	∥, ⊥

C_{3v}	A_1	A_2	E
A_1	∥	—	⊥
A_2	—	∥	⊥
E	⊥	⊥	∥, ⊥

C_3	A	E
A	∥	⊥
E	⊥	∥, ⊥

3. Molecular Orbital Theory

3.1 Introduction

An understanding of chemical bonds and of the electronic structures of crystals and minerals is of fundamental importance for understanding mineral matter in general.

Atoms in geochemistry (in solutions and melts as in minerals) never occur in a free state; their properties in compounds are not additive. Because of this, properties in compounds are not additive. Because of this, properties of free atoms (charges, ionization potentials, average orbital radii) are of significance only as far as they enter into the calculations of electronic states of compounds. In these calculations they are used as effective values, possibly values determined specifically for a given compound; they are used also together with other data which do not represent properties of the elements (for example, exchange integrals). Such calculations, even approximate ones, are made by much more complicated methods than additive considerations.

Considerations of the geochemical distribution of the elements, mechanisms of inorganic reactions, different properties of oxygen and sulfur compounds, forms of transfer of ore components, nature of hydrothermal solutions, changes in mantle properties viewed essentially as changes in the chemical bonding in minerals at high pressures, and all non-thermodynamic aspects of chemical processes find a basis in data on the nature of the chemical bond.

Crystal structures are no more than empirically determined geometrical arrangements of atoms. Structural crystallography in its very principles does not contain the laws of chemical bonding. These laws are supplied by quantum chemistry. They must provide explanations of the crystal structures themselves: why one or another coordination and valency of atoms should occur, why certain isomorphous substitutions should be encountered while others are not permitted, why interatomic distances vary for the same pair of atoms in different structures, why the radiacls are formed in oxygen compounds and sulfosalts, what forces of cohesion act between atoms in crystals and between sheets in sheet silicates, what is the temperature- and pressure-dependence of crystal structures, and so on.

Properties of minerals depend not only on the nature of constituent atoms and their arrangements in a structure; the crystal structure does not explain properties directly. It is certainly a necessary prerequisite for understanding the properties, but knowledge of electronic structure is just as necessary. Optical, magnetic, electric, chemical, mechanical properties result from electronic properties governed by chemical bonding. Studies of the electronic structure lead to the interpretation of properties and parameters describing them.

Introduction

One can distinguish three periods in the development of ideas on chemical bonding, connected in fact with the development of ideas about the atoms thermselves: (1) before Bohr's theory of the atom (Berzelius, Butlerov, Werner); (2) on the basis of Bohr's concept of atom but before quantum mechanics (Kossel, Lewis); (3) quantum mechanical, (a) general concepts and simple systems (Heitler and London, Hund, Mulliken), (b) molecular orbital and band calculations, spectroscopy measurements of crystals.

Vestiges of each of these periods remain in contemporary ideas, where there thus coexist with concepts at very different levels of sophistication.

Already in the beginning of the 19th century, Berzelius (1812), based on the phenomenon of electrolysis as understood at that time, suggested that compounds are composed of positively (metal) and negatively (oxygen, sulfur, halogens) charged particles; that chemical forces of cohesion were reduced to the electrostatic interaction between these particles. This suggestion proved sufficient for the calculation of lattice energies of ionic crystals in the Madelung-Born model which followed more than a hundred years later (see Chap. 7.2).

Considerations of directional chemical forces and valencies, then led to the first models of the stereochemistry of organic compounds as valent diagrams (Butlerov, Kuper, Kekulé) and to structural models of inorganic complexes commonly having six-fold coordination (Werner).

Immediately after Bohr's explanation of the structure of the atom, the first two electronic models of the chemical bond, suggested by Kossel and Lewis, appeared simultaneously.

In Kossel's model: (1) the completely filled eight-electron (s^2p^6) and eighteen-electron ($s^2p^6d^{10}$) shells are most stable; (2) during interactions some of the atoms give up the surplus electrons beyond these complete shells, thus becoming the positively charged ions, whereas other atoms accept these electrons in their incomplete shells, thus becoming the negatively charged ions; (3) these positively and negatively charged ions are bonded by electrostatic interactions.

In the Lewis model the electron is not transferred completely from one atom to another, but the unpaired electron of every atom entering in the bonding becomes paired and the bonding is effected by these two electrons. This common pair of electrons explained the connecting lines in structural chemical formulas.

The insuperable restriction of these models of chemical bonding is the fact that they are based on the Bohr (orbital) model of atom. As a consequence of this, the depiction of the chemical bond was far from an adequate description of interaction of atoms and did not suggest in principle the possibility of calculations of bonding in specific compounds.

Thus as quantum mechanics (1926) and then quantum chemistry (1927–1930) became established and tested with the simplest molecules, and as molecular orbital theory was developed and applied to complexes and crystals, the Kossel-Lewis ideas found wide-spread acceptance in inorganic chemistry and geochemistry. It was from this basis that the systems of ionic radii were developed, and the lattice energies of ionic crystals calculated. Their influence on geochemistry and mineralogy was enormous; their persistence is still evident today. At present the Kossel-Lewis models have only historical significance. They did not enter into quantum mechanical concepts, which were developed from new principles.

The third period in the development of ideas on the chemical bond began with the Schrödinger equation (1926) and pertinent calculations of the hydrogen atom and later the helium atom. In 1927, using the Schrödinger equation, Heitler and London calculated molecule hydrogen ion H_2^+ and hydrogen molecule H_2; this study initiated quantum chemistry. Already in these studies the types of interaction that constitute the mechanism of the chemical bond were revealed, the new nonclassical type of interaction, that of exchange interaction, was derived, the Coulomb, exchange and overlap integrals were written, and the order of the approximate solution of the Schrödinger equation for the molecule was suggested. This approach was called the Valence Bond method (VB).

In 1929–1930, Hund and Mulliken repeated the calculation by the Molecular Orbital method (MO).

With these calculations, quantum mechanical theory of the chemical bond began. The VB and MO methods were developed as two ways of representing the wave function and two mathematical treatments of the problem of the interaction of atoms.

The contemporary state of the chemical bond theory can be characterized as follows.

1. Even the presentation of the chemical bond problem is multifarious. Its mineralogical and geochemical aspects are reduced to chemical concepts and theories. It should be emphasized that the general theory of chemical bonding in minerals does not differ in any way from that in inorganic crystals. Its chemical aspects (valency, ionic radii, electronegativity, and the complicated and versatile concept of the chemical bond) are reduced to physical concepts (electronic structure as the physical equivalent of chemical bond, types of interactions of atoms and electrons, forces and energies, electron density distribution). All these physical parts of the problem are expressed by quantum-mechanical concepts (wave functions, molecular orbitals, operator of energy, Schrödinger equation). The quantum-mechanical problems lead to mathematics: secular equation, overlap and exchange integral, spherical harmonics etc., which here take the physical meaning of the bond characteristics (and vice versa: from the mathematical treatment to quantum chemistry, to physical pictures, to chemical concepts, to mineralogical and geochemical interpretations and applications).

2. In mineralogy, as in chemistry, it is becoming increasingly clear that the introduction of these equations, operators, and integrals is inevitable, in spite of difficulties connected with the assimilation of complicated quantum mechanical descriptions of the chemical bond and its mathematical expression. It is now the only way of considering the nature of the chemical bond and its features in different compounds. All chemical concepts and ideas concerning the interactions of atoms in compounds must be considered from the point of view of their correspondence to the quantum-mechanical description.

3. In the initial stages the problem was mainly to understand the general nature of the chemical bond in its quantum-mechanical expression. Now the molecular orbital, bond orbital, energy band calculations have been made for many specific crystals and minerals, not as type schemes but with numerical data correlated with the experimental (spectroscopic) observations.

Introduction

4. The possibilities of chemical bonding calculations are radically increased and simplified owing to use of computers and to the availability of standard and special programs, tabulated data on the wave functions, overlap integrals and other parameters.

5. The possibilities of experimental representation of chemical bonding parameters are enormously increased owing principally to the development in the fifties of solid state spectroscopy, i.e., X-ray and electron spectroscopy, Mössbauer spectroscopy, electron paramagnetic resonance, nuclear magnetic and nuclear quadrupole resonance. The electronic structure diagrams are directly related to some of these spectra. Each of the methods of solid state spectroscopy gives the parameters needed to characterize features of chemical bonding directly, not only in general terms, but also in specific compounds.

6. Along with all this, it is natural in this time of transition to expect the coexistence of concepts and ideas that have arisen in different periods and that now coexist with contemporary concepts. This is connected with the fact that all basic principles used by mineralogists and geochemists in descriptions of the chemical bond (electronegativity, lattice energy, ionic radii, valency and charges) are now in a state of radical revision.

7. It is of the outmost importance that by the sixties the two methods of quantum-mechanical treatment of the chemical bond – the Molecular Orbital (MO) and Valence Bond (VB) – became sufficiently developed to describe different aspects of chemical bonding in many types of compounds. This transformed them from methods of calculation of the simplest molecules to a well-grounded approach to the chemical bond in general, with systems of logically and mathematically interrelated concepts and methods of calculation. It proved then that in the early calculations of simple molecules both MO and VB methods were considered as complementary to each other, their further development showed quite definitely the predominance of the MO method and the insufficiency of the VB method.

However, since the MO and VB methods are not only schemes of calculation, but also ramified systems of concepts, definitions, and parameters, all the considerations of the chemical bond theory inherit the features of the initial models of these two methods. That is why, in considering each problem of chemical bonding, one must specify carefully the system of reference, either the MO or VB method by which the problem is being treated and discussed.

It should be noted also that historically the preceding decades, which were the formative years of modern chemistry, saw the definite prevalence of the VB method, whereas the MO method exerted almost no influence on its development. It is natural that in this period, such a fundamental monograph on the chemical bond as that of Pauling [113] is completely based on the VB method.

Solid state spectroscopy promoted the contemporary development of the MO method and resultant changes in chemical bond concepts that began in the sixties. The approaches are now trending to bond-orbital method and energy band calculations.

There are now monographs and reviews considering the MO method at different levels [78, 79, 83, 86, 88, 90, 91, 94–96, 104, 106, 110, 111, 114, 116], the VB method [87, 89, 97, 113], the general problem of the chemical bond [82, 87, 89,

93, 94, 99, 103, 113, 118–120, 124], and results of the calculation of minerals, complexes, and molecules [108, 109, 423, 424, 426–428, 431, 432, 449, 451, 452, 460, 477–480].

3.2 General Theory of the Chemical Bond; Molecular Orbital Method; Valence Bond Method

3.2.1 Description and Systematics of Molecular Orbitals

In order to understand the concept of molecular orbitals (MO) obtained usually as combinations of atomic orbitals, let us consider them in the same order as atomic orbitals (see Chaps. 1.3–1.8), because all the notions discussed earlier for atomic orbitals have a logical and mathematical development applied to molecules.

Wave Function. Just as one-electron wave functions, which describe the behavior of each electron in an atom (i.e., atomic orbitals, AO) and represent a model for electron structure of atom, molecular orbitals are one-electron wave functions describing the electron behavior in the field of two or more nuclei and represent quantum-mechanical model for the electron structure of molecule or complex, i.e., a chemical bond model.

Molecular orbitals are obtained as the linear combination of atomic orbitals (MO LCAO):

$$\psi_{MO} = \psi_A \pm \psi_B.$$

Near the atom A the electron behavior is approximated by the atomic orbitals ψ_A, while near the atom B by ψ_B. The wave functions ψ_A and ψ_B are here the atomic orbitals 1s, 2s, 2p, 3p and so on. For example, for the H_2 molecule (the hydrogen AO is 1s):

$$\psi_{MO} = 1s_A \pm 1s_B.$$

For the CO molecule (the carbon AOs are $2s^2$, $2p^3$, the oxygen AOs are $2s^2 2p^4$) some molecular orbitals are:

$$\psi'_{MO} = 2s_C \pm 2s_O; \quad \psi''_{MO} = 2p_C \pm 2p_O \quad \text{and others.}$$

Normalization and molecular orbital coefficients. Wave function ψ_{MO} has the same physical meaning as atomic orbitals (see Chap. 1.3): its square ψ^2_{MO} gives the probability for finding an electron in element volume dv. In order to normalize to unity the probability of finding an electron in all the space of a given wave function one introduces the normalizing constant N (as for AO; see Chap. 1.4): $N^2 \int \psi^2_{MO} dv = 1$.

Thus the MO can be written:

$$\psi_{MO} = N(\psi_A \pm \psi_B).$$

Description and Systematics of Molecular Orbitals

To take into account the contribution of every AO (ψ_A and ψ_B) in MO one uses coefficients c_i:

$$\psi_{MO} = N(c_1\psi_A \pm c_2\psi_B)$$

or designating $c_1/c_2 = \lambda$:

$$\psi_{MO} = N(\lambda\psi_A \pm \psi_B).$$

We obtain the expression for the normalizing constant N:

$$N^2\int\psi_{MO}^2 dv = 1; \quad N^2\int(c_1\psi_A + c_2\psi_B)^2 dv = 1;$$

$$N^2[c_1^2\int\psi_A^2 dv + c_2^2\int\psi_B^2 dv + 2c_1c_2\int\psi_A\psi_B dv] = 1.$$

Since the atomic orbitals ψ_A and ψ_B are used already normalized then $\int\psi_A^2 dv = 1$ and $\int\psi_B^2 dv = 1$.

The integral $\int\psi_A\psi_B dv$ is denoted by the letter S and is called the overlap integral (see below). Then

$$N^2[c_1^2 + c_2^2 + 2c_1c_2 S] = 1; \quad N = \frac{1}{\sqrt{c_1^2 + c_2^2 + 2c_1c_2 S}};$$

$$\psi_{MO} = N(c_1\psi_A + c_2\psi_B) = \frac{1}{\sqrt{c_1^2 + c_2^2 + 2c_1c_2 S}}(c_1\psi_A + c_2\psi_B).$$

For homonuclear molecules (H_2, O_2 etc.): $N = 1/\sqrt{2c^2 + 2c^2 S}$. If one neglects overlapping, i.e. taking $S = 0$, then $c_1^2 + c_2^2 = 1$; $2c^2 = 1$ and $c = 1/\sqrt{2}$. Thus $\psi_{MO} = 1/\sqrt{2}(\psi_A + \psi_B)$. For the general case of heteronuclear molecules, coefficients c_1, and c_2 have to be calculated (see below).

σ- and π-Molecular Orbitals; Bonding, Antibonding and Nonbonding Molecular Orbitals. Picture representation of atomic orbitals (of angular part of wave function) is given as the boundary surfaces (see Chap. 1.4) and for electrons in all types of atoms there are only four kinds of these surfaces: s, p, d, f (Fig. 10).

It is natural to propagate this model for the case of combination of atoms by the MO LCAO method. This obviously requires the consideration of all possible overlapping between the s, p, and d orbitals. By this procedure, therefore, one obtains boundary surfaces for corresponding molecular orbitals.

It emerges that all possible types of pair combination for s, p and d orbitals are restricted by the two types: σ and π molecular orbitals; each of them can be bonding (σ^b, π^b) and antibonding (σ^*, π^*).

The overlapping of atomic orbitals by lobes gives the σ molecular orbitals (Fig. 38) which are symmetric with respect to rotation about the bond direction (z-axis). The σ orbitals are made of overlapping of $s-s$, $s-p_z$, $s-d_{x^2-y^2}$, $s-d_{z^2}$ atomic orbitals (i.e., the s orbitals give only σ molecular orbitals) and also of overlapping p_z-p_z, $p_z-d_{z^2}$, $p_z-d_{x^2-y^2}$.

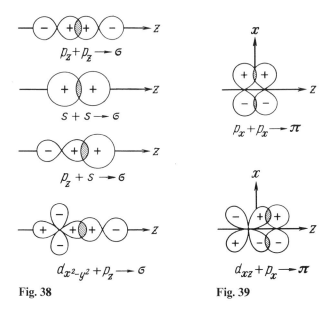

Fig. 38. Formation of σ molecular orbitals from atomic s, p, d orbitals

Fig. 39. Formation of π molecular orbitals from atomic p and d orbitals

The π orbitals arise from overlapping of p_x-p_x, p_y-p_y, p_x-d_{xz} (and similar) atomic orbitals and are not symmetric with respect to rotation about the bond direction (Fig. 39). The π bonds are always far less stable than σ bonds.

Antibonding σ* and π* molecular orbitals are formed from the same pairs of atomic orbitals not, however, as a result of their addition, but of subtraction. For example, for H_2 molecules (Fig. 40):

$$\sigma^b = c_1 1s_A + c_2 1s_B,$$
$$\sigma^* = c_1 1s_A - c_2 1s_B.$$

The atomic orbital overlapping by the lobes of the same sign leads to a bonding molecular orbital, while that of opposite sign to an antibonding.

In bonding orbitals the electron density increases in the region between the two nuclei, the molecular orbital energy is less than that of the atomic orbitals, the molecule is stable, the overlap integral $S > 0$, the electron spins are antiparallel.

In antibonding orbitals the electron density between two nuclei diminishes to zero, the electrons with parallel spins repulse, the molecule energy is higher than the AO energies and the molecule is unstable, the overlap integral $S < 0$.

Every pair of atomic orbitals forming the molecular orbital gives not one but always two molecular orbitals: bonding and antibonding. This is reflected in the formation of the two energy levels: a lower level corresponding to σ^b or π^b, and a higher one corresponding to σ^* or π^*. In the ground state, the bonding orbitals are usually completely occupied, while the antibonding orbitals are empty or partially occupied by d electrons.

Fig. 40a–d. Bonding (σ^b) and antibonding (σ^*) σ molecular orbitals. **a** Overlapping of s orbital boundary surfaces of two atoms; **b** overlapping of the wave functions; **c** contours of constant electron density of σ^b and σ^* molecular orbitals; **d** corresponding molecular orbital scheme

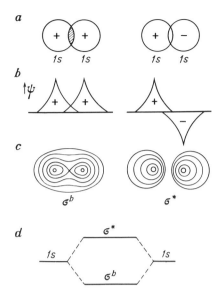

Even (g – gerade) and uneven (u – ungerade) molecular orbitals. The σ^b molecular orbitals made from identical atomic orbitals have a center of symmetry and are referred to as even, and often abbreviated as σ_g. The antibonding σ^* orbitals without a center of symmetry are uneven (σ_u^* or simply σ_u). On the contrary, π^b bonding orbitals are uneven (π_u) while π^* antibonding are even (π_g^* or simply π_g). The parity in the MO theory, as in the crystal field theory, determines some of the selection rules for optical transitions.

For the molecular orbitals mode from different atomic orbitals there is no classification on the g and u orbitals.

Compare now the different notation for the molecular orbitals:

$\sigma^b - \sigma_g - \sigma$ and $\pi^b - \pi_u - \pi$ – bonding,

$\sigma^a - \sigma_u - \sigma^*$ and $\pi^a - \pi_g - \pi^*$ – antibonding.

Nonbonding Orbitals. Ligand atomic orbitals not having a type of orbital among the metal atomic orbitals with which they could overlap are called nonbonding orbitals. They do not form molecular orbitals. The possibility of molecular orbital formation is strictly stipulated by symmetry conditions. Symmetry type of nonbonding ligand orbitals has no analogs among the central atom orbitals. The overlap integral in this case is equal to zero (Fig. 41).

Energy Level Diagrams, Electron Configurations, and Terms. Just as in atomic energy level diagrams representing the $1s$, $2s$, $2p$... atomic orbital energies and the number of electrons populating them, the molecule energy level diagrams show the energies of the molecular orbitals made from the every possible combination of the atomic orbitals.

To achieve this, one constructs the energy levels for each of the atoms and then forms from them the molecular orbitals under the following conditions: (1) the

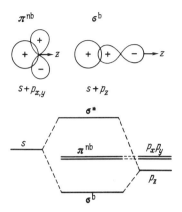

Fig. 41. Nonbonding and bonding orbitals. (Summary overlapping of s and p_x atomic orbitals is nill; these orbitals are of different symmetry with respect to axis of the molecule)

atomic orbitals should have a comparable energy, (2) they must overlap appreciably, (3) they must have the same symmetry type, (4) each pair of atomic orbitals forming a bonding molecular orbital must give at the same time an antibonding molecular orbital, (5) each of the atomic orbitals contributes in greater or lesser degree to all the molecular orbitals of the same symmetry type.

For molecules, as for atoms, the electron distribution in the orbitals is written as the electron configuration. For example, for the oxygen atom: $1s^2, 2s^2 2p^4$; for the oxygen molecule (with 12 valent electrons):

$$[1s^2][1s^2](\sigma_s^b)^2(\sigma_s^*)^2(\sigma_z^b)^2(\pi_x^b,\pi_y^b)^4(\pi_x^*)^1(\pi_y^*)^1.$$

The electron configurations of MO are obtained by the same Aufbau principle as for atoms (see Chap. 1.7): electrons are filled in successively from the lowest molecular orbitals upwards with no more than two electrons with paired spins in each nondegenerate molecular orbital (according to Pauli's exclusion principle).

The electronic state of a molecule as a whole is represented by a set of all possible molecular orbitals and is described by a term obtained analogously as a term of the atoms (see Chaps. 1.8 and 2.2). Term symbols are designated:

for atoms: ^{2S+1}L with $L=0$, 1, 2, 3, 4 ...
                                                          S, P, D, F, G ...

for linear molecules: $^{2S+1}\Lambda$ with $\Lambda = 0$, 1, 2, 3, 4 ...
                                                          Σ, Π, Δ, Φ, Γ ...

for nonlinear molecules and complexes: $^{2S+1}\Gamma$ where Γ is the symmetry type (A, B: one-dimensional, nondegenerate type; E: two-dimensional, doubly degenerate; T: three-dimensional, triply degenerate, see below and cf. also p. 68).

For the molecules and complexes with closed electron shells or with all paired electrons in the molecular orbitals, both the angular momentum and spin are equal to zero and thus the term is $^1\Sigma$ for linear molecules and $^1A_1$ for nonlinear molecules.

For molecules with one of the molecular orbitals having one unpaired electron ($S=1/2$ and $2S+1=2$) the term is for $[...]\sigma^1$ molecular orbital $^1\Sigma$, for $[...]\pi^1$ $^1\Pi$, for $[...]a^1$ ^2A, for $[...]b^1$ ^2B, for $[...]e_g^1$ ^2E_g, for $[...]t_{2g}^1$ $^2T_{2g}$ etc. For an electron configuration with two and more unpaired electrons there are several terms: ground state term and terms of the one or more excited states. For example, for $[...]\pi^2$ the ground state term is $^3\Sigma$ and the excited ones are $^1\Sigma$, $^1\Delta$ (see [22], p. 723–734).

Molecular Orbital Diagrams for A_2, AB_1, AB_2 Linear Molecules and Molecule Ions. The sequence for discussion of the molecular orbital for these types of molecule may be as follows:

diatomic:

homonuclear A_2 molecules with 2, 4...14 electrons and A_2 molecule ions with 1, 3...15 electrons;

heteronuclear AB molecules with 4, 6...14 electrons and AB molecule ions with 3, 5...15 electrons;

triatomic:

linear AB_2 molecules (and bent AB_2 molecule ions; see below).

The molecular orbital diagram for diatomic A_2 molecules and molecule ions is shown in Figure 42; it is the same for all the A_2 compounds. One discerns further the A_2 molecule by a number of electrons taking into account only valent electrons. For example, the electron number is 5 for N atom ($2s^2 2p^3$), 6 for O atom ($2s^2 2p^4$), 7 for F atom ($2s^2 2p^5$) and 10 for N_2 molecule, 12 for O_2, 14 for F_2 (with

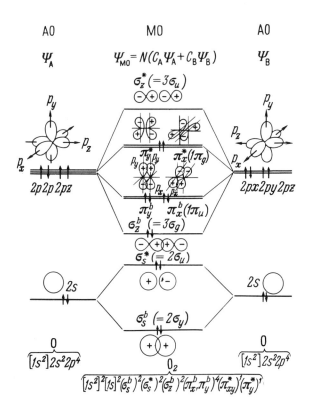

Fig. 42. Molecular orbital scheme for A_2-type molecules

even electron numbers) and 11 for N_2^-, 13 for O_2^-, 15 for F_2 molecule ions (with uneven electron numbers).

The electron configurations for these molecules and molecule ions are obtained from the same MO scheme (Fig. 42) filled by the respective number of the electrons. The A_2 molecules, except O_2, have all the filled molecules orbitals with paired electrons and are thus characterized by a single term $^1\Sigma_g$. In the 12-electron molecule O_2 the electrons in the $(\pi_x^*)^1 (\pi_y^*)^1$ molecular orbital are unpaired and the O_2 molecule is thus paramagnetic and is characterized by the three terms: $^3\Sigma_g$ (the ground state), $^1\Delta_g$ and $^1\Sigma_g$.

The A_2 molecule ions (called also radical ions) with an uneven number of electrons are paramagnetic with a single unpaired electron in last molecular orbital, its state designating the very radical type:

11-electron $(\pi_2^*)^1$ radicals (for example, N_2^-):
$$(\sigma_1)^2 (\sigma_2)^2 (\pi_1)^4 (\sigma_3)^2 (\pi_2^*)^1;$$
13-electron $(\pi_2^*)^1$ radicals (O_2^-, S_2^- ...):
$$(\sigma_1)^2 (\sigma_2)^2 (\pi_1)^4 (\sigma_3)^2 (\pi_2^*)^3;$$
15-electron $(\sigma_2^*)^1$ radicals (F_2^-, Cl_2^-, O_2^{3-}, S_2^{3-} ...):
$$(\sigma_1)^2 (\sigma_2)^2 (\pi_1)^4 (\sigma_3)^2 (\pi_2^*)^4 (\sigma_2^*)^1.$$

For diatomic AB molecules (heteronuclear) the same MO scheme can be used and there are the same types of radical ions:

11-electron $(\pi_2^*)^1$ radicals: NO, CO^-, SO^+, PF^+ ...;
13-electron $(\pi_2^*)^3$ radicals: SO^-, ClO, PF^- ...;
15-electron $(\sigma_1^*)^1$ radicals: FCl....

There is no classification of AB molecules into even and uneven types of molecular orbitals existing in A_2 molecules.

In linear AB_2 molecules (for example, $CO_2 = O-C-O$, see Fig. 43) as distinct from A_2 and AB molecules, one starts with formation of oxygen (or other B_2 atoms) O...O group orbitals which are obtained in the same way as the molecular orbitals for the O_2 molecule (Fig. 42). One then determines the σ or π character for carbon atomic orbitals: s is always of σ type and always even thus $s \to \sigma_g$; p orbitals are always uneven and $p_z \to \sigma_u$, p_x and $p_y \to \pi_u$. When this is done, one obtains the molecular orbitals for CO_2 as the linear combinations of carbon atomic orbitals and O...O group orbitals (Fig. 43).

Molecular Orbital Classification According to Symmetry Types; the MO Scheme for the Bent Molecule Ion CO_2^-. MO classification into σ and π types is sufficient only for simple diatomic and linear triatomic molecules. This is related also to symmetry, not, however, to point group symmetry but to rotation symmetry: σ orbitals are cylindrically symmetrical about the molecule axis while π orbitals are not.

In bent AB_2 molecule ions as well as in AB_3, AB_4, AB_6 complexes MO, formation is determined by the point group of symmetry of these molecule ion or complexes and the molecular orbitals are labeled not σ and π but by the symmetry-type designations (for example, a_1, b_1, a_2, b_2 in bent CO_2^-).

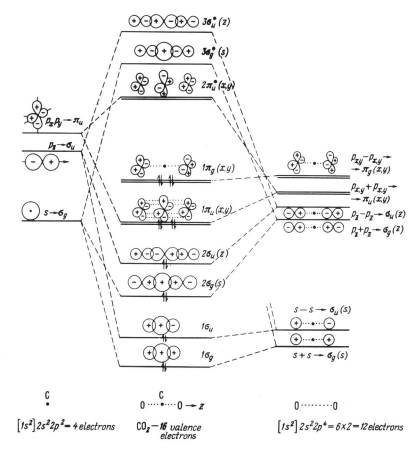

Fig. 43. Molecular orbital scheme for linear three-atomic molecule AB_2 (for example, $CO_2 = O-C-O$)

The MO diagrams for nonlinear molecules are constructed in the following manner (for CO_2^- taken as an example, see Fig. 44).

1. Determine the symmetry point group for the molecule, radical or complex. For CO_2^- it is C_{2v}. Write the symmetry elements for this point group: C_{2z} (twofold axis coinciding with the axis of the molecule ion), $\sigma_v(xz)$ (vertical plane of symmetry coinciding with xz coordinate plane) and $\sigma_v(yz)$.

2. Consider the action of the corresponding symmetry operations (a) upon each of the central atom (carbon here) atomic orbitals and (b) upon each of the ligands (O...O here) group orbitals. For that mark off only two kinds of the behavior: an orbitals is symmetrical with respect to the given symmetry operation (i.e., positive and negative lobes of atomic orbitals do not change under rotation about C_2 axis or reflection in σ_v plane) or antisymmetrical (the plus and minus lobes change). The results are marked respectively as $+1$ and -1. For example, one obtains for the p_z carbon orbital $+1, +1, +1$ with respect to $C_{2z}, \sigma_v(xz), \sigma_v(yz)$, for $p_x-1, +1$, -1 etc. There is in the C_{2v} point group in all four, a type of behavior labeled a_1, b_1,

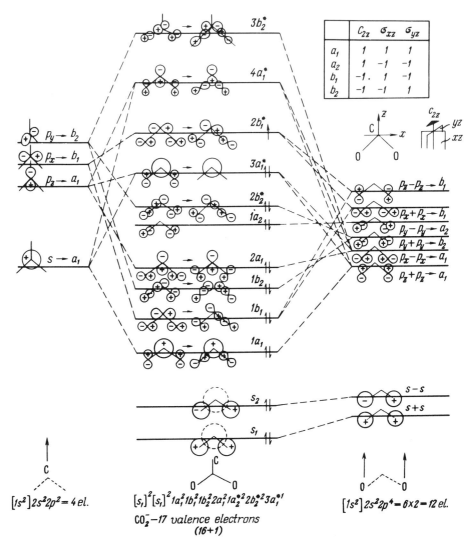

Fig. 44. Molecular orbital scheme for bent three-atomic molecule AB_2 (for example, CO_2^-)

a_2, b_2. The types symmetrical with respect to C_{2z} are designated a, those antisymmetrical b, symmetrical with respect to $\sigma_v(xz)$ are designated by the subscript 1 ($a_1 b_1$), antisymmetrical with respect to $\sigma_v(xz)$ and therefore symmetrical with respect to $\sigma_v(xz)$ are designated by the subscript 2 ($a_2 b_2$).

This procedure is completely analogous to the consideration of the action of the symmetry elements upon the atomic orbitals discussed in crystal field theory (see Chap. 2.2).

Two consideration are new here: (1) not only d orbitals but also s and p orbitals of a central atom are considered, and (2) by means of group orbitals one considers also ligand atoms. For example, two atomic orbitals p_z and p_z of the two oxygen

Description and Systematics of Molecular Orbitals

atoms of CO_2^- transform due to their addition into a_1 type and by substraction (in the case of different relative position of plus and minus lobes of the two p_z orbitals; see Fig. 44) into b_1 type.

Thus one obtains for the carbon atom $C(2s^2 2p^4)$ in CO_2^-:

	C_{2z}	$\sigma_v(xz)$	$\sigma_v(yz)$	
s	+1	+1	+1	a_1
p_x	−1	+1	−1	b_1
p_y	−1	−1	+1	b_2
p_z	+1	+1	+1	a_1

For two oxygen atoms O...O in CO_2^-:

	C_{2z}	$\sigma_v(xz)$	$\sigma_v(yz)$	
$s_1 + s_2$	+1	+1	+1	a_1
$s_1 - s_2$	−1	+1	−1	b_1
$p_x - p_x$	−1	+1	−1	b_1
$p_x + p_x$	+1	+1	+1	a_1
$p_y + p_y$	−1	−1	+1	b_2
$p_y - p_y$	+1	−1	−1	a_2
$p_z + p_z$	+1	+1	+1	a_1
$p_z - p_z$	−1	+1	−1	b_1

3. Obtain the molecular orbitals as the combination of the atomic orbitals of the central atom and of the ligand atom group orbitals, taking into account the following bilateral considerations: (a) only atomic orbitals with the same symmetry-type can combine to form the molecular orbital; the molecular orbitals are designated by the same symmetry-type symbols as the atomic orbitals from which they are constructed; (b) all atomic orbitals with the same symmetry interact giving a contribution to the corresponding molecular orbital; the amount of the atomic orbital contribution of each to a molecular orbital depends on the overlapping and on the AO energy difference; if there are several molecular orbitals of the same type, they are numbered from the lowest MO upwards: $1a$, $2a$, $3a$, and so on.

In Figure 44 each type of molecular orbital, for instance the a_1 molecular orbitals ($1a_1$, $2a_1$, $3a_1$, $4a_1$) should be shown as constructed from all the a_1 carbon orbitals and all the a_1 oxygen group orbitals. However, in this figure, for simlicity, the interactions are shown between the atomic and group orbitals of most similar energies, since the contribution from more remote orbitals is rather weak.

The crystal or molecular structure elements corresponding to a molecular orbital are represented by the whole CO_2^- molecule or by complexes AB_3, AB_4, AB_6, that is one-electron molecular orbitals describe each electron behavior in the field of all the nuclei of the complex. (Two electrons which can occupy the molecular orbital are distinguished only by spin direction).

Electron configuration lists the molecular orbital symmetry types and the numbers of electrons in these molecular orbitals. For example, for CO_2^-: $1a_1^2\ 1b_1^2\ 1b_2^2\ 2a_1^2\ (1a_2^*)^2\ (2b_2^*)^2\ (3a_1^{*1}$. The classification into bonding, antibonding and nonbonding orbitals is retained in these descriptions: a_1, a_1, a_2 hare here bonding orbitals, a_2^*, b_2^*, a_1^* antibonding.

Molecular orbital terms are designated by the same symmetry types as in the crystal field theory (see Chap. 2.2); for CO_2^- the term is $^2A_1$.

3.2.2 Molecular Orbital Energy and Coefficient Calculations (the H_2^+ Molecule Ion Example)

The consideration of this calculation pursues far-reaching aims: it reveals the heart of the quantum-mechanical approach to chemical bond: the whole deduction is from the Schrödinger equation (for molecules) as all thermodynamics is deduced from two basic laws. Here also the nature and types of chemical bond forces are manifested (already in the Hamiltonian operator expression) which were in principle inaccessible for prequantum approaches. In the course of the solution of the Schrödinger equation, the terms are introduced which both qualitatively and quantitatively characterize contributions of the different types of bond forces: the integrals of quantum chemistry, molecular orbital energies and coefficients. A choice of the wave function substituted in the Schrödinger equation determines the choice of the chemical bond method: molecular orbitals or valence bonds. Without knowledge of the secular equation (which is also used in many sections of solid state spectroscopy) it is difficult to follow the thread of modern bond theories.

It is expedient to discern in all stages of the calculation the three aspects: mathematical expression, physical meaning and chemical significance. It is extremely important to consider constantly, the integrals, the operators, the functions only as the expressions (as the symbols) for the corresponding physical phenomena.

Schrödinger equation for the molecule, operator of energy, interaction forces. In calculating electron behavior in the molecule as in atoms (see Chap. 1.3) and crystals (see Chap. 4.1), begin with the Schrödinger equation:

$$H\psi = E\psi.$$

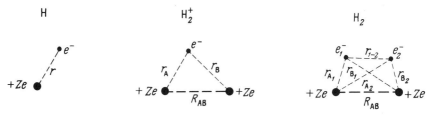

Fig. 45. Types of interactions for the hydrogen atom (H), hydrogen molecule ion (H_2^+) and hydrogen molecule (H_2)

Let us consider some comments to the meaning of the operator H and the wave function ψ. The Hamiltonian operator H is here, the same operator as in the calculations of free atoms (see Chap. 1.3) but it takes into account in addition the interactions depending on the fact that the electron in the molecule occurs in the field of two or more nuclei.

Let us write the Hamiltonian operators for the hydrogen atom (H), for the hydrogen molecule ion (H_2^+) and for the hydrogen molecule (H_2) (see Fig. 45):

(H) $\quad H = \dfrac{h^2}{8\pi^2 m}\nabla^2 - \dfrac{e^2}{r}; \quad \psi_A = \psi_{1s};$

(H_2^+) $\quad H = \dfrac{h^2}{8\pi^2 m}\nabla^2 = \dfrac{e^2}{r_A} - \dfrac{e^2}{r_B}; \quad \psi_{MO} = c_1\psi_A + c_2\psi_B;$

(H_2) $\quad H = \dfrac{h^2}{8\pi^2 m}(\nabla_1^2 + \nabla_2^2) - \dfrac{e^2}{r_{A_1}} - \dfrac{e^2}{r_{B_1}} - \dfrac{e^2}{r_{A_2}} - \dfrac{e^2}{r_{B_2}} + \dfrac{e^2}{r_{1-2}} + \dfrac{e^2}{R_{AB}};$

$\psi_{MO} = \psi_{MO_1} \cdot \psi_{MO_2} = (c_1\psi_{A_1} + c_2\psi_{B_1})(c_3\psi_{A_2} + c_4\psi_{B_2}).$

The first term in all three expressions gives the kinetic energy and the other terms give the potential energy of Coulomb interaction between two charges: electron – electron, electron – nuclear, nuclear – nuclear separated by the distances r_A, r_B, r_{1-2}, R_{AB} (see Fig. 45).

In the hydrogen atom it is simply the interaction between the nuclear charge $+Ze$ with the electron charge $-e$ with the electron radius, r, i.e., $+e^2/r$. This expression corresponds to the point charges model (as in the old quantum theory, see Chap. 1.2) and describes the interaction between the point electron and the point nucleus. However, since the electron density is distributed in a volume of a charge cloud described by wave function (ψ^2), all actions in quantum mechanics are made with wave functions and are written in the operator form: $H\psi = \left(\dfrac{h^2}{2m}\nabla^2 - \dfrac{e^2}{r}\right)\psi$ (see Chap. 1.3). By substitution of the wave function analytical expression (for hydrogen $1s\psi = \exp(+ar)$), multiplying of each operator term by this wave function and integrating over all the space in which the electron is distributed, we obtain the numerical value for the energy. In this quantum mechanical way[3] we obtain the same value as by the semiclassical method earlier (see Chap. 1.2).

The energy operator for the hydrogen molecule ion H_2^+ includes the electrostatic interaction between the electron and already with the two nuclei (and the interaction between the two nuclei). For the hydrogen molecule H_2 (with two electrons) it includes the interactions between each electron and its own and other nuclei, between two electrons and between two nuclei.

Thus the interaction forces determining the chemical bond between atoms are of one single origin: they are always forces of the electrostatic electron and nuclear interactions. However, since the interactions occur not between point charges but

---

3 A very readable account of the quantum mechanical calculation of the hydrogen ground state energy may be found in [89], pp. 140–142.

between the charges described by wave functions, the nonclassical interaction forces appear (see below).

The wave function ψ which is substituted in the Schrödinger equation is represented in the MO method for the H_2^+ molecule ions by the molecular orbital $\psi_{MO} = c_1 \psi_A + c_2 \psi_B$ constructed as a linear combination of the identical atomic orbitals of the two hydrogen atoms A and B ($\psi_A = 1s$ and $\psi_B = 1s$). For the H_2 hydrogen molecule having two electrons, the wave function ψ is a product of two identical molecular orbitals for each of the two electrons (ψ_{MO_1} and ψ_{MO_2}) each constructed as linear combination of atomic orbitals.

Molecular orbital energy and the variation principle. The energy expression can be obtained from the Schrödinger equation by multiplying both parts by ψ integrating over all the space of variables and taking E outside the integral sign:

$$H\psi = E\psi; \quad \int \psi H \psi d\tau = \int \psi E \psi d\tau; \quad \int \psi H \psi d\tau = E \int \psi^2 d\tau;$$

$$E = \frac{\int \psi H \psi d\tau}{\int \psi^2 d\tau}$$

where $d\tau$ is the volume element.

Thus, if the two-point charge interaction energy is $E = e^2/r_{1-2}$, the interaction energy expressed by the operator $H = e^2/r_{1-2}$ and for two electrons characterized by the wave functions ψ_1 and ψ_2 is given by:

$$E = \int \psi_1 H \psi_2 d\tau = \int \psi_1 \frac{e^2}{r_{1-2}} \psi_2 d\tau.$$

The energy value E is determined by utilizing the variation principle: one obtains the true energy E by minimizing the wave function ψ with respect to each of the coefficients c_i, i.e., $\dfrac{\partial E}{\partial c_1} = 0$ and $\dfrac{\partial E}{\partial c_2} = 0$.

Secular equation. Without unfolding for the time being the operator H term, let us substitute the ψ_{MO} expression in the energy equation:

$$E = \frac{\int \psi H \psi \cdot d\tau}{\int \psi^2 d\tau}; \quad \psi_{MO} = c_1 \psi_A + c_2 \psi_B;$$

$$E = \frac{\int (c_1 \psi_A + c_2 \psi_B) H (c_1 \psi_A + c_2 \psi_B) d\tau}{\int (c_1 \psi_A + c_2 \psi_B)^2 d\tau}$$

$$= \frac{\int (c_1 \psi_A H c_1 \psi_A + c_1 \psi_A H c_2 \psi_B + c_2 \psi_B H c_1 \psi_A + c_2 \psi_B H c_2 \psi_B) d\tau}{\int (c_1^2 \psi_A^2 + 2 c_1 c_2 \psi_A \psi_B + c_2^2 \psi_B^2) d\tau}$$

$$= \frac{c_1^2 \int \psi_A H \psi_A d\tau + 2 c_1 c_2 \int \psi_A H \psi_B d\tau + c_2^2 \int \psi_B H \psi_B d\tau}{c_1^2 \int \psi_A^2 d\tau + 2 c_1 c_2 \int \psi_A \psi_B d\tau + c_2^2 \int \psi_B^2 d\tau}.$$

These transformations are rather clumsy but simple.

Molecular Orbital Energy and Coefficient Calculations

It follows from this expression that the energy determination is here reduced to the estimation of the four types of integral important to quantum chemistry. The following symbols are introduced for these integrals: $\int \psi_A H \psi_A d\tau = H_{AA}$ and $\int \psi_B H \psi_B d\tau = H_{BB}$ are referred to as the Coulomb integrals; $\int \psi_A H \psi_B d\tau = \int \psi_B H \psi_A d\tau = H_{AB}$ are the exchange integrals; $\int \psi_A \psi_B d\tau = S_{AB}$ or simply S is the overlap integral $\int \psi_A^2 d\tau = S_{AA}$ and $\int \psi_B^2 d\tau = S_{BB}$ are the normalization integrals. If the ψ_A and ψ_B function are already normalized as is usual (see Chap. 1.4) then the normalization integrals are equal to zero.

Without unfolding here these integrals (it will be done below) let us write the energy expression:

$$E = \frac{c_1^2 H_{AA} + 2c_1 c_2 H_{AB} + c_2^2 H_{BB}}{c_1^2 + 2c_1 c_2 S + c_2^2}.$$

To obtain the minimum energy the expression is differentiated partially with respect to c_1 and then E/c_1 is set equal to zero. There are c_1 both in the numerator and denominator in the E expression.

To differentiate this expression with respect to c_1: (1) multiply the denominator by the derivative of the numerator with respect to c_1, (2) obtain the product of the numerator and the derivative of the denominator with respect to c_1, (3) substract the latter from the former, (4) divide the result by the square of the denominator:

$$\frac{\partial E}{\partial c_1} = \frac{(c_1^2 + 2c_1 c_2 S + c_2^2)(2c_1 H_{AA} + 2c_2 H_{AB}) - (c_1^2 H_{AA} + 2c_1 c_2 H_{AB} + c_2^2 H_{BB})(2c_1 + 2c_2 S)}{(c_1^2 + 2c_1 c_2 S + c_2^2)^2} = 0.$$

Since E/c_1, is equal to zero, then:

$$(c_1^2 + 2c_1 c_2 S + c_2^2)(2c_1 H_{AA} + 2c_2 H_{AB})$$
$$= c_1^2 (H_{AA} + 2c_1 c_2 H_{AB} + c_2^2 H_{BB})(2c_1 + 2c_2 S).$$

On dividing both parts by $(c_1^2 + 2c_1 c_2 S + c_2^2)$ one obtains on the right part the E value:

$$(2c_1 H_{AA} + 2c_2 H_{AB}) = \frac{(c_1^2 H_{AA} + 2c_1 c_2 H_{AB} + c_2^2 H_{BB})}{(c_1^2 + 2c_1 c_2 S + c_2^2)} (2c_1 + 2c_2 S);$$

$$(2c_1 H_{AA} + 2c_2 H_{AB}) = E(2c_1 + 2c_2 S) \quad \text{or}$$

$$c_1 (H_{AA} - E) c_2 (H_{AB} - SE) = 0.$$

On differentiating with respect to c_2 and taking $\partial E/\partial c_2 = 0$ equal to zero one obtains similarly:

$$c_1 (H_{AB} - SE) + c_2 (H_{BB} - E) = 0.$$

Thus obtain the two linear equations which can be written in the determinant form:

$$c_1 c_2 \begin{vmatrix} H_{AA}-E & H_{AB}-SE \\ H_{AB}-SE & H_{BB}-E \end{vmatrix} = 0.$$

Since the coefficients c_1 and c_2 are not equal to zero, then:

$$\begin{vmatrix} H_{AA}-E & H_{AB}-SE \\ H_{AB}-SE & H_{BB}-E \end{vmatrix} = 0.$$

This is widely used in quantum mechanics and in spectroscopy secular equation. (This name is retained from celestial mechanics.)

Solution of the secular equation gives the electron energy value E in a given molecular orbital ψ_{MO} and with knowledge of the E value, one determines the coefficients c_1 and c_2 from the above linear equations. For this divide each term by $H_{AB}-SE$ (taken $H_{AA}=H_{BB}$ because A and B here are the same $1s$ orbital of the hydrogen atom):

$$\begin{vmatrix} \dfrac{H_{AA}-E}{H_{AB}-SE} & 1 \\ 1 & \dfrac{H_{AA}-E}{H_{AB}-SE} \end{vmatrix} = 0,$$

that is, we obtain the equation $\begin{vmatrix} x & 1 \\ 1 & x \end{vmatrix} = 0$ type from which $x^2 - 1 = 0$; $x = -1$; $x = +1$.

Then

$$\frac{H_{AA}-E}{H_{AB}-SE} = -1 \quad \text{and} \quad E_1 = \frac{H_{AA}+H_{AB}}{1+S};$$

$$\frac{H_{AA}-E}{H_{AB}-SE} = +1 \quad \text{and} \quad E_2 = \frac{H_{AA}-H_{AB}}{1-S}.$$

The two energy values E_1 and E_2 correspond to bonding and antibonding molecular orbitals.

From the linear equation one obtains:

$$c_1(H_{AA}-E) + c_2(H_{AB}-SE) = 0$$

$$c_1 = -c_2 \left(\frac{H_{AB}-SE}{H_{AA}-E} \right).$$

For a homonuclear molecule (with $H_{AA}=H_{BB}$) one obtains[4]: $(H_{AB}-SE) = (H_{AA}-E)$ and hence $c_1 = -c_2$, i.e., in homonuclear molecules the coefficients in $\psi_{MO} = c_1\psi_A + c_2\psi_B$ are obviously equal.

---

4 From the secular equation: $(H_{AA}-E)(H_{BB}-E) = (H_{AB}-SE)^2$ but if $H_{AA}=H_{BB}$ then $(H_{AA}-E)^2 = (H_{AB}-SE)^2$ and $(H_{AA}-E) = (H_{AB}-SE)$.

Molecular Orbital Energy and Coefficient Calculations

The c_1 and c_2 values are determined from the normalization requirement:

$$\int \psi_{MO}^2 d\tau = 1; \quad \int (c_1\psi_A + c_2\psi_B)^2 d\tau = 1;$$

$$(c_1^2\psi_A^2 + 2c_1c_2\psi_A\psi_B + c_2^2\psi_B^2)d\tau = 1;$$

$$c_1^2\int\psi_A^2 d\tau + 2c_1c_2\int\psi_A\psi_B d\tau + c_2^2\int\psi_B^2 d\tau = 1.$$

Since the ψ_A and ψ_B wave function are already normalized (that is $\int\psi_A^2 d\tau = 1$ and $\int\psi_B^2 d\tau = 1$) and $\int\psi_A\psi_B d\tau = S$ and $c_1 = \pm c_2$ thus

$$c_1^2 \pm 2c_1c_2 S + c_1^2 = 1; \quad c_1^2(2 \pm 2S) = 1;$$

$$c_1 = c_2 = \frac{1}{\sqrt{2+2S}} \quad \text{and} \quad c_1 = -c_2 \frac{1}{\sqrt{2-2S}}.$$

Quantum Chemistry Integrals: Coulomb, Exchange and Overlap Integrals. There remains only to find numerical values of the molecular orbital energy (E_1 and E_2) and coefficients (c_1 and c_2). The above expressions for them are simple enough and one needs only to calculate the three integrals: H_{AA}, H_{AA} and S. Consider these calculations for the molecule ion H_2^+ taken as an example (see Fig. 45).

Unfold the operator H expression for H_2^+ and substitute it in the expressions of H_{AA} and H_{AB}

$$H_{AA} = \int\psi_A H\psi_A d\tau = \int\psi_A\left(-\frac{\hbar^2}{2m}\nabla^2 - \frac{e^2}{r_A} - \frac{2}{r_B}\right)\psi_A d\tau.$$

Recall the Schrödinger equation for the free hydrogen atom (see Chap. 1.3):

$$H\psi_A = E\psi_A \quad \text{or} \quad \left(-\frac{\hbar^2}{2m}\nabla^2 - \frac{e^2}{r_A}\right)\psi_A = E_A\psi_A.$$

On substituting it for H_{AA} one obtains:

$$H_{AA} = \int\psi_A\left(E_A - \frac{2}{r_B}\right)\psi_A d\tau = E_A - \int\psi_A \frac{e^2}{r_B}\psi_A d\tau = E_A + I;$$

$$H_{AA} = E_A + I.$$

Thus the H_{AA} integral represents the interaction energy of the electron with its own nuclear and with the other nuclei. It can be represented as a decrease of the hydrogen atom energy E_A caused by the electrostatic attraction (I) of the electron charge cloud with a density $e\psi_A^2$ to the other nuclear of the H_2^+ molecule ion.

On substituting for $\psi_A = \psi_{1s}$ its analytical expression one obtains:

$$-I = \int \psi_A \frac{e^2}{r_B} \psi_A d\tau = \int \psi_{1s} \frac{e^2}{r_B} \psi_{1s} d\tau = \int e^{-r} \frac{e^2}{r_B} e^{-r} d\tau$$

$$= e^2 \left[\frac{1}{R_{AB}} - \frac{e^{-2R}}{R_{AB}} (1 + R_{AB}) \right].$$

Similarly, on substituting the same operator H expression in the expression of the H_{AB} integral:

$$H_{AB} = \int \psi_A H \psi_B d\tau = \int \psi_A \left(-\frac{\hbar^2}{2m} \nabla^2 - \frac{e^2}{r_A} - \frac{e^2}{r_B} \right) \psi_B d\tau$$

$$= \int \psi_A \left(E_A - \frac{e^2}{r_B} \right) \psi_B d\tau = E_A \int \psi_A \psi_B d\tau - \int \psi_A \frac{e^2}{r_B} \psi_B d\tau$$

$$= E_A S + K.$$

The integral K referred to as *exchange (or resonance) integral* has no classical analog. It reflects the fact that the molecular orbital ψ_{MO} is constructed from both the atomic orbitals ψ_A and ψ_B and that the electron is distributed in both nuclei regions, its properties being partially those of ψ_A and ψ_B. For two electron hydrogen molecules it could be deceptively easily represented as an electron exchange between the two nuclei occurring as a result of the indiscernibility of electrons or of resonance of two structures with different electron distribution. However, the exchange integral exists also in the one-electron system (as in hydrogen molecule ion H_2^+) and its meaning is connected with the statistical and probabilistical picture of electron density distribution, in this case in the region around the two nuclei.

Overlap integral S can be calculated analytically or can be taken directly from the tables [34].

$$S = \int \psi_A \psi_B d\tau = \int \psi_{1s}^2 d\tau = \frac{1}{\pi} \int e^{-r_A + r_B} d\tau \quad \text{(in atomic units)}.$$

Using elliptical coordinates

$$\mu = \frac{r_A + r_B}{R_{AB}} \quad \text{and} \quad \nu = \frac{r_A - r_B}{R_{AB}}$$

and on integrating, one obtains

$$S = e^{-R}(1 + R_{AB} + 1/3 R_{AB}^2).$$

A comparison of the final expressions for I, K and S shows that for s electrons (we considered the 1s hydrogen atom orbital and the molecular orbitals constructed

Fig. 46. Energy variation as function of internuclear distance R_{AB} in H_2^+

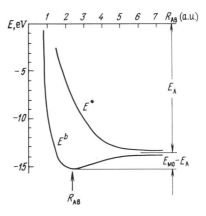

from it) these integrals depend only on the internuclear distance R_{AB}. For other electrons these integrals are presented in the tables ([84] and others) as functions of elliptical coordinates μ and ν.

Potential energy variation with internuclear distance. Substitute H_{AA} and H_{AB} thus obtained in the energy expressions E_1 and E_2 found by the secular equation solution:

$$E_1 = \frac{H_{AA} + H_{AB}}{1+S} = \frac{(E_A + I) + (E_A S + K)}{1+S} = \frac{E_A(1+S) + I + K}{1+S}$$

$$= E_A + \frac{I+K}{1+S};$$

$$E_2 = \frac{H_{AA} - H_{AB}}{1-S} = E_A + \frac{I-K}{1-S}.$$

However, in order to obtain the total potential energy add to these values of electron interaction energy the internuclear repulsion energy e^2/R_{AB}:

$$E_1 + e^2/R_{AB} \quad \text{and} \quad E_2 + e^2/R_{AB}.$$

Each of these values and their sum are functions of internuclear distance R_{AB}. In Figure 46 the two curves, one for a bonding state with $E_{MO} = E_1 + e^2/R_{AB}$ and other for an antibonding with $E_{MO} = E_2 + e^2/R_{AB}$, tend at $R_{AB} \to \infty$ to a limit with $E_A = E_B = 0.5 e^2/a_0 = -13.6 \text{ eV}^5$, i.e., to the hydrogen atom 1s is state energy.

The minimum on the curve of bonding state energy (E^b in Fig. 46) corresponds to the equilibrium internuclear distance R_{AB} and shows that the hydrogen molecule ion H_2^+ is stable, that the ψ_{MO} molecular orbital energy is less than the free atom energy E_A; the energy difference between E_{MO} and E_A corresponds to the molecule dissociation energy. The values calculated within the limits of this simple

---

5 The I and K values are measured in e^2/a_0 units (a_0 is the Bohr radius); $e^2/a_0 = 27.2 \text{ eV}$, i.e., is equal to the doubled energy of the hydrogen atom ground state; I and K vary from 1 to 0 (with R_{AB} respectively 0 and ∞). S is dimensionless.

theory are $E_{MO} - E_A = 1.76$ eV and $R_{AB} = 1.32$ Å. The improved calculations give results very close to the experimental values 2.79 eV and 1.06 Å.

3.2.3 Molecular Orbital Calculation for Octahedral and Tetrahedral Complexes of Transition Metal and Nontransition Element Ions

The general scheme of consideration of these complexes of molecular orbitals is similar to that of the simple molecules like H_2^+ discussed earlier.

Let us first construct a qualitative molecular orbital diagram for these complexes and then consider methods of calculation.

Octahedral complexes (band on example of CrF_6^{3-}).

A. The Molecular Orbital Diagram Construction. Basis atomic orbitals used to construct molecular orbitals are: (a) that of a central ion in nontransition element complexes: ns, np and empty nd; for instance, for Al $3s^2 3p^1$ and empty $3d$ (the same type AO and in the same order can be written for other nontransition elements); (b) that of a central ion in transition metal complexes: nd, $(n+1)s$, $(n+1)p$; for instance, for Cr $3d^5 4s^1$ and empty $4p$ (the same for other transition metals: $3d^n 4s^2 4p^0$ or $4d^n 5s^2 5p^0$); (c) that of ligand ions: ns, np; for instance, for oxygen: $2s^2 2p^4$, for fluorine $2s^2 2p^5$, for sulfur $3s^2 3p^4$, for chlorine $3s^2 3p^5$.

Thus the discussion of the CrF_6^{3-} complex is of more general significance. Variations in qualitative MO diagrams depend on atomic orbital relative energies, on a number of electrons in the molecular orbital, and also on distortions of coordination polyhedra leading to deviation of the symmetry from the octahedral one.

The energy differences between the $2s$ and $2p$ orbitals are rather large and, moreover, the oxygen, fluorine etc., $2s$ orbitals are still lower with respect to the $3s$, $3p$, $3d$ cation orbitals, hence one considers usually the bond formation between the cation $3s$, $3p$, $3d$ orbitals and the $2p$ orbitals of the six oxygen or fluorine ions.

1. Consider the transformation of these atomic orbitals into the symmetry types in the octahedral complex (Fig. 47).

The metal s, p, d atomic orbital transformation into the symmetry types for the octahedral O_h point group can be taken from Table 14.

The concept of symmetry type and the derivation of these types for atomic orbitals were considered at an example of the C_{3h} point group earlier (see Fig. 30).

Here, in Figure 47, only final results of these transformations for O_h point group are shown as follows:

$$s \to a_{1g}; \quad p \to t_{1u}; \quad d_{xy}, d_{yz}, d_{xz} \to t_{2g};$$
$$d_{x^2-y^2}, d_{z^2} \to e_g.$$

(Using Table 14 one can obtain the AO transformations into the symmetry types for the all important point groups.)

The fluorine (or oxygen etc.) ligand ions are considered not separately but all together because they are equivalent in the octahedron, and molecular orbitals are delocalized over all the complex. Hence ligand ions form the group orbitals made

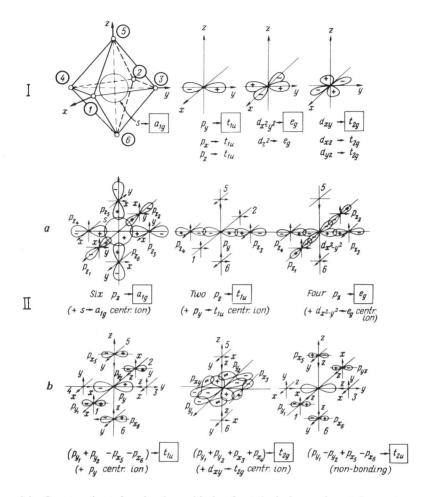

Fig. 47a and b. Construction of molecular orbitals of octahedral complex. I Symmetry types of atomic s, p, d orbitals of central ion in octahedron. II Symmetry types of group orbitals of six ligand oxygen ions formed from p_z, p_y, p_x orbitals (shown together with central ion orbitals of the same symmetry). **a** p_z orbitals forming σ bonds with $s \to a_{1g}$, $p \to t_{1u}$, $d \to e_g$ central ion orbitals; **b** p_x and p_y orbitals forming π bonds with $p \to t_{1u}$ and $d \to t_{2g}$ central ion orbitals and nonbonding orbitals

as a combination of the several ions atomic orbitals (see the construction of group orbitals for CO_2 and CO_2^- in Figs. 42 and 43).

All three p_x, p_y, p_z metal orbitals in the center of a complex contribute to one of the molecular orbitals (see Fig. 47) while the six p_x, six p_y and six p_z atomic orbitals from the six fluorine ions in the vertices of the octahedron form several group orbitals which belong to different symmetry types but are of the same energy (i.e., these group orbitals are degenerate). In Figure 47 they are shown together with the central ion atomic orbitals, hence one can conceive intuitively that the related group orbitals are of the same symmetry.

To designate these group orbitals, one numbers the ligand ions as in Figure 47 and constructs for each of them the coordinate system with the z axis directed to the central ion.

There are two types of ligand p orbitals. The six p_z orbitals directed to the central ion s, p, d orbitals form σ bonds while the p_x and p_y orbitals of the six ligand ions form π bonds.

2. On combining the central ion atomic orbitals with the ligand group orbitals having the same symmetry, one obtains the molecular orbitals for an octahedral complex:

σ molecular orbitals:

$$a_{1g} = (s \to a_{1g})_{centr} + [1/\sqrt{6}(p_{z_1} + p_{z_2} + p_{z_3} + p_{z_4} + p_{z_5} + p_{z_6}) \to a_{1g}(\sigma_p)],$$
$$t_{1u} = (p_x \to t_{1u})_{centr} + [1/\sqrt{2}(p_{z_1} - p_{z_2}) \to t_{1u}\sigma_p],$$
$$= (p_y \to t_{1u})_{centr} + [1/\sqrt{2}(p_{z_3} - p_{z_4}) \to t_{1u}\sigma_p],$$
$$= (p_z \to t_{1u})_{centr} + [1/\sqrt{2}(p_{z_5} - p_{z_6}) \to t_{1u}\sigma_p],$$
$$e_g = (d_{x^2-y^2} \to e_g)_{centr} + [1/\sqrt{2}(p_{z_1} - p_{z_2} + p_{z_3} - p_{z_4}) \to e_g\sigma_d],$$
$$= (d_{z^2} \to e_g)_{centr} + [1/2\sqrt{3}(2p_{z_5} + 2p_{z_6} - p_{z_1} - p_{z_2} - p_{z_3} - p_{z_4}) \to e_g\sigma_d)].$$

π molecular orbitals:

$$t_{1u} = (p_x \to t_{1u})_{centr} + [1/2(p_{x_3} + p_{y_5} - p_{x_4} - p_{y_6}) \to t_{1u}\pi_p],$$
$$= (p_y \to t_{1u})_{centr} + [1/2(p_{y_1} + p_{x_5} - p_{y_2} - p_{x_6}) \to t_{1u}\pi_p],$$
$$= (p_z \to t_{1u})_{centr} + [1/2(p_{x_1} + p_{y_3} - p_{x_2} - p_{y_4}) \to t_{1u}\pi_p],$$
$$t_{2g} = (d_{xy} \to t_{2g})_{centr} + [1/2(p_{y_1} + p_{y_2} + p_{x_3} + p_{x_4}) \to t_{2g}\pi_d],$$
$$= (d_{xz} \to t_{2g})_{centr} + [1/2(p_{x_1} + p_{x_2} + p_{y_5} + p_{y_6}) \to t_{1u}\pi_d],$$
$$= (d_{yz} \to t_{2g})_{centr} + [1/2(p_{y_3} + p_{y_4} + p_{x_5} + p_{x_6}) \to t_{2g}\pi_d].$$

Nonbonding ligand group orbitals:

$$t_{2u} = 1/2(p_{x_3} - p_{y_5} - p_{x_4} + p_{y_6}),$$
$$= 1/2(p_{x_1} - p_{x_5} - p_{y_2} + p_{y_6}),$$
$$= 1/2(p_{y_1} - p_{y_3} - p_{x_2} + p_{y_4}),$$
$$t_{1g} = 1/2(p_{y_1} - p_{y_5} + p_{x_2} - p_{y_4}),$$
$$= 1/2(p_{y_5} - p_{x_5} + p_{y_4} - p_{x_6}),$$
$$= 1/2(p_{x_1} - p_{x_3} + p_{y_2} - p_{x_4}).$$

Since there are no central ion orbitals which could be transformed into t_{2u} and t_{1g} symmetry types, the corresponding ligand group orbitals remain nonbonding.

Thus there are in all for the octahedral complex (both of transition metal and nontransition elements) three σ bonding molecular obitals a_{1g} (made from s + six p_z), t_{1u} (p + two p_z), e_g ($d_{x^2-y^2}$, d_{z^2} + four p_z), two π bonding molecular orbitals t_{1u} (p + two p_x and two p_y), t_{2g} (d_{xy}, d_{xz}, d_{yz} + two p_x and two p_y) and corresponding antibonding molecular orbitals and also two nonbonding ligand orbitals.

With the ordering of the central ion atomic orbitals $3s - 3p - 3d$ one obtains a MO diagram for nontransition element complexes, while for the $3d - 4s - 4p$ ordering one obtains a MO diagram for transition metal complexes (Fig. 48).

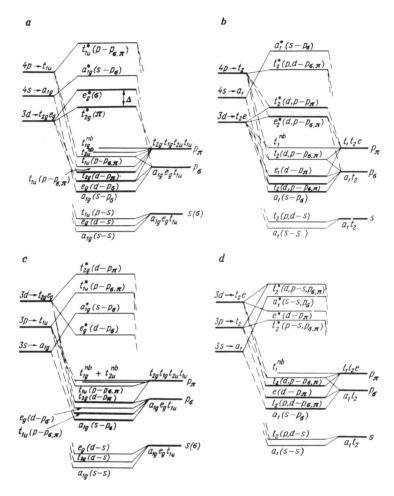

Fig. 48a–d. Molecular orbitals schemes for transition metal and third-row elements in octahedral and tetrahedral complexes: **a** transition metal in octahedron; **b** the same in tetrahedron; **c** third row element in octahedron; **d** the same in tetrahedron

B. Molecular Orbital Energy Levels and Coefficient Calculation. Outline of the Calculation Procedure. The data of departure for MO calculation are: the valent atomic orbitals (for Cr $3d$, $4s$, $4p$, for F $2s$, $2p$), the coordination and the interatomic distances, the local point symmetry.

Where can the actual numerical values for the atomic orbitals be taken from?

One ought to know for the MO calculation (1) the energies of these atomic orbitals and (2) the electron density radial distribution (for the overlap integral calculation).

The AO energies can be taken from the experimental atomic spectra (from the atomic energy levels tables, see [38]) as the ionization energies for $3d$, $4s$ and $4p$ electrons of a given atom (Cr in this case).

However, it must be taken into account that the atom in the compound occurs not in the form of Cr^{3+} but with a fractional effective charge, for instance, in the form of $Cr^{+0.77}$, and its electron configuration is not $3d^5 4s 4p^0$ but corresponds to the fractional effective configuration, for instance $3d^{4.73} 4s^{0.21} 4p^{0.29}$.

In order to obtain the ionization energies for these fractional charge and electron configuration take from atomic spectra data the energies of ionization of these electrons from the states with integral charge and electron configuration which occur for the free atoms. The actual fractional charges and electron configurations of an atom in a compound have some intermediate values between the free atom integral charge values $M^0 - M^+ - M^{2+}$ and between the integral electron configuration $3d^n 4s - 3d^{n-1} 4s^2 - 3d^{n-1} 4s 4p$ and so on.

The interpolation procedure for the determination of these intermediate values will be discussed below. However, these are not the only complications. Those effective charge ($Cr^{+0.72}$) and electron configurations ($3d^{4.73} 4s^{0.21} 4p^{0.29}$) for which the ionization energy must be determined can themselves be estimated only as a final result of the calculation.

This is why the solution results from successive cycles of calculation (iteration): in assuming reasonable values of the effective charge and electron configuration one estimates the molecular orbital energies and coefficients from which one evaluates the new (calculated) effective charge and electron configuration. Then using these calculated values of charge and configuration as input, one adjusts them by an iterative procedure until input and output agree, at which point the solution is considered self-consistent.

Thus, the MO parameters represent a result and an expression of the self-consistency between electron properties of atoms with a crystal or complex structure: its coordination, local symmetry, and interatomic distances.

Consider now the MO calculation for an octahedral complex of a transition metal (CrF_6^{3-}) [78, 126].

1. Write the molecular orbitals (see Fig. 48) formed from the four Cr atomic orbitals $3d$ (t_{2g}, e_g), $4s$ (a_{1g}), $4p$ (t_{1u}) and the seven F group orbitals $2p_\sigma$ (a_{1g}, t_{1u}, e_g), $2p_\pi$ (t_{1u}, t_{2g}, t_{2u}, t_{1g}):

$$\psi(a_{1g}) = c_1 \psi(a_{1g} 4s) \pm c_2 \psi(a_{1g} 2p_\sigma),$$
$$\psi(e_g) = c_3 \psi(e_g 3d) \pm c_4 \psi(e_g 2p_\sigma),$$
$$\psi(t_{2g}) = c_5 \psi(t_{2g} 3d) \pm c_6 \psi(t_{2g} 2p_\pi),$$
$$\psi(t_{1u}) = c_7 \psi(t_{1u} 4p) \pm c_8 \psi(t_{1u} 2p_\sigma) \pm c_9 \psi(t_{1u} 2p_\pi),$$
$$(t_{1g}) = \psi(t_{1g} 2p_\pi) - \text{nonbonding},$$
$$(t_{2u}) = \psi(t_{2u} 2p_\pi) - \text{nonbonding}.$$

Here $\psi(a_{1g})$ etc. are the molecular orbitals, $\psi(a_{1g} 4s)$ etc. are the Cr $4s$ atomic orbitals transformed in the octahedron into the a_{1g} symmetry type, $\psi(a_{1g} 2p_\sigma)$ etc. are the F σ group orbitals made from the F $2p$ atomic orbitals and transformed in the octahedron into the a_{1g} symmetry type (cf. p. 118).

There are in all eleven molecular orbitals of the six types. This set of molecular orbitals written above is the same for all octahedral complexes.

Molecular Orbital Calculations

If one also takes into account the F 2s atomic orbitals, one obtains three more group orbitals: $2s \rightarrow a_{1g}, e_g, t_{1u}$; then the number of the molecular orbitals will be equal to fourteen (see Fig. 48).

2. For each of the available four types of molecular orbital (without the nonbonding t_{1g} and t_{2u}), i.e., $\psi(a_{1g}), \psi(e_g), \psi(t_{2g}), \psi(t_{1u})$, write the secular equation

$$\begin{vmatrix} H_{AA} - E & H_{AB} - GE \\ H_{AB} - GE & H_{BB} - E \end{vmatrix} = 0.$$

Here, as in the case of the hydrogen molecule, the diagonal terms contain Coulomb integrals H_{AA}, H_{BB} and the off-diagonal ones contain exchange integrals H_{AB} and group overlap integral G (instead of S, see below).

For the molecular orbital $\psi(a_{1g})$ made from the Cr 4s orbital and F $2p_\sigma$ group orbitals the secular equation is written as

$$\begin{array}{c|cc} & 4s & 2p_\sigma \\ \hline 4s & H_{4s4s} - E_{a_{1g}} & H_{4s2p} - G_{4s2p}E_{a_{1g}} \\ 2p_\sigma & H_{4s2p_\sigma} - G_{4s2p}E_{a_{1g}} & H_{2p2p} - E_{a_{1g}} \end{array} = 0.$$

For the molecular orbital $\psi(e_g)$ made from the Cr 3d orbital and the F $2p_\sigma$ group orbitals:

$$\begin{array}{c|cc} & 3d & 2p_\sigma \\ \hline 3d & H_{3d3d} - E_{e_g} & H_{3d2p} - G_{3d2p} \\ 2p_\sigma & H_{3d2p} - G_{3d2p}E_{e_g} & H_{2p2p} - E_{e_g} \end{array} = 0.$$

For the $\psi(t_{1u})$ made from Cr 4p and F $2p_\sigma, 2p_\pi$:

$$\begin{array}{c|ccc} & 4p & 2p & 2p \\ \hline 4p & H_{4p4p} - E_{t_{1u}} & H_{4p2p} - G_{4p2p}E_{t_{1u}} & H_{4p2p} - G_{4p2p}E_{t_{1u}} \\ 2p & H_{4p2p} - G_{4p2p}E_{t_{1u}} & H_{2p2p} - E_{t_{1u}} & H_{2p2p} - G_{2p2p}E_{t_{1u}} \\ 2p & H_{4p2p} - G_{4p2p}E_{t_{1u}} & H_{2p2p} - G_{2p2p}E_{t_{1u}} & H_{2p2p} - E_{t_{1u}} \end{array} = 0.$$

3. Solutions of these secular equations give the molecular orbital energy values:

$$E_{MO} = \frac{H_{AA} + H_{BB} - 2H_{AB}G_{AB} \pm \sqrt{(H_{AA} - H_{BB})^2 + 4[H_{AB}^2 + H_{AB}^2 G_{AB}^2 - H_{AB}G_{AB}(H_{AA} + H_{BB})]}}{2(1 - G_{AB}^2)}.$$

Two roots of the equations for $\psi(a_{1g}), \psi(e_g), \psi(t_{2g})$ give each two values E_1 and E_2 of a bonding and an antibonding molecular orbital; three roots of the $\psi(t_{1u})$ equation give the energy values for three t_{1u} molecular orbitals. To evaluate these energies one must determine: the Coulomb integrals H_{AA} and H_{BB}, i.e., H_{3d3d}, H_{4s4s}, H_{4p4p} for Cr and H_{2s2s}, H_{2p2p} for F; the exchange integrals H_{AB}, i.e., H_{4s2p} (for a_{1g}), H_{3d2p} (for e_g), H_{3d2p} (for t_{2g}), H_{4p2p} and H_{4p2p} (for t_{1u}); the group overlap integrals G_{AB}, i.e., G_{4s2p} and so on.

4. Calculate the Coulomb integrals H_{3d3d}, H_{4s4s}, H_{4p4p} for Cr in CrF_6^{3-}.

In calculating the hydrogen molecule ion H_2^+ it was shown that the Coulomb integral $H_{AA} = E_A + I$, i.e., H_{AA} represents the electron energy for the hydrogen atom ground state (i.e., ionization energy) plus the energy of the interaction with the other nucleus of the molecule ion H_2^+. (For an isolated hydrogen atom $H_{AA} = E_A =$ ionization energy; if rejecting I one loses the term accounting for interactions of an electron in molecule).

For more complex systems as it is CrF_6^{3-}, the H_{AA} are approximated as the negative values of Valence State Ionization Energies (VSIE) or Valence State Ionization Potential (VSIP) (recall that potential is energy divided by electron charge) [128, 155–157].

Determine the valence state ionization energies (VSIE) for 3d electron from the configurations d^n, $d^{n-1}s$, $d^{n-1}p$ (i.e., $3d^6$, $3d^54s$, $3d^54p$) of M^0, M^+, M^{2+} atoms with integral charges $q = 0, +1, +2$ (i.e., Cr^0, Cr^+, Cr^{2+}); for 4s electron from the configurations $d^{n-1}s$, $d^{n-2}s^2$, $d^{n-2}sp$ ($3d^54s$, $3d^44s^2$, $3d^44s4p$) of M^0, M^+, M^{2+} atoms; for 3p electron from the configuration $d^{n-1}p$, $d^{n-2}sp$, $d^{n-2}p^2$ ($3d^54p$, $3d^44s4p$, $3d^44p^2$) of M^0, M^+, M^{2+} atoms, i.e., nine VSIE values for each electron.

The VSIE contrary to ionization energy (IE) takes into account interelectron (orbital) interaction, i.e., the fact that there can be several states for an initial (for instance, M^0) and a final (M^+) electron configuration described by different terms (Fig. 49):

$$VSIE = IE - T_{ground} + T_{ioniz.},$$

where IE is the ionization energy of a given electron from a given electron configuration, i.e., the energy difference between ground terms of M^0 and M^+ ions (or M^+ and M^{2+} and so on).

T_{ground} is the excitation energy of the lowest in a given configuration term for the initial M^0 ion (or M^+, M^{2+}...) [84] or the difference between weighted average energy of all the initial configuration terms [79];

$T_{ioniz.}$ is the same for the M^+ ion electron configuration with ionized electron (see also [128]).

The VSIE values are already computed and are available in the tables [78, 82, 84] from which [78] we can take them for Cr (see Table 28).

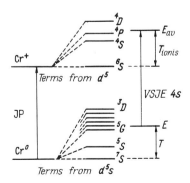

Fig. 49. Relation between ionization potential (IP) and valence state ionization energy (VSIE). E_{av} is average energies of ground state terms and of ionized state terms

Table 28. Valence state ionization energies (VSIE) for Cr electrons for integer d^n, $d^{n-1}s$, $d^{n-1}p$ electronic configurations and for integer charges $q=0, +1, +2$ (i.e., Cr^0, Cr^+, Cr^{2+}; in $cm^{-1} \cdot 10^{-3}$ units)

	$Cr^0 (Cr^0 \to Cr^+)$		$Cr^+ (Cr^+ \to Cr^{2+})$		$Cr^{2+} (Cr^{2+} \to Cr^{3+})$	
VSIE of 3d electron						
d^n	$d^6 \to d^5$	35.1	$d^5 \to d^4$	124.6	$d^4 \to d^3$	243.6
$d^{n-1}s$	$d^5s \to d^4s$	57.9	$d^4s \to d^3s$	163.6	$d^3s \to d^2s$	288.8
$d^{n-1}p$	$d^5p \to d^4p$	67.7	$d^4p \to d^3p$	174.4	$d^3p \to d^2p$	—
VSIE of 4s electron						
$d^{n-1}s$	$d^5s \to d^5$	53.2	$d^4s \to d^4$	118.8	$d^3s \to d^3$	200.5
$d^{n-2}sp$	$d^4sp \to d^4p$	63.3	$d^3sp \to d^3p$	138.2	$d^2sp \to d^2p$	—
$d^{n-2}s^2$	$d^4s^2 \to d^4s$	74.7	$d^3s^2 \to d^3s$	143.2	$d^2s^2 \to d^2s$	—
VSIE of 4p electron						
$d^{n-1}p$	$d^5p \to d^5$	28.4	$d^4p \to d^4$	83.2	$d^3p \to d^3$	152.5
$d^{n-2}sp$	$d^4sp \to d^4s$	37.8	$d^2sp \to d^3s$	—	$d^2sp \to d^2s$	—
$d^{n-2}p^2$	$d^4p^2 \to d^4p$	38.1	$d^3p^2 \to d^3p$	98.2	$d^2p^2 \to d^2p$	—

The VSIE of integral charges ($q=0, +1, +2...$) being determined, one may evaluate, the VSIE of a fractional effective charge $Cr^{+0.77}$ sought for the same integral electron configurations d^n, $d^{n-1}s$, $d^{n-1}p$ and so on. It can be done graphically (Fig. 50) or analytically by using the expression:

$$VSIE = Aq^2 + Bq + C,$$

where q is the charge, A, B, C are parameters of a given electron configuration and a given electron type available in the VSIE tables [78, 82]. They are listed for Cr in Table 29. The VSIE (in 10^3 cm^{-1}) for the charge $q = +0.77$ ($Cr^{+0.77}$) according to $Aq^2 + Bq + C$:

3d from $d^n = 101.41$
$d^{n-1}s = 137.56$
$d^{n-1}p = 148.13$

4s from $d^{n-1}s = 102.28$
$d^{n-2}sp = 119.54$
$d^{n-2}s^2 = 126.02$

4p from $d^{n-1}p = 69.31$
$d^{n-2}sp = 82.79$
$d^{n-2}p^2 = 82.09$.

Finally, estimate the VSIE of 3d, 4s, 4p electrons not only for the fractional effective charge $q = +0.77$ but also for the fractional effective electron con-

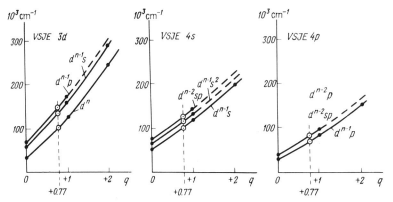

Fig. 50. Valence state ionization energies (VSIE) of $3d$, $4s$ and $4p$ electrons from different electronic configurations of Cr as function of charge q

Table 29. Constants A, B and C for Cr

Initial electron configuration		A	B	C
d	d^n	14.75	74.75	35.1
	$d^{n-1}s$	9.75	95.95	57.9
	$d^{n-1}p$	9.75	96.45	67.7
s	$d^{n-1}s$	8.05	57.55	53.2
	$d^{n-2}s^2$	8.05	66.85	63.3
	$d^{n-2}sp$	8.05	60.45	74.7
p	$d^{n-1}p$	7.25	47.55	28.4
	$d^{n-2}p^2$	7.25	52.85	37.8
	$d^{n-2}sp$	7.25	52.85	38.1

For $q=0$ (M^0) VSIE $=C$; for $q=+1$ (M^+) VSIE $=A+B+C$.

figuration $3d^{4.73}4s^{0.21}4p$ which namely represents the Coulomb integral H_{3d3d}, H_{4s4s}, H_{4p4p} values. For this:

a) Represent the effective electron configuration $d^{q(d)}s^{q(s)}p^{q(p)}$ (i.e., $3d^{4.73}4s^{0.21}4p^{0.29}$) as a linear combination of the integral electron configuration for:

$$H_{dd}: d^{q(d)}s^{q(s)}p^{q(p)} = ad^n + bd^{n-1}s + cd^{n-1}p;$$
$$H_{ss}: d^{q(d)}s^{q(s)}p^{q(p)} = ad^{n-1}s + bd^{n-2}s^2 + cd^{n-2}sp;$$
$$H_{pp}: d^{q(d)}s^{q(s)}p^{q(p)} = ad^{n-1}p + bd^{n-2}p^2 + cd^{n-2}sp.$$

Designate: m is the electron number (for Cr $m=6$); p is the effective charge on Cr ($q=+0.77$); $n=m-q=5.23$; $q_d=4.73$; $q_s=0.21$; $q_p=0.29$. For example, H_{dd}(Cr): $d^{4.73}s^{0.21}p^{0.29} = ad^{5.23} + bd^{4.23}s + cd^{4.23}p$.

b) Determine in these expressions the coefficients a, b, c:

$$s \to a \cdot 0 + b \cdot 1 + c \cdot 0 = q_s; \quad b = q_s;$$
$$p \to a \cdot 0 + b \cdot 0 + c \cdot 1 = q_p; \quad c = p_p;$$
$$d \to a \cdot n + b(n-1) + c \cdot (n-1) = q_d; \quad a = 1 - q_s - q_p,$$

($q_d = n - q_s - q_p = n - b - c$; on substituting it in the expression for d one obtains $a = 1 - q_s - q_p$). In the same way obtain the coefficients a, b, c for H_{ss} and H_{pp}.

c) Determine finally the values H_{dd}, H_{ss}, H_{pp}, on substituting the coefficients a, b, c values and the VSIE for $q = 0.77$ and the integral configurations:

$$H_{dd} = (1 - q_s - q_p)[d\text{-VSIE from } d^n] + q[d\text{-VSIE from } d^{n-1}s]$$
$$+ q_p[d\text{-VSIE from } d^{n-1}p],$$

$$H_{ss} = (2 - q_s - q_p)[s\text{-VSIE from } d^{n-1}s] + (q_s - 1)[s\text{-VSIE from } d^{n-2}s^2]$$
$$+ q_p[s\text{-VSIE from } d^{n-2}sp],$$

$$H_{pp} = (2 - q_s - q_p)[p\text{-VSIE from } d^{n-1}p] + (q_p - 1)[p\text{-VSIE from } d^{n-2}p^2]$$
$$+ q_s[p\text{-VSIE from } d^{n-2}sp].$$

For Cr utilizing the VSIE value for $q = +0.77$ (from Table 28) and substituting the coefficients ($q_s = 0.21$; $q_p = 0.29$):

$$H_{dd} = 0.50 \cdot 101.41 + 0.21 \cdot 137.56 + 0.29 \cdot 148.13 = 122.7 \cdot 10^3 \text{ cm}^{-1};$$
$$H_{ss} = 1.50 \cdot 102.28 - 0.79 \cdot 119.54 + 0.29 \cdot 126.02 = 95.6 \cdot 10^3 \text{ cm}^{-1};$$
$$H_{pp} = 1.50 \cdot 69.31 - 0.71 \cdot 82.79 + 0.21 \cdot 83.09 = 62.5 \cdot 10^3 \text{ cm}^{-1}.$$

Therefore, the Coulomb integrals for Cr in CrF_6^{3-} are determined as $H_{3d3d} = 122{,}700 \text{ cm}^{-1}$, $H_{4s4s} = 95{,}600 \text{ cm}^{-1}$, $H_{4p4p} = 62{,}500 \text{ cm}^{-1}$.

5. Evaluate the Coulomb integrals H_{2p2p} (and H_{2s2s}) for F in CrF_6^{3-}. Two methods are used for evaluating H_{2p2p}: (a) by determining H_{2p2p}, H_{2s2s} as the VSIE (as above for the Cr atom): construct curves (or obtain an analytical expression) showing dependence of the VSIE of F (or oxygen etc.) on the charge for the different integral electron configuration (see the VSIE tables for elements of II and III periods [84]); using these curves one obtains the VSIE for a given effective charge and a given electron configuration which need to be adjusted to the metal charge and configuration; (b) by assigning a single fixed value to the H_{2p2p} evaluated approximately from the comparison of the ionization energy of free atoms, hydrides and other compounds, using also experimental estimations of ligand ion effective charge in different compounds; this assumption is permissible in this case because of very much lower variations of the ligand VSIE with the charge than for the metal atoms.

Let us assume [78] for F in $CrF_6^3 H_{2p2p_\sigma} = -160{,}400 \text{ cm}^{-1}$, $H_{2p2p_\sigma} = -150{,}500 \text{ vm}^{-1}$.

6. Evaluate the overlap integrals and the exchange integrals related to them.

Firstly, find the standard diatomic overlap integrals $S = \int \psi_M \psi_F d\tau$, i.e., $S(4s2p_\sigma)$, $S(4p2p_\sigma)$, $S(3d2p_\sigma)$, $S(4p2p_\pi)$, $S(3d2p_\pi)$ (see Fig. 48 and the MO types on p. 120). They could be calculated using analytical expressions for the corresponding wave functions but now they can be taken directly from overlap integral tables [80]. Here the interatomic distances are required because S is a function of interatomic distance (Cr–F in CrF_6^{3-} is 1.93 Å).

Since the molecular orbitals are constructed from metal atomic orbitals and ligand group orbitals one then finds the group overlap integrals $G = \int \psi_M \psi_L d\tau$ connected with the diatomic overlap integrals S by simple expressions. The conversion coefficients depend on coordination; for the octahedron they are (the numerical values are given for CrF_6^{3-}):

$$G_{a_{1g}} = \sqrt{6} \cdot S(4s, 2p_\sigma) = 0.2673,$$

$$G_{e_{g\sigma}} = \sqrt{3} \cdot S(3d, 2p_\sigma) = 0.2321,$$

$$G_{t_{2g\pi}} = \sqrt{4} \cdot S(3d, 2p_\pi) = 0.1728,$$

$$G_{t_{1u\sigma}} = \sqrt{2} \cdot S(4p_\sigma, 2p_\sigma) = 0.1427,$$

$$G_{t_{1u\pi}} = \sqrt{4} \cdot S(4p_\pi, 2p_\pi) = 0.2788.$$

7. Exchange integrals H_{AB} (the off-diagonal elements of the secular determinant) are proportional to overlap integrals and expressed usually as

$$H_{AB} = K \left(\frac{H_{AA} + H_{BB}}{2} \right) G_{AB} \quad \text{or} \quad H_{AB} = K \sqrt{H_{AA} \cdot H_{BB}} \cdot G_{AB},$$

where $k = 1.67$ for octahedron and $k = 2$ for tetrahedron.

8. Thus we have the H_{AA}, H_{AB} and G_{AB} values and can write numerically the secular equation for each MO type, for example, for $\psi(e_g)$:

$$\begin{vmatrix} H_{3d3d} - E_{e_g} & H_{3d2p} - G_{3d2p} E_{e_g} \\ H_{3d2p} - G_{3d2p} E_g & H_{2p2p} - E_{e_g} \end{vmatrix} = 0;$$

$$\begin{vmatrix} (-122.98 - E) & (-65.12 - 0.2321 E) \\ (-65.12 - 0.2321 E) & (-160.4 - E) \end{vmatrix} = 0.$$

The solutions of this secular equation give two roots:

$$E = \frac{H_{dd} + H_{pp} - 2H_{dp} G_{dp} \pm \sqrt{(H_{dd} - H_{pp})^2 + 4[H_{dp}^2 + H_{dp}^2 G_{dp}^2 - H_{dp} G_{dp}(H_{dd} + H_{pp})]}}{2(1 - G_{dp}^2)};$$

$$E_1 = -163{,}480 \text{ cm}^{-1}; \quad E_2 = -92{,}520 \text{ cm}^{-1}.$$

One can use also the simple approximate expressions [82]:

$$E_1 = H_{pp} - \frac{(H_{dp} - H_{pp} G_{dp})^2}{(H_{dd} - H_{pp})}; \quad E_2 = H_{dd} + \frac{(H_{dp} - H_{dd} G_{dp})^2}{(H_{dd} - H_{pp})}.$$

Molecular Orbital Calculations

Table 30. Energies and coefficients of CrF_6^{3-} molecular orbitals

MO	$E(10^3 cm^{-1})$	c_1	c_2	c_3
$1a_{1g}$	−174.68	$c_{4s}=0.2260$	$c_{2p_\sigma}=0.9156$	—
$1e_g$	−169.73	$c_{3d}=0.4387$	$c_{2p_\sigma}=0.8028$	—
$1t_{2g}$	−163.48	$c_{3d}=0.3917$	$c_{2p_\sigma}=0.8549$	—
$1t_{1u}$	−162.86	$c_{4p}=0.0826$	$c_{2p_\sigma}=0.8664$	$c_{2p_\pi}= 0.4240$
$2t_{1u}$	−150.06	$c_{4p}=0.0809$	$c_{2p_\sigma}=0.4905$	$c_{2p_\pi}=-0.8670$
t_{2u}	−149.15	—	—	$c_{2p_\pi}= 1.0000$
t_{1g}	−145.62	—	—	$c_{2p_\pi}= 1.0000$
$3t_{2g}^*$	−105.93			
$2e_g^*$	− 92.52			
$2a_{1g}^*$	− 73.12			
$3t_{1u}^*$	− 43.05			

In a similar way one obtains the other molecular orbital energies listed in Table 30.

9. Determine the molecular orbitals coefficients c_i using the two equations

$$c_1(H_{AA}-E)+c_2(H_{AB}-G_{AB}E)=0,$$

$$c_1(H_{AB}-G_{AB}E)+c_2(H_{BB}-E)=0,$$

and the normalization condition

$$c_1^2+c_2^2+2c_1c_2G_{1,2}=1.$$

For example, for the $\psi(e_g)$ molecular orbital:

$$c_1(H_{3d3d}-E_g)+c_2(H_{3d2p}-G_{3d2p}E_{e_g})=0,$$

$$c_{3d}(-122680+169770)+c_{2p}(65130+0.2321 \cdot 169770)=0,$$

$$c_{3d}=0.547 c_{2p};$$

using the normalization condition and on substituting on it the value $G_{3d2p}=0.2321$ one obtains: $c_{3d}=0.4387$ and $c_{2p}=0.8028$.

Thus the $\psi(e_g)$ molecular orbital is written as the 3d and 2p combination:

$$\psi(e_g)=c_1\psi(3d)+c_2\psi(2p)=0.4387 \cdot \psi(3d)+0.8028 \cdot \psi(2p).$$

Similarly, one calculates the other MO coefficients.

MO energy and coefficients tables (such as Table 30) represent the final result of MO calculations.

10. These values are used for calculation of the effective charge and effective electron configuration assuming them as initial value for a new cycle of calculation repeated until the resulted charge and configuration (output) agree with the initial ones (input). The whole process is programmed to iterate automatically to a self-consistent charge and configuration and run on a computer.

Transition Metal Tetrahedral Complexes (Based on Example of MnO_4^-). General procedure for tetrahedral (and other symmetry type complexes) MO calculation is the same as for the octahedral CrF_6^{3-}, the differences depend on a different symmetry of MnO_4^-.

A. The Molecular Orbital Diagram Construction. Basis atomic orbitals of Mn ($3d^5 4s^2$) are the same as those of Cr ($3d^5 4s$), i.e., 3d, 4s, 4p, but in tetrahedral symmetry they transform differently: $s \to a$, $p \to t_2$, $d \to t_2$, e (the signs g and u indicating parity are absent here because of noncentro-symmetricity of the tetrahedron).

Basis ligand group orbitals, though formed from the same 2s and 2p oxygen (or fluorine) atomic orbital, transform in tetrahedral symmetry differently:
the 2s orbitals of the four oxygen atoms in tetrahedral complexes (having 8 electrons) give σ group orbitals a_1 (with 2 electrons) and t_2 (with 6 electrons); the $2p_x$, $2p_y$, $2p_z$ orbitals of the four oxygen atoms (i.e., 12 oxygen atomic orbitals with a possibility of accommodating 24 electrons) give σ group orbitals a_1 and t_2 (with 2 and 6 electrons) and π group orbitals e, t_1 and t_2 (4, 6 and 6 electrons).

In combining the central atom atomic orbitals and ligand group orbitals of the same symmetry one obtains the follow molecular orbitals for tetrahedral complex:

MO   AO (of Mn)   group orbitals of four oxygens

$$\psi(a_1) = c_1 \psi(a_1 4s) \pm c_2 \psi(a_1 2s) \pm c_3 \psi(a_1 2p_\sigma),$$
$$\psi(e) = c_4 \psi(3de) \pm c_5 \psi(2p_\pi e),$$
$$\psi(t_2) = c_6 \psi(3dt_2) \pm c_7 \psi(4pt_2) \pm c_8 \psi(2p_\sigma t_2) \pm c_9 \psi(2p_\pi t_2),$$
$$\psi(t_1) = \psi(2p_\pi t_1) - \text{nonbonding.}$$

There are in all (taking into account group orbitals formed from 2s and 2p oxygen orbitals) 11 molecular orbitals: three of a_1, two of e, five of t_2 and one of t_1.

These 11 molecular orbitals of the four types are the same for all tetrahedral complexes: transition metal as nontransition elements. It is only the molecular orbital ordering and coefficients which vary.

In the tetrahedral complex MO diagram in Figure 48b, the basis central atom orbitals are positioned in the order 3d, 4s, 4p for transition metals and therefore the positioning of the 11 resulting molecular orbitals is typical for all transition metal tetrahedral complexes.

B. Molecular Orbital Energy and Coefficients Calculation. For the three MO types: a_1, e, t_2 (t_1 is nonbonding) one writes (as for CrF_6^{3-}) three secular equations. Their solutions give: three roots for $\psi(a_1)$ constructed from three orbitals, two roots for $\psi(e)$ constructed from two orbitals and five roots for $\psi(t_2)$ from five orbitals.

To obtain numerical solution of these equation evaluate the Coulomb integrals (as VSIE). For MnO_4^- [126] they are equal to:

Mn: $H_{sd3d} = 121{,}280\ cm^{-1}$,

$\qquad H_{4s4s} = 93{,}410\ cm^{-1}$,

$\qquad H_{4p4p} = 58{,}400\ cm^{-1}$,

$Mn^{+0.66}(3d^{5.82}4s^{0.18}4p^{0.34})$

O: $H_{2s2s} = 260{,}800\ cm^{-1}$,

$\qquad H_{2p2p} = 101{,}780\ cm^{-1}$;

Group overlap integrals for the tetrahedral complex are:

$Gt_2 = 2/\sqrt{3} \cdot S(3d_\sigma 2p_\sigma)$,

$Gt_2 = 2/\sqrt{2/3} \cdot S(3d_\pi 2p_\pi)$,

$Gt_2 = 2/\sqrt{3} \cdot S(3d_\sigma 2s)$,

$Gt_2 = 2/\sqrt{3} \cdot S(4p 2p_\sigma)$,

$Gt_2 = 2/\sqrt{3} \cdot S(4p 2s)$,

$Gt_2 = 2/\sqrt{2}/\sqrt{3} \cdot S(4p 2p_\pi)$;

$Ge = 2/\sqrt{2}/\sqrt{3} \cdot S(3d_\pi 2p_\pi)$,

$Ga_1 = 2 \cdot S(4s 2s)$,

$Ga_1 = 2 \cdot S(4s 2p_\sigma)$.

Having evaluated exchange integrals H_{AB} from the expression $H_{AB} = k \cdot \sqrt{H_{AA} H_{BB}} \cdot G_{AB}$, substitute the H_{AA}, H_{BA}, H_{AB}, G_{AB} in the secular equations for each type of the molecular orbitals and find the molecular orbital energies and coefficients.

Nontransition Elements in Tetrahedral Complexes. These are such important oxyanions as SiO_4^{4-}, PO_4^{3-}, SO_4^{2-} (and also SiF_4, PF_4^+, SF_4^{2+} etc.).

Basis atomic orbitals, for example, of Si $(3s^2 3p^2)$ transform in the tetrahedron by the same manner as $3d$, $4s$, $4p$ in MnO_4^-, i.e., $s \to a_1$, $p \to t_2$, $d \to t_2$, e but are positioned in another order: $3s - 3p - 3d$ (while in MnO_4^- $3d - 4s - 4p$).

Group orbitals of the four oxygen atoms are the same as in MnO_4^-. Thus for SiO_4^{4-} (and for other nontransition tetrahedral complexes) one writes the same four types of molecular orbital: $\psi(a_1)$, $\psi(e)$, $\psi(t_2)$, $\psi(t_1)$, in all 11 molecular orbitals (see above the MO of MnO_4^-). Different ordering of the basis central atom orbitals leads to different ordering of the molecular orbitals and to different MO diagram (see Fig. 48a).

The results of the molecular orbital energies and coefficients calculation are given below in a discussion of the chemical bond in silicates (see Chap. 8.2).

3.2.4 Valence Bond Method; Hybrid Atomic Orbitals

Consider the general model of valence bond method (VB) and its immediate consequences; its comparison with the MO method helps to estimate more completely the meaning of the both methods.

Wave Function in MO and VB Methods, Delocalized Orbitals and Localized Electron Pairs. Wave function in the MO method is represented by molecular orbital $\psi_{MO} = c_1\psi_A + c_2\psi_B$; near the A nucleus the electron properties are approximated by an atomic ψ_A orbital and near the B nucleus by ψ_B orbital, but here there are no two atomic orbitals: each electron is distributed over all the space near the two nuclei.

In the VB method initial wave function is taken as the product of atomic orbitals: $\psi_{VB_1} = \psi_{A_1} \cdot \psi_{B_2}$; electron 1 occurs near the A nucleus (ψ_{A_1}) and electron 2 near the B nucleus (ψ_{B_2}).

However, because of the exchange of the electrons, the same molecule state can be described by the other wave function: $\psi_{VB_2} = \psi_{A_2} \cdot \psi_{B_1}$ with electron 2 on A and electron 1 on B. Then the trial function ψ_{VB} is expressed as a linear combination of the two product functions.

Compare the expression for the wave functions in the MO and VB methods for the H_2 molecule:

MO method

$$\psi_{MO_1} = c_1\psi_{A_1} + c_2\psi_{B_1}$$
$$\psi_{MO_2} = c_3\psi_{A_2} + c_4\psi_{B_2}$$

$$\psi_{MO} = \psi_{MO_1} \cdot \psi_{MO_2}$$
$$= (c_1\psi_A + c_2\psi_B)(c_3\psi_A + c_4\psi_B)$$

VB method

$$\psi_{VB_1} = \psi_{A_1} \cdot \psi_{B_2}$$
$$\psi_{VB_2} = \psi_{A_2} \cdot \psi_{B_1}$$

$$\psi_{VB} = c_1\psi_{VB_1} + c_2\psi_{VB_2}$$
$$= c_1(\psi_{A_1}\psi_{B_2}) + c_2(\psi_{A_2}\psi_{B_2}).$$

One can choose also the wave function with both electrons on A (i.e., A^-B^+) and both electrons on B (A^+B^-). Then:

$$\psi_{V_\theta} = \psi_{V_{\theta_1}} + \psi_{V_{\theta_2}} + \psi_{A^+B^-} + \psi_{A^-B^+} = \psi_{A_1B_2} + \psi_{A_2B_1} + \psi_{B_1B_2}.$$

Electron delocalization in molecular orbitals (i.e., electron distribution over the total space of many nuclei of a molecule or complex) is one of the most important features of the MO method. It is naturally connected with the quantum mechanical concept of electron density distribution and is confirmed by direct experimental (spectroscopic) observations.

The choice of the formulation of the wave function leads, as already mentioned, to far-reaching consequences. ψ_{MO} is a one-electron function (as it is AO) while ψ_{VB} is two-electron; ψ_{MO} can be not only a two-nuclear but also many-nuclear function. For example, for the tetrahedral AB_4 complex $\psi_{MO} = c_1\psi_A + c_2\psi_B + c_3\psi_B + c_4\psi_B + c_5\psi_B$ is a many-nuclear and one-electron wave function describing the state of an electron in the field of the many nuclei.

Delocalized Molecular Orbitals in the MO Method and Localized Electron Pairs in the VB Method. The state of the whole molecule is composed from as many ψ_{MO} as there are electrons in this molecule. On the contrary, ψ_{VB} is always the two-center (two-nuclei) function; in many-atomic molecules the bonding between every two atoms is described separately and the bond state of the whole molecule is represented by the sum of the individual two-atomic binding. In this way, in the AB_4 molecule one considers the four localized bindings A–B. This is why the VB method is also called the localized electron pairs method.

Because of electron delocalization in molecular orbitals, the symmetry of the molecule or complex determines immediately the symmetry of the molecular orbital. The atomic orbitals of the same symmetry transformation type compose the molecular orbitals designated only by the transformation type for a given point group of symmetry (of the molecule or complex): a_1, b_1, a_2, b_2, t_1, t_2, e, etc.

In the VB method the symmetry type classification is inapplicable since in this method one considers the two-center binding only.

Basic Features of VB Method. The whole system of concepts in the VB method ensues logically and indispensably from the choice of the wave function in the form of the product of atomic orbitals:

1. All the bindings are considered as two-center ones (not only for two-atomic molecules but also in the case of many-atomic complexes).

2. The bindings are represented by the localized electron pairs with paired (opposite directed) spins. However, it is necessary only for s electrons not for p (and other type) electrons which can be distributed in p_x, p_y, p_z orbitals.

3. The localized two-center bindings are considered as directed binding in AB_3 complex (for instance, in CO_3^{2-}) directed from A to three B atoms, in AB_4 (for instance, in CH_4) from A to four B, etc. In these cases the angles in the plane complex AB_3 are 120°, in the tetrahedral AB_4 are 109°28' while the atomic orbitals p_x, p_y, p_z are directed at an angle of 90° to each other. Thus, the introduction by Pauling of the concept of hybrid atomic orbitals which are obtained by the mixing of the different atomic orbitals of the atom was a natural consequence: for example, one obtains from s and p orbitals of the atom the hybrid sp orbital of the same atom (see below).

4. The assumption of the localized bonds demanded not only an explanation of the angles between the bond directions (it was performed with the introduction of hybrid orbitals) but also co-ordination with the valencies of atoms. Every two-center bond corresponds to the valency stroke of stereochemistry. In the CO_3^{2-} molecule, for example, the carbon is four-valent and in the valent scheme of this molecule it must be four such strokes departing from carbon (Fig. 51). However,

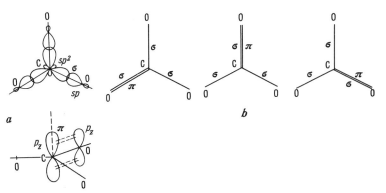

Fig. 51a and b. Valence schemes of CO_3^{2-}: **a** localized σ orbitals (formed from sp^2 carbon orbitals and sp oxygen orbitals) and π orbitals (from p_z orbitals of C and O); **b** valent structures resonance of which describes electronic structure of CO_3^{2-} according to valence bond method

with the three oxygen atoms in CO_3^{2-}, it was necessary to introduce the concept of bond multiplicity: in CO_3^{2-} there are two single σ bonds and one double σ, π bond.

5. However, because all three oxygen atoms in CO_3^{2-} are identical, the π bond belongs equally to each of the C–O directions and, since delocalization of the π bond would be contradictory to the concept of localized bonds, the electronic structure of CO_3^{2-} is described not by one but by the three valent schemes. Each of these three states is not realizable separately; they represent a kind of "end member" of the bond the superposition or resonance of which describes the actual electronic structure of the molecule. The concept of the resonant valent structures is the indispensable element of the description of many atomic molecules in the model of localized bonds of the VB method.

Such is the logical scheme of the rigidly interrelated concepts and ways of bond descriptions which compose the very basis of the VB method. The concepts of the general theory of the chemical bond, many of which are connected inseparably with the VB method, will be considered below.

Hybrid Orbitals. Consider the electron configuration of carbon in the molecule CH_4 with four equivalent bonds C–H. Such a geometrical form of this molecule needs (1) four equivalent electrons and (2) directed bonds with the tetrahedral angles of 109°28′. In the free carbon atom ($1s^2 2s^2 2p^2$) there are among the valent electrons two paired s electrons and two unpaired p electrons with the spherical boundary surfaces for s electrons and the boundary surfaces directed at the angle of 90° to each other for p electrons.

This is why one assumes that the electron configuration of C in compounds varies with the coordination (Figs. 52 and 53). In tetrahedral coordination there occur: (1) unpairing and promotion of the p electrons and excitation of one of the s electrons in the p state ($s^2 p^2 \rightarrow sppp$), (2) mixing (hybrdization) of the s and p states leading to the formation of the four equivalent hybrid orbitals each consisting of a quarter of an s orbital and three quarters of a p orbital:

$$\psi_{sp^3}^2 = 1/4\psi_s^2 + 1/4\psi_{px}^2 + 1/4\psi_{py}^2 + 1/4\psi_{pz}^2.$$

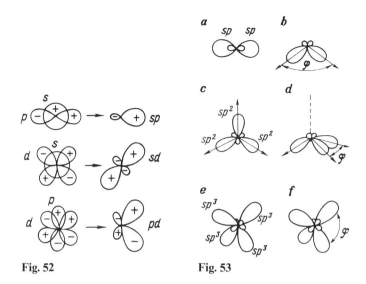

Fig. 52. Hybrid sp, sd, and pd orbitals

Fig. 53a–f. Hybrid sp^n orbitals: **a** two hybrid sp orbitals; $s^2p^2 \to (sp)(sp)p_\pi p_\pi$; $c_p^2/c_s^2 = 1/2$; $\varphi = 180°$; **b** hybrid sp-orbitals with various c_p^2/c_s^2 ratios and $\varphi < 180°$ (for example, carbon in bent molecular ion CO_2^-); **c** sp^2; $\varphi = 120°$ (carbon in plane radical CO_3^-); **d** sp^n; $\varphi < 120°$ (carbon in pyramidal radical ion CO_3^-); **e** sp^3; $\varphi 109°28'$ (tetrahedral angle); carbon in CH_4; **f** sp^n; $\varphi \neq 109°28'$; Si in SiO_4^{4-}

In three-fold coordination (CO_3^{2-}) there are three hybrid sp^2 orbitals and the remaining p_z orbital forms the π bonds.

In two-fold coordination (CO_2), besides two hybrid sp orbitals, the two p orbitals remain which take part in the π bonds.

Finally, in the two-atomic molecule CO the same orbitals are formed as in CO_2 but one of the sp hybrid orbitals remains non participating in bonding.

The forms of sp, sp^2, sp^3 hybrid orbitals are similar, their number and directions are determined by coordination (Fig. 53).

In the same way the sp^n hybrid orbitals form in other atoms with s and p valent electrons.

Hybrid orbitals can arise also in the cases of mixing of other atomic orbitals: s–d, p–d, s–p–d. The systematic deduction of all the possible hybrid orbitals for different coordinations can be found in many works [96, 97, 100, 113] and others. Formation, for instance, of sp^3 or sd^3 hybrid orbitals depends on the relative energy differences of the s–p and s–d atomic orbitals. The concept of hybrid atomic orbitals is the most characteristic feature of the VB model. The very statement of the problem, including the certain number of directed equivalent bonds each with one cation electron to form electron pairs, corresponds to the principles of the VB method.

A hybrid orbital is an atomic orbital but not one of the free atom (there is no atomic orbital hybridization in a free atom) but that of the atom in compounds where the type of the hybrid orbital is determined first of all by the coordination:

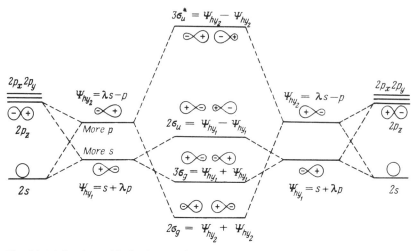

Fig. 54. Molecular orbital scheme with hybrid atomic orbitals

sp^3 or sd^3 for all atoms in tetrahedral coordination, d^2sp^3 in octahedral coordination etc.

In the MO method there is no necessity for atomic orbital hybridizations. In some cases they are used also in constructing MO schemes, but only as a nonobligatory element, because the mixing of states is described in the MO method by the atomic orbital coefficients in molecular orbitals (Fig. 54).

The mixing of states is considered in the MO as in the VB methods. However, in the VB method one considers first the mixing of all valent orbitals of the atom resulting in the formation of the hybrid orbitals of every atom and then the formation of the chemical bonds between atomic hybrid orbitals.

In the MO method each atomic orbital participates in the formation of the molecular orbital; different atomic orbitals of the same atoms can contribute in a given molecular orbital if it is allowed by the symmetry of the complex. (But $s \to a_1$ and $p \to t_1$ in the tetrahedron and thus having different symmetry they do not mix in the MO model). The whole set of molecular orbitals describes the mixing of states in the bond formation.

3.3 Analysis of the MO Scheme: Information Obtained from MO and Basic Concepts of the Theory of the Chemical Bond

The general model of the chemical bond in the MO method is the MO scheme (similar, for example, to Fig. 48) with its MO energies and coefficients and also with the intermediate terms of calculations: Coulomb (H_{AA}), exchange (H_{AB}) and overlap (G_{AA}) integrals. This scheme can be considered as a concise expression of the whole system of concepts and parameters composing efficient formalism in the description of the chemical bond.

Fig. 55. Bonding and antibonding molecular orbitals. Bonding molecular orbitals are constructed mainly from ligand atomic orbitals *(bottom)*. Antibonding molecular orbitals are constructed mainly from central ion atomic orbitals *(top)*; they are empty in nontransition element complexes but are occupied by d electrons in transition metal complexes

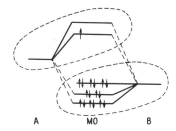

It follows from the MO scheme, first of all, that the chemical bond is realized not by the transfer of an electron from cation to anion, and not by common possession of the electron pair, but through the redistribution of all valent electrons of the central ion and the ligand with the participation of all valent and empty orbitals (for example, $3s3p3d$ for the elements of $3d$ period or $3d4s4p$ for the transition metals). The distribution of the valent electrons of molecule or complex occurs in the form of shells round two or more nuclei, i.e., there are a number of molecular orbitals each representing the distribution of one, two or several electrons.

For example, the 11 molecular orbitals describe the bond in tetrahedral complexes, the 14 molecular orbitals exist in octahedral complexes etc.

All the molecular orbitals are enumerated in order of increasing energies with the indication of the number of electrons in each orbital; they are written in the form of the electron configuration of the molecule (or complex). For example, for SiO_4^{4-}: $(1a)^2 (1t_2)^6 (2a)^2 (2t_2)^6 (1e)^4 (3t_2)^6 (t_1)^6 (4t_2^*)^0 (3a^*)^0 (5t_2^*)^0 (2e^*)^0$, i.e., there are in all 32 electrons distributed in the seven bonding and nonbonding molecular orbitals and four anti-bonding molecular orbitals. For each of these molecular orbtials ($1a, 1t_2, 2a$ and so on) one indicates their composition: from which atomic orbitals of the central ion and which group orbital it is made and what are their relative amounts, for example: $\psi(1a) = c_1 \psi(3s) + c_2 \psi(2p_\sigma)$. Overlapping and relative energies of the molecular orbitals show the amounts of their contribution in the total energy of the bond.

The distribution of each electron in the molecular orbital is characterized by its delocalization, representing one of the most important features of the MO method. The electron is distributed not in a space between two atoms but over the whole complex whose atomic orbitals enter in the given molecular orbital, i.e., over the whole part of the space described by the boundary surfaces of the atomic orbitals which compose the given molecular orbital (see Fig. 47).

It is expedient to discern in the MO schemes (see Fig. 48) the types of molecular orbital existing irrespective of coordination, symmetry and other peculiarities of complexes.

A. Molecular orbitals made mainly from ligand group orbitals (Figs. 55, 56).

1. Weakly bonding, made mainly from the deep-lying valent group orbitals of the ligands ($2s$ for oxygen, fluorine, $3s$ for sulfur, chlorine).

2. Strongly bonding, made from s, p, d or d, s, p atomic orbitals of the central atom and $2p$ or $3p$ ligand group orbitals.

3. Weakly bonding, made from the higher lying and often empty atomic orbitals of the central atom and the ligand group orbitals of the same symmetry.

Fig. 56. Relative position of strong bonding ($c_A \ll c_B$), weak bonding ($c_A \ll c_B$) and nonbonding ($c_A = 0$, $c_B = 1$; $\psi_{MO} = \psi_B$) molecular orbitals

4. Nonbonding orbitals, the ligand orbitals having no counterparts of the same symmetry among the atomic orbitals of the central atom.

B. Antibonding molecular orbitals made mainly from the atomic orbitals of the central atom (Fig. 55).

5. (a) With d electrons (in the case of transition metals); (b) empty molecular orbitals.

The actual MO schemes are determined further by the coordination, symmetry point group of the complex and relative positions of the atomic orbitals (s–p–d or d–s–p).

Consider now: (1) features of MO schemes reflecting the general idea of the chemical bond; (2) the chemical meaning of the each term of the MO schemes (not only E_i and C_i but also intermediate terms H_{AA}, H_{AB}, G_{AB}); (3) the relation of these MO schemes to the base concepts already formed in the course of practical experience of inorganic chemistry as also to the new general concepts arising from this scheme; (4) relation to the MO schemes of the experimental parameters which can be read from it immediately or calculated (thus acquiring the meaning of the characteristics of certain features of the chemical bond); (5) the ways of using the MO schemes in each consideration of the chemical bond features in particular groups of compounds and in actual compounds.

3.3.1 Coulomb Integrals H_{AA} in the MO Method-Ionization Potentials-VSIE; Deep Meaning of the Self-Consistency; Electronegativity

During the past 40 years in geochemistry and mineralogy, as well as in chemistry, the most common concept of the chemical bond theory was electronegativity (with the closely related potentials of ionization, electron affinities of atoms and lattice energies). However, these are lacking in the MO diagram. The related terms in the MO method are Coulomb integrals H_{AA} determined mostly as valence state ionization energies (VSIE). The procedure of the determination of H_{AA} as VSIE has been described earlier (see p. 122).

Let us consider the quantum-mechanical meaning of H_{AA}–VSIE and, thereupon, compare it with the mentioned chemical concepts.

What are the relations between H_{AA}–VSIE and the ionization energy of atoms, and why is it impossible to use the ionization potentials directly in the MO calculations?

There are three ways of defining H_{AA}:

$H_{AA} = \int \psi_A H \psi_A d\tau$;

$H_{AA} = E_A + I = IE_A + I$ (see p. 113);

$H_{AA} = VSIE\ M^{0,n} d^d s^s p^p$ (see p. 122).

a) The energy operator H includes all the interactions of the electron in the molecular orbital $\psi_{MO} = c_1\psi_A + c_2\psi_B$, but it is related only to the part of the wave function ψ_{MO} (i.e., it is $H\psi_A^2$). In the energy expression (see p. 109), each of the H_{AA}, H_{BB}, H_{AB} takes into account only a part of the interactions.

b) H_{AA} represents the interaction energy of the electron with its nucleus (ionization energy IE_A) plus the energy of the interaction of the electron with another atom's nucleus and other electrons. The same may be expressed as the energy of the interaction of the electron with the nucleus of its atom (ionization energy), the latter having acquired the effective charge and effective electron configuration and thus here replacing the interaction with another atom, i.e., it is the interaction with its atom in the valence state.

c) The valence state concept (VSIE) implies a hypothetical state, in which a free atom can have an effective charge and an electron configuration similar to that in the molecule but without participating in the chemical bonding, i.e., the state of an atom in a molecule with other atoms removed adiabatically (i.e., with no electron rearrangement). It is evident that such a state is not realizable either in an atom or in a molecule.

Hence, one can understand that the Coulomb integrals H_{AA} represent only the intermediate quantities in the calculations, as they are only one of the terms and have by themselves no real chemical meaning.

Indeed, the MO energies and the c_i coefficients describing the electron distribution in the molecular orbitals depend on the H_{AA}, H_{AB}, G_{AB} values. Though determined as "valence state energies", the Coulomb integrals do not represent the ionization potentials either of the atom in a molecule, or of the whole molecule, as the loss of the electrons (ionization) arises from the molecular orbital, whose energy is determined not only by H_{AA}, but also by H_{AB} and G_{AB} (see below).

Of great importance is one further aspect of the H_{AA}–VSIE concept: the final values of H_{AA} for a given compound, which must be the initial values, are obtained when the calculations are completed as the result of the iteration i.e., as a result of the self-consistency between H_{AA}, the effective charge, the effective electron configuration, the overlap integrals and the interatomic distances (for a given coordination). However, these initial values (including the interatomic distances obtained in principle as the equilibrium distances in the potential curve dependence) as well as all other values in the MO diagram, are obtained as a result of calculations i.e., of mutual adjustment.

None of the atom properties remains constant. In every compound all values acquire in principle the individual properties inherent to this particular compound.

The MO–LCAO–SCF Method is a Means for the Self-Consistency of the Properties of the Atoms to Enter Chemical Bonding. The Chemical Bond is the Self-Consistency of the Atom Properties in a Compound. At the same time H_{AA} is not only an intermediate term but it represents also the corrected (self-consistent) value which permits calculation of all the values represented in the MO schemes. It is the part of the energy which is contributed from the each atomic orbital (from $3d$, $4s$, $4p$ etc.) into all the molecular orbitals of the same symmetry. It is only H_{AA} in the MO schemes which represents the properties of the given atom entering

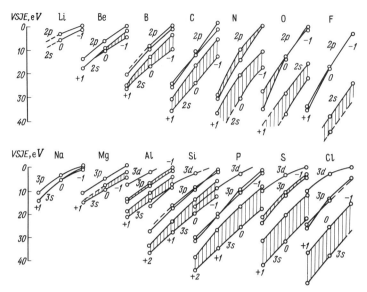

Fig. 57. Valence state ionization energies (VSIE) as function of charge

complexes represented by the molecular orbitals. It is the energy which would be attributed to the atomic orbital being completely nonbonding.

The tables of ionization potential of atoms which usually can be found in textbooks of chemistry and geochemistry are of rather limited use, because for consideration of molecules and complexes one needs a continuous series of VSIE values which vary with the variation of charges and electron configuration from which the ionization of the different electrons occur. Such values of the VSIE taken from [84] are represented in Figure 57.

From all these characteristics of H_{AA}–VSIE there follows logically the estimation of the concept of electronegativity suggested in 1932 by Pauling. Combined with the practical experience of chemistry and the most rational features of the ideas about electropositive and electronegative atoms, the concept of electronegativity was widely used in systematics of the empirical data of chemistry and formed the essential part of chemical, mineralogical and geochemical approaches. It was used as the chemical characteristic of atoms, allowing an estimation of their properties in compounds.

However, the inherent contradictoriness of electronegativity, its discrepancy with experimental observations and some principal objections led to revising it and renouncing its use [123].

The number of publications on electronegativity is enormous but most of them are of minor importance; as principal works can be quoted [81, 113, 115, 122, 125].

Pauling's determination of electronegativity [113]. In order to understand better the method of obtaining electronegativity according to Pauling, one needs to realize that this determination was done within the framework of the VB method (see above). Electron structure in this method is described by the superposition, or resonance, of the hypothetic pure ionic structure $\psi_{ion}(A^+B^-)$ and of the hypothetic pure covalent structure $\psi_{cov}(A-B)$.

Then the total energy of the bond E_{AB} can be written as the sum of the energies of these hypothetic structures:

$$E_{AB} = E_{ion}(A^+B^-) + E_{cov}(A-B),$$

or in the form of the energies of dissociation:

$$D_{AB} = D(A^+B^-) + D(A-B).$$

The D_{AB} value is taken as the experimental value of the dissociation energy of the molecule AB, the $D(A-B)$, i.e., the covalent component of the dissociation energy, is estimated as half of the sum of the dissociation energies of the covalent molecules A_2 and B_2, i.e., $1/2(D_{AA} + D_{BB})$, or as their geometric average $\sqrt{D_{AA} \cdot D_{BB}}$. Then the ionic part of the dissociation energy is determined as the difference

$$D(A^+ + B^-) = \Delta = D_{AB} - 1/2(D_{AA} + D_{BB}).$$

For example ([87], p. 151), $D_{HF} = 135.1 \text{ kcal mol}^{-1}$; $1/2(D_{H_2} + D_{F_2}) = 1/2(103.4 + 37.7) = 70.6 \text{ kcal mol}^{-1}$; $\Delta_{HF} = D_{HF} - 1/2(D_{H_2} + D_{F_2}) = 64.5 \text{ kcal mol}^{-1}$.

These differences between the total energy of heteroatomic bond $A-B$ and the average energy of the homoatomic covalent bonds $A-A$ and $B-B$ equal to the ionic part of the bond energy is taken as a measure of charge separation in the ionic–covalent molecule AB, that is, as a measure of the difference of the electronegativities $(X_A - X_B)$ of the atoms A and B, or as a measure of the energy of the resonance. Here $(X_A - X_B)$ is proportional to Δ equal to $D(A^+B^-)$.

However, one needs two more steps to determine electronegativity.

First, the comparison of the Δ values for several compounds showed that they are not additive; for example, $\Delta_{HF} + \Delta_{FCl} \neq \Delta_{HCl}$. Then as a measure of electronegativity the value $\sqrt{\Delta}$ (i.e., "square root from resonance energy") was chosen as showing approximate additivity for some simple molecules (such as HCl, HF, ClF). Thus finally the electronegativity difference $X_A - X_B$ was equated with this value:

$$X_A - X_B = \sqrt{\Delta}.$$

Here both parts of the expression are represented either in kcal mol^{-1} or in eV; if Δ is in kcal mol^{-1} while X_A and X_B have to be in eV then $(X_A - X_B)$ eV $= \sqrt{\Delta/23.06} = 0.208\sqrt{\Delta}$ kcal mol^{-1} (because 1 eV = 23.06 kcal mol^{-1}) and $(X_A - X_B)$ kcal mol^{-1} = $23.06\sqrt{\Delta}$ eV.

Secondly, this expression determines only the difference of the electronegativities and not their absolute value. For one of the chemical elements, electronegativity has then to be taken as an arbitrary value. Assuming for fluorine, as the most electronegative element, an electronegativity value equal to 4, Pauling obtained the electronegativity scale for the elements of the periodic system [4, 11, 79, 105, 113 and others].

The scale is constructed in such a way that the difference of the electronegativities of two elements gives $\sqrt{\Delta}$ in eV. For example: $X_F = 4.0$; $X_H = 2.1$; $X_F - X_H = 1.9 = \sqrt{\Delta}$; $\Delta = 3.6$ eV $= 82.8$ kcal mol^{-1} (cf.: $\Delta_{HF} = D_{H^+F^-} = D_{HF} - 1/2(D_{H_2} + D_{F_2}) = 135.1 - 1/2(103.4 + 37.7) = 64.5$ kcal mol^{-1}.

The principal objections to this determination of electronegativity are as follows:

a) The whole method of the consideration corresponds to the VB system of ideas and thus the concept of electronegativity being determined in this way inherits, naturally, the inadequacy of the method.

The values $\Delta = D_{A^+B^-}$ and D_{A-B} are not observable; they correspond to the end members of valent schemes, pure ionic and pure covalent, neither of which is realized. However, differences in electronegativities still have physical meaning in the framework of the VB method as the energy of the ionic structure, i.e., as conventional ionic part of the total energy (i.e., as abstract intermediate value).

However, electronegativities X_A and X_B themselves remain here predominantly conventional values which are far from the real physical meaning. Their definition as "the property of atoms to attract electrons" does not correspond to their actual meaning. There is also no physical meaning in the unit measure $X_A - X_B = \sqrt{\Delta}$, i.e., square root of energy but not energy itself.

Electronegativity represents a summary property of the atom and does not reflect electron distribution in certain molecular orbitals. No aspects of this definition of electronegativity are compatible with any aspects of the MO method.

b) In this definition electronegativity has been considered as a constant of the each element, as its chemical characteristic. However, it proves correct neither from the point of view of theory nor of experimental observations.

Even in the framework of the VB method it soon became clear that the electronegativity value depends on the valency, on the state and degree of hybridization, on the properties of other atoms entering the chemical bonding and other factors.

However, the repudiation from the attribution of the constant electronegativity value to each element immediately deprives the determination of electronegativity of its numerical basis because in $\Delta = D_{AB} - 1/2(D_{AA} - D_{BB})$ the covalent part of energy D_{A-B} can be obtained only as an average value from D_{AA} and D_{BB} and the total energy D_{AB} is taken from the data for diatomic molecules AB, while the energies of the bonding A−B in polyatomic compounds do not have unequivocal determinations.

There can, therefore, be no scale of constant values of the electronegativities for elements, as it is impossible to obtain experimental data by this method for the construction of an adjustable scale. Numerous attempts to obtain an improved electronegativity scale are impossible by the very way in which the problem is presented.

c) There thus exists an understandable inconsistency between the electronegativities determined from Δ and those calculated from experimental data.

The value Δ taken as the difference of the tabulated electronegativities $X_A - X_B$ does not generally coincide with Δ_{exp} calculated as the difference of the experimental D_{AB} and average D_{A-B}. For example [122], for NaCl $\Delta = 101$ kcal mol^{-1}, $\Delta_{exp} = 60.8$ kcal mol^{-1}; for LiF $\Delta = 207$ kcal mol^{-1}, $\Delta_{exp} = 105.4$ kcal mol^{-1} etc.

There is no additivity, not only for the Δ values but also for $\sqrt{\Delta}$ even for the simple diatomic molecules. For example [122]: $(\sqrt{\Delta_{HCl}} + \sqrt{\Delta_{ClF}} = 192.3$ kcal mol$^{-1}) \neq (\sqrt{\Delta_{HF}} = 184.2$ kcal mol$^{-1})$ or $(\sqrt{\Delta_{HJ}} + \sqrt{\Delta_{JCl}} = 66.9$ kcal mol$^{-1}) \neq (\sqrt{\Delta_{HCl}} = 108.1$ kcal mol$^{-1})$.

Empirical Determinations of Electronegativity. Because the measure of electronegativity difference according to Pauling's determination is not observable experimentally (Δ) is only the measure of the energy of resonance), numerous attempts have been undertaken to construct electronegativity scales based on different measurable properties which can be connected with electron distribution: formation heats, dipole moments, interatomic distance, force constants, resonance frequencies of nuclear quadrupole resonance, different vibration modes in infrared spectra, energy gaps etc. The relations were represented by empirical equations.

In some cases this property of a compound was correlated with the electronegativity of the atom entering in the compound. In other cases it was related to the properties of the covalent simple compounds, while the electronegativity difference had to reflect the deviation of the chemical bond from the pure covalent. For example:

$$r_{AB} = r_A - r_B - 0.09(X_A - X_B),$$

where r_{AB} is interatomic distance in AB; r_A and r_B are interatomic distances in simple compounds A and B; $(X_A - X_B)$ is the electronegativity difference [81]; or $\Delta E_{AB} = \Delta E_A + \Delta E_B + a(X_A - X_B) - bn_{AB}$, where ΔE_{AB}, ΔE_A, ΔE_B are the energy gaps in the compounds AB, A and B respectively, n is the principal quantum number, a and b are the empirical constants [81].

Similar equations were used both for the calculations of the corresponding properties of compounds (for example, interatomic distances in the compounds composed by any elements for which the electronegativity value is known) and for the construction of electronegativity scales based on the measurements of these properties.

It is evident that these empirical equations do not provide any new material for determining electronegativity. The principal reason for applying such equations is the empirically adjusted coefficients and the fact that the variation of a given measured value in different compounds reflects the periodicity of variation of properties in the Mendeleev system of elements. The theories of the relation of properties, e.g., the spectroscopic parameters, to the properties of atoms are of a much more complicated character.

All electronegativity scales (empirical as are also those based on the ionization potentials discussed below) can be reduced to the Pauling scale by simple division or multiplication by some empirical factor (though electronegativities can be expressed in these scales in different unit measures: in the Pauling scale it is represented by the square root of energy while in that of Mulliken by energy).

Mulliken's determination of electronegativity and its evolution to VSIE. Since the tendency of atoms towards formation of positive or negative ions can be associated with the energy of the valent electron (or ionization potential of the atom) and with the energy of attraction of an additional electron by atom (electron affinity), Mulliken (1934) suggested the determination of electronegativity based on the following considerations.

In order to estimate which atom, A or B, is more electronegative, i.e., which of them in compound AB forms a positive ion and which negative, let us compare the energy:

$$A+B \to A^+B^-, \quad \text{i.e.,} \quad IP_A - F_B,$$
$$A+B \to B^+A^-, \quad \text{i.e.,} \quad IP_B - F_A,$$

where IP is the ionization potential and F is the electron affinity of atoms A and B.

If $IP_A - F_B < IP_B - F_A$, it means that it is easier to separate the electron from A and transfer it to B, i.e., $A+B \to A^+B^-$, than to obtain $A+B \to B^+A^-$. That is why the compound A^+B^- is formed and A is less electronegative than B. Or if $IP_A + F_A < IP_B + F_B$ (grouping together the parameters of each atom), then A is less electronegative than B. Thus the electronegativity according to Mulliken was taken as

$$EN = 1/2(IP + F).$$

For electropositive atoms the electron affinities are close to zero and their electronegativities are simply equal to the ionization potentials. Mulliken's scale of electronegativities represents essentially the ionization potential values for metals and semimetals and the values of the ionization potential plus electron affinity for the remaining elements.

However, this leads logically to the evolution of the Mulliken scale: the ionization potentials and consequently, the electronnegativities depend on charge and valency; they differ for different electrons (i.e., one distinguishes the ionization potentials of s, p, d electrons) and differ for ionization from different electron configurations. Then the average ionization potentials of atoms as well as the electronegativities determined from them lose their sense [115]; for example, $EN = \Sigma I_n/n + F$, where ΣI_n is the sum of the ionization potentials, n is valency and F is the electron affinity.

Without discussing the further development of Mulliken's determination of electronegativity [81] note that the consideration of effective charges and effective electron configuration leads from ionization potentials to VSIP (valence state ionization potentials), or VSIE, to which the Coulomb integrals H_{AA} are equated in the MO method. (The sum of ionization potential and electron affinity has no physical meaning because these values are related to the different electrons and different orbitals).

Thus the most rational determination of electronegativity, i.e., that of Mulliken, leads in its logical development to H_{AA}–VSIE which, however, do not represent the extent of the ability of the atom in a molecule to attract electrons, and which are the intermediate value in calculations used to understand the order of calculation.

Then, however, the original significance of electronegativity as constant, summary and additive characteristic of the chemical properties of elements is lost.

3.3.2 LCAO Coefficients c_i and Electronic Population Analysis; Ionicity – Covalency of Chemical Bonding and Effective Charge; Valence and Charge

Formation of chemical bonding involves redistribution of electron density (partial transfer of electron and related transfer of charge from a less electronegative atom to more electronegative one). In a molecule, for instance, LiH the Li valent electron 2s occurs more often near H, so that Li has a partially positive charge in this molecule; it is partially positive since a for some part of the time the electron still occurs near Li. Effective charges in the LiH molecule are not the charges of free atoms Li^0 and H^0 and not Li^{+1} H^{-1} as they would be in the case of a complete transfer of the electron; the actual distribution of charge obtained from experimental and calculated data is $Li^{+0.8}$ $H^{-0.8}$.

Degree of charge separation is described by the concepts of ionicity – covalency. Compounds are covalents when valent electrons are shared equally by the atoms; these can be only homoatomic molecules (H_2, O_2 and so on) and crystals (diamond, sulfur etc.). Pure ionic states would correspond to complete transfer of the electron from one atom to another, and they represent extreme states which never occur. All heteroatomic molecules and all inorganic crystals, including most parts of minerals, correspond to the intermediate ionic-covalent type of electron (charge) distribution.

The notion of ionicity – covalency has differing content and differing physical meaning in different theoretical approaches to the chemical bond and in theories describing relations between different experimental parameters (first of all spectroscopic parameters) and electron distribution. Hence these determinations of ionicity – covalency are not comparable directly and they describe different aspects of a complex phenomenon of chemical bond.

In the valence bond method which treats chemical bonding as the resonance of hypothetical limit states (pure covalent AB and pure ionic A^+B^- with a small contribution of A^-B^+) the notion of ionicity is closely related to the notion of electronegativity (see above) and moreover, with respect to latter it is the initial one.

In the expression $\psi_{AB} = a\psi_{ion}(A^+B^-) + b\psi_{cov}(AB)$ the coefficients a and b give the fractions of an ionic and a covalent structure in the presentation of the bonding state, and the ionic resonance energy, or dissociation energy, $D(A^+B^-) = \Delta = D_{AB} - 1/2(D_{AA} + D_{BB})$, is a measure of the ionicity of the chemical bonding. It is precisely this measure of ionicity that is taken as a value of difference of electronegativities whose absolute values are not determined but are counted from an arbitrarily chosen magnitude. By means of this determination of electronegativity from the degree of ionicity (i.e., from electronegativity difference) and taking a scale of electronegativity, one determines the degree of ionicity for concrete compounds from the electronegativity difference using diagrams constructed on the basis of a few points from experimental data (see below) for some simple molecules with a known degree of ionicity [142].

The molecular orbital method presents a much more detailed picture of electron density distribution in chemical bonding formation; this is obtained as a result of molecular orbital population analysis.

For a bonding molecular orbital: $\psi^b = N(c_A\psi_A + c_B\psi_B)$. Taken $c_A/c_B = \lambda$ we obtain $\psi^b = N(\lambda\psi_A + \psi_B)$. If $\lambda = c_A/c_B = 1$ the molecular orbital is equally occupied by electrons of the central atom and ligand; it is a case of the covalent bond. The value $\lambda = c_A/c_B = 0$ corresponds to the extreme case of pure ionic bond for bonding orbitals or to the nonbonding molecular orbital $\psi_{MO} = \psi_B$.

The values of c_i and γ are connected by the condition of normalization to unity:

$$c_A^2 + c_B^2 + 2c_Ac_BG_{AB} = 1 \quad \text{or} \quad 1 + \gamma^2 + 2\gamma G_{AB} = 1.$$

The coefficients c_i in the tables of MO calculations have not yet any real significance. They represent only initial data for population analysis of molecular orbitals, while the physical meaning is connected with the squares of these coefficients c_i^2, corresponding to the electron population of the molecular orbital (in the same way as the square of wave function ψ^2 but not wave function ψ itself corresponds to electron density distribution).

Consider the determinations given by Mulliken (see [1]) illustrated by the $3d$ electrons of Mn in $\psi(1e)$ molecular orbital of MnO_4^-: $c_{3d(1e)}$ – coefficient c_{3d} in the molecular orbital $\psi(1e) = c_{3d}\psi_{3d} + c_{2p}\psi_{2p}$; N_{1e} – number of electrons in $(1e)$ molecular orbital equal 4 for $\psi(1e)$; $N_{1e} \cdot c_{3d(1e)}^2$ – net atomic population, i.e., contribution of $3d$ electrons of Mn atom in $\psi(1e)$ MO; $N_{1e} \cdot 2c_{3d(1e)} \cdot c_{2p(1e)} G_{1e(3d2p)}$ – overlap population for $3d$ and $2p$ electrons in $\psi(1e)$ MO; (from the expression $c_1^2 + c_2^2 + 2c_1c_2G_{1,2} = 1$ the overlap population is equal $2c_1c_2G_{1,2} = 1 - c_1^2 - c_2^2$); $n_{3d(1e)} = N_{1e} \cdot c_{3d(1e)}^2 + N_{1e} \cdot 1/2(2c_{3d(1e)} \cdot c_{2p(1e)}) \cdot G_{1e(3d,2p)}$ gross atomic population, i.e, contribution of $3d$ electrons of Mn in $\psi(1e)$ MO consisted of the net atomic population and half of the overlap population; $\Sigma(n_{3d(1e)} + n_{3d(1t_2)} + n_{3d(2t_2)}) = q_d$ – total $3d$ electrons fraction on Mn atom in all molecular orbitals of MnO_4^-; q – effective atomic charge: $q = m - (q_d + q_s + q_p)$, where m is number of valent electrons in free Mn atom ($3d^54s^2$, i.e., $m = 7$).

For the molecular orbital $\psi(1t_2)$ composed of two atomic orbitals and two group ligand orbitals, the overlap population is

$$N_{1t_2}(2c_{3d}c_{2p\sigma} + 2c_{3d}c_{2p\pi} + 2c_{4p}c_{2p\sigma} + 2c_{4p}c_{2p\pi}),$$

and the gross atomic populations are:

$$n_{3d(1t_2)} = N_{1t_2} \cdot c_{3d(1t_2)}^2 + N_{1t_2} \cdot 1/2(2c_{3d(1t_2)} \cdot c_{2p\sigma})$$
$$+ N_{1t_2} \cdot 1/2(2c_{3d(1t_2)} \cdot c_{2p\pi});$$

$$n_{4p(1t_2)} = N_{1t_2} \cdot c_{4p(1t_2)}^2 + N_{1t_2} \cdot 1/2(2c_{4p(1t_2)} \cdot c_{2p\sigma})$$
$$+ N_{1t_2} \cdot 1/2(2c_{4p(1t_2)} \cdot c_{2p\pi}).$$

Table 31. Molecular orbital coefficients c_i and energies

MO	E,	c_A		c_B	
$1a_1$	−19.94	$c_{4s}=$	0.478	$c_{2p_\sigma}=$	0.613
$1t_2$	−18.4	$c_{3d}=$	0.637	$c_{2p_\sigma}=$	0.539
		$c_{4p}=$	−0.054	$c_{2p_\pi}=$	0.24
$1e$	−11.8	$c_{3d}=$	0.583	$c_{2p_\pi}=$	0.626
$2t_2$	−11.21	$c_{3d}=$	−0.143	$c_{2p_\sigma}=$	0.527
		$c_{4p}=$	0.244	$c_{2p_\pi}=$	−0.737
t_2	−9.76	—		$c_{2p_\pi}=$	1.000
$3t_2$	−7.13	$c_{3z}=$	0.715	$c_{2p_\sigma}=$	−0.605
		$c_{4p}=$	0.401	$c_{2p_\pi}=$	−0.439
$2e$	−3.73	$c_{3d}=$	0.912	$c_{2p_\pi}=$	−0.815
$4t_2$	−2.60	$c_{3d}=$	−0.189	$c_{2p_\sigma}=$	0.119
		$c_{4p}=$	1.02	$c_{2p_\pi}=$	0.458
$2a_1$	−0.03	$c_{4s}=$	1.32	$c_{2p_\sigma}=$	1.03

Table 32. Group overlap integrals G_{AB} for MnO_4^-

MO	G_{AB}
a_1	$G_{4s2p_\sigma}=0.67$
e	$G_{3d2p_\pi}=0.36$
t_2	$G_{3d2p_\sigma}=0.252$
t_2	$G_{3d2p_\pi}=0.207$
t_2	$G_{4p2p_\sigma}=0.022$
t_2	$G_{4p2p_\pi}=0.26$

Table 33. Overlap population $N_i 2c_A c_B G_{AB}$ calculation with using data of Tables 31 and 32

MO	$4s2p_\sigma$	$3d2p_\pi$	$3d2p_\pi$	$4p2p_\sigma$	$4p2p_\pi$	$\sum n\sigma$	$\sum n\pi$	$\sum(n\sigma+n\pi)$
$1a_1$	0.7804	—	—	—	—	0.7804	—	0.7804
$1t_2$	—	1.073	0.386	−0.008	0.042	1.0650	0.428	1.493
$1e$	—	—	1.048	—	—	—	1.048	1.048
$2t_2$	—	−0.228	0.262	0.034	0.566	−0.194	0.828	0.634
						1.6514	2.304	3.9554

For the ligand atoms the gross atomic population is:

$$n_{2p(1e)} = N_{1e} \cdot c^2_{2p(1e)} + N_{1e} \cdot 1/2(2c_{3d(1e)} \cdot c_{2p(1e)} \cdot G_{1e(3d2p_\pi)}).$$

Electron population analysis is given for the MnO_4^- in Tables 31–35 [78, 126].

This analysis gives the following information: (1) the effective atomic charges: as the calculations (see Tables 31–35) indicate they show to what extent electrons belong to the central atom and to ligand atoms; they represent those parts of electron density (of charges) which remain after the deduction of the net atomic

Table 34. Atomic populations $n = Nc_A^2 + 1/2 N \cdot 2c_A c_B G_{AB}$, effective charge and effective electronic configuration of Mn in MnO_4^-

AO	MO	Nc_A^2 (c_A from Table 31)	$1/2(N \cdot 2c_A c_B G_{AB})$ (from Table 33)		n	
4s	$1a_1$	$2 \cdot 0.478^2 = 0.45$	$4s2p_\sigma$	0.39		
					0.84	
						$q_s = 0.84$
4p	$1t_2$	$6 \cdot 0.054^2 = 0.017$	$4p2p_\sigma$	-0.004		
			$4p2p_\pi$	-0.021		
				0.017	0.03	
	$2t_2$	$6 \cdot 0.244^2 = 0.35$	$4p2p_\sigma$	0.017		
			$4p2p_\pi$	0.283		
				0.300	0.65	
						$q_p = 0.68$
3d	$1e$	$4 \cdot 0.583^2 = 1.36$	$3d2p_\pi$	0.524	1.88	
	$1t_2$	$6 \cdot 0.637^2 = 2.40$	$3d2p_\sigma$	0.536		
			$3d2p_\pi$	0.193		
				0.729	3.13	
	$2t_2$	$6 \cdot 0.143^2 = 0.12$	$3d2p_\sigma$	-0.114		
			$3d2p_\pi$	0.131		
	$\sum Nc_A^2 = 4.69$			0.017	0.14	
				$q_d = 5.1$		

Total number of electrons on Mn is $\sum(q_d + q_s + q_p) = 0.67$. Effective electronic configuration of Mn in MnO_4^- is $3d^{5.15} 4s^{0.84} 7p^{0.68}$. Effective charge on Mn ($3d^5 4s^2$) is $7 - 6.67 = 0.33$, i.e., $Mn^{+0.33}$.

Table 35. Effective charge and electronic configuration of oxygen in MnO_4^-

AO	MO	$N \cdot c_B^2$ (c_B from Table 31)	$1/2(N \cdot 2c_A c_B G_{AB})$ (from Table 33)		n
$2p_\sigma$	$1a_1$	$2 \cdot 0.613^2 = 0.75$	$2p_\sigma 4p$	0.39	1.14
	$1t_2$	$6 \cdot 0.539^2 = 1.74$	$2p_\sigma 4p$	-0.004	
			$2p_\sigma 3d$	0.536	
				0.532	2.27
	$2t_2$	$6 \cdot 0.527^2 = 1.67$	$2p_\sigma 4p$	0.017	
			$2p_\sigma 3d$	-0.114	
				-0.097	1.57
			$\sum 2p_\sigma = 4.98:4 = 1.24$		
$2p_\pi$	$1t_2$	$6 \cdot 0.24^2 = 0.35$	$2p_\pi 4p$	0.021	
			$2p_\pi 3d$	0.193	
				0.214	0.56
	$2t_2$	$6 \cdot 0.737^2 = 3.26$	$2p_\pi 4p$	0.282	
			$2p_\pi 3d$	0.132	
				0.414	3.67
	$1e$	$4 \cdot 0.626^2 = 1.57$	$2p_\pi 3d$	0.524	2.09
	t_1	$6 \cdot 1.000^2 = 6.00$		0	6.00
		$\sum Nc_B^2 = 9.34 \, (\varrho_{e3}t_1)$	$\sum 2p_\pi = 12.32:4 = 3.08$		

Total number of electrons on four oxygens is $= 17.30$. Effective electron configuration of O in MnO_4^- is $2p^{1.24} 2p^{3.08}$. Effective charge on $O(2p^4)$ is $4 - (1.24 + 3.08) = 0.32$, i.e., $O^{-0.32}$.

population and the overlap population of molecular orbitals; (2) the overlap population (the overlap density) are a measure of the bond strength for a given molecular orbital and at the same time a measure of the bond covalency; in pure covalent compounds with $c_A = c_B$ the overlap densities can be of different magnitude; (3) the atomic populations c_A^2 and c_B^2 as $c_A^2 + 1/2(2c_A c_B G_{AB})$ and $c_B^2 + 1/2(2c_A c_B G_{AB})$; the relation c_A^2/c_B^2 indicates a charge separation in a given molecular orbital, while the relation $\Sigma c_A^2/\Sigma c_B^2$ for all electrons in all the molecular orbitals of a complex gives the best approach to the concept of ionicity – covalency. For the MnO_4^- complex (see Tables 34, 35) $\Sigma c_A^2/\Sigma c_B^2 = 4.69/9.34 \approx 0.50$.

Note that none of these characteristics taken separately is sufficient to describe electron distribution in chemical bonding formation, but all of them need to be used. Indeed, effective charges as determined in Tables 34 and 35 do not reflect the electron distribution in the molecular orbitals described by c_i^2 coefficients. Effective charges determined from the MO calculation do not immediately give a measure of ionicity: from the relation q_{eff} to maximum ionic charge (equal to number of valent electrons) one obtains too low a value of ionicity for multivalent atoms; for example, for $Mn^{+0.33}$ in MnO_4^- (with $Mn\ 3d^5 4s^2$, i.e., 7 valent electrons) ionicity would be equal in this case $0.33/7 \approx 0.05$.

Ionicity could probably be determined by the relation $(q + \Sigma c_A^2)$, i.e., effective charge plus atomic population, to valence. For $Mn^{+0.33}$ in MnO_4^- $(q + \Sigma c_A^2) = 0.33 + 4.69 = 5.02$ (see Table 34), then the ionicity is $5/7 \approx 0.70$.

The effective charges enter in expressions determining experimental parameters in a number of methods and in this way they can be obtained for certain compounds.

One of the simple methods for estimating the effective charges is measurement of dipole moments of molecules. The dipole moment $\mu = q_{eff} R$, where q_{eff} is the atomic effective charge in the molecule and R is the interatomic (internuclear) distance.

A relation of the measured dipole moment to the pure ionic one with the atomic charge equated with electron charge $\mu_{ion} = eR = 4.8 \cdot 10^{-10}\ ESU \cdot 10^{-8} R(Å) = 4.8R$ (in Debye units, $D = 10^{-18}\ ESU$) is taken as a measure of the bond ionicity in a molecule.

For example, in the HCl molecule $\mu_{meas} = 1.03 D$; $\mu_{ion} = eR = 4.8 \cdot 1.28 = 6.07 D$; $\mu_{meas}/\mu_{ion} = 1.03/6.07 = 0.17$; i.e., the degree of ionicity is 17%.

With a knowledge of the interatomic distances in molecules, one can obtain for them μ_{ion}, and taking experimental values for μ_{meas} [112], we can obtain effective charges and degrees of ionicity for the molecules. The ionicity values for the HCl, HBr, HI molecules determined by this method were taken by Pauling [113] for the calibration of the curve in the diagram connecting electronegativitiy with degree of ionicity.

Experimental values of effective atomic charges in crystals are determined by all the methods of solid state spectroscopy, from dielectric properties and others. Let us consider here merely some general remarks concerning these methods.

Atomic effective charge determination in compounds can be done only in terms of electron density (electron charge) distribution in the compounds. Empirical effective charges are the values calculated from different experimental parameters and thus depend on the specific aspects of theories elaborated for these methods and also on the particular mechanisms of the differing interactions, essentially composing these methods. Empirical effective charge values are not correlated directly with that obtained in the course of molecular orbital calculations. For example, the effective charge of Si in SiO_2 from X-ray spectroscopy data is $+1.97$, while that of Si in SiF_4, resulting from MO calculation is $+0.18$.

Parameters of some other methods from which effective charges are calculated (dipole moments, dielectric constants) are not of quantum mechanics origin.

A method of solid state spectroscopy such as EPR gives the characteristics for electron density distribution only for one of the molecular orbitals (antibonding in the case of EPR of transition metal ions), i.e., one obtains c_A^2/c_B^2 for one of the MO but not $\Sigma c_A^2/\Sigma c_B^2$ and not $q = m - (q_s + q_p + q_d)$ (see Tables 34 and 35). Moreover even the different EPR parameters: g-factor, hyperfine structure, superhyperfine structure are related to the different mechanisms of electronic interactions. Thus without further reductions one cannot expect coincidence in the effective charge values obtained by measurement with different methods.

However all types of calculation and measurement show that the atomic effective charges in inorganic compounds are not great and are much less than pure ionic ones [101].

An anion charge does not exceed the value -1, while that of cation does not exceed $+2$. Multicharge ions do not exist. For example, there are no Si^{4+} ions in compounds, but $Si^{+1.97}$ exists in quartz, there are no Al^{3+} but one observes $Al^{+1.38}$ in orthoclase and so on. The MO calculations give still lower values for effective charges (see Tables 34 and 35), thus confirming the electroneutrality principle suggested by Pauling [113].

Valence and Charge. The definition of valency by classical chemistry as a capacity of atoms to attach a certain number of atoms of another element was interpreted at earlier stages of the chemical bond theory from the point of view of the concepts of ionic and covalent compounds. For ionic compounds valency was identified with a number of electrons lost by a cation or gained by an anion and also with atomic charges in compounds (Na^+, Mg^{2+}, Al^{3+}... as singly, doubly, triply charged ions). For covalent compounds it was equated to a number of shared electron pairs (Cl_2 – one shared electron pair, diamond – four pairs and so on).

In the valence bond method, the idea of multicharge ions with shared electron pairs leads necessarily to the concept of valence structure resonance.

Both experiments and calculation have shown for certain that valence cannot be identified with charge. In $Al_2^{3+}O_3^{2-}$, for instance, the Al atomic charge is $+1.53$ and that of oxygen is -1.02. This corresponds to the delocalization of each valence electron over all the atoms of a molecule or complex. Electron distribution among the atoms involved in a molecular orbital is determined by fractional values of the coefficients c_i^2. In the final data of MO schemes there are only fractional values of the effective charges (for example, $Mn^{+0.33}$ in MnO_4^-) and the fractional electron configurations ($Mn\ 3d^{5.15}4s^{0.84}4p^{0.68}\ O\ 2p^{2.24}2p^{3.08}$). The electrons lost by Mn occur mostly near the oxygen atom (the coefficients c_B are greater than c_A in nearly all bonding and of course, nonbonding molecular orbitals).

Sometimes, rejecting with good reason the identity of valence and charge, one denies the very notion of valency and tries to use instead the fractional values of charge ($Si^{+1.97}$, $O^{-1.02}$ etc.). However, parallel with the continuous (fractional) characteristics of electron density and charge distribution there exist also their discrete characteristics.

Solid state spectroscopy being the principal experimental method estimation of chemical bond parameters (effective charges included) can discern unambiguously the valence states of atoms.

For example, the Fe^{2+} and Fe^{3+} valence states are sharply different in any type of compound with any amount of covalent character of a bond according to EPR spectra, optical absorption spectra (in both of them the completely different energy levels schemes correspond to the spectra), finally, simply according to the colors of their compounds. They are different according to their chemical and geochemical behavior: in the oxidation zone of ore deposits only Fe^{3+} and not Fe^{2+} exists in all minerals with atomic charges; on the contrary, in lunar minerals Fe^{2+} and only the small quantities of Fe^{3+} were observed (from EPR data).

Luminescence spectra of TR^{3+} and TR^{2+} are also principally different. In crystals with electron-hole centers the Pb^+ and Pb^{3+}, O^- and other centers are clearly distinguished.

Thus it is necessary to find in the MO scheme a consistent definition of the notion of valency for transition metal-ions and nontransition elements, for positive and negative ions, covalent and ionic compounds and to retrace its relation to coordination number, to formulas of compounds, to a number of attached atoms.

In the MO scheme the valency can only be the genealogical concept, since the final results of MO calculations lead to effective charges and effective configurations. However, the MO scheme itself has genealogical features as to the construction of molecular orbitals from the atomic orbitals (as their linear combination – LCAO). The number of electrons (integer) in molecular orbitals is determined from basis atomic orbitals. For example, in the molecular orbitals of MnO_4^- there are 24 electrons: $(1a_1)^2 (1t_2)^6 (1e^4) (2t_2)^6 (t_1)^6$. They are obtained from 4 electrons of each of 4 oxygen atoms (oxygen electron configuration $2p^4$), from 1 charge electron of MnO_4^- complex (i.e., from the neighbor Mn atoms) and from 7 electrons of Mn ($3d^5 4s^5$). Here Mn^{+7} is sevenvalent with the charge $Mn^{+0.33}$ ($q = +0.33$) oxygen O^{2-} is divalent with the charge $O^{-0.33}$ ($q = -0.33$).

In MnO_4^{2-} 24 electrons are obtained: 16 from O_4^{2-}, 6 from Mn^{6+}, 2 from MnO_4^{2-} complex. Here 6 electrons of Mn are in the bonding molecular orbitals but the 7th Mn electron must occupy the antibonding $3t_2^*$ molecular orbitals, thus determining the electron configuration of the Mn ion here as $3d^1$.

The definition of valency in the framework of the MO scheme can be given as follows.

The valency of positive ions is equal to the number of electrons supplied by the atom on bonding molecular orbitals. For negative ions it is equal to the number of holes left by the atom on bonding and nonbonding molecular orbitals.

In crystals the determination of the valency must related to the crystal structure. For example, in $FeCl_3$, in the octahedral complex $(Fe^{3+}Cl_6^-)^{3-}$ there can be 36 electrons in the bonding molecular orbitals (see Fig. 48): 30 electrons from Cl_6 ($2p^5 \cdot 6$), 3 electrons from the charge of the $FeCl_6^{3-}$ complex (i.e., from the neighbor Fe atoms) and 8 electrons from Fe ($3d^6 4s^2$) of which only 3 electrons occupy the bonding molecular orbitals, while 5 electrons occur in the antibonding orbitals. Thus the total number of electrons is 41 with 36 in bonding MO and 5 in antibonding ones.

The valency of Fe is here 3 (Fe^{3+}), 5 electrons occur in antibonding molecular orbitals (Fe^{3+} electron configuration is $3d^5$), the effective charge is $Fe^{+0.n}$, the coordination number is six ($FeCl_6^{3-}$) and the chemical formula $FeCl_3$.

All the nontransition ions in nondefect positions in crystal structures have valencies equal to the numbers of the outer electrons (or to the number of holes in the outer subshells) and corresponding to their position in the periodic table of elements (Na^+, Mg^{2+}, Al^{3+}, Si^{4+}...).

For transition elements of the same valency, for example, $Ti^{3+}(3d^1)$ $V^{3+}(3d^2)$, $Cr^{3+}(3d^3)$... after occupation of the bonding orbitals by a certain number of electrons (3 electrons in these cases), the remaining electrons further occupy the antibonding orbitals.

The integer number of electrons in molecular orbitals (2 in a, 4 in e, 6 in t molecular orbitals) explains the stability wellknown in chemistry of the 2-, 8-, 18-electron configurations.

A special reference must be made to oxygen valency. Free ions O^{2-} do not form because of the negative value of the affinity to the second electron. However, in compounds, oxygen is always divalent, owing to the placing of two electrons of a central atom in the oxygen nonbonding orbitals, or in the bonding molecular orbitals which are mostly oxygen orbitals. In a defect position the hole center O^- can be formed only in crystals: $O^{2-}(2p^6) + e^+$ (i.e., the hole e^+ trapping or the loss of electrons $e^-) \rightarrow O^-(2p^5)$. This center is readily identified by means of EPR spectra.

Ionicity-Covalency in the MO Scheme. The relative position of the molecular orbitals with respect to the energies corresponding to the Coulomb integrals – VSIE (H_{AA} = VSIE of $A^{+0.n}$ and H_{BB} = VSIE of $B^{-0.n}$, see above) is determined by the difference between H_{AA} and H_{BB} (or the electronegativity difference) and by the overlap integral G_{AB} value. Exchange integral H_{AB} can be obtained as $k \cdot G_{AB} \cdot (H_{AA} + H_{BB})/2$.

It is the position of the molecular orbitals and values of these integrals which in the MO scheme determine the amount of ionic-covalent character of a bond described by the c_i coefficients and the overlap density (Fig. 58).

In the case of a covalent bond (Fig. 58a):

$$E_b = \frac{H_{AA} + H_{AA}}{1 - G}; \quad E^* = \frac{H_{AA} + H_{AA}}{1 + G}; \quad c_A = c_B.$$

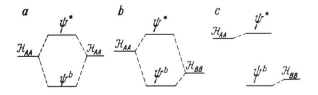

Fig. 58a–c. Relative position of molecular orbital in dependence of ionicity–covalency of the bond

In the case of an ionic-covalent bond (Fig. 58b):

$$E_b = H_{BB} - \frac{(H_{AB} - H_{BB}G_{AB})^2}{(H_{AA} - H_{BB})}; \quad H_{AB} = k \cdot G_{AB}\left(\frac{H_{AA} + H_{BB}}{2}\right);$$

$$E^* = H_{AA} + \frac{(H_{AB} - H_{AA}G_{AB})^2}{(H_{AA} - H_{BB})}; \quad c_A = c_B\left(\frac{H_{BB} - E}{H_{AB} - G_{AB}E}\right).$$

In the case of an ionic bond (Fig. 58c):

$$E_b \approx H_{BB}; \quad E^* \approx H_{AA}; \quad c_A \ll c_B.$$

Ionization Potentials and Electron Affinities of Molecules and Complexes in the MO Scheme; Oxydation – Reduction Potentials in the MO Scheme. In geochemistry and mineralogy, ionization potentials of atoms were used in many characteristics of atomic behavior: properly as ionization potentials [70, 76], in the determination of electronegativities (see above), as a main contribution to lattice energy (see Chap. 7.2), as a measure of acidity–basicity of the minerals etc. In all these cases, one used the free atom values with the charges equal to the valences [105].

However, only free ions can be multicharge (see p. 148) while in compounds (molecules, crystals, solutions, melts) multicharge ions do not exist and the effective charges are rather less than the valences. Atoms in compounds, behave not as Si^{4+}, Al^{3+}, Mg^{2+} but rather as $Si^{+0.n}$, $Al^{+0.n}$, $Mg^{+0.n}$.

Let us compare the ionization potentials of Si^{4+} and Si^+. The first ionization potential $Si^0 \to Si^+$ is 187.7 kcal mol^{-1}, while the sum of the ionization potentials $Si^0 \to Si^+ \to Si^{2+} \to Si^{3+} \to Si^{4+}$ is equal to 6256.10 kcal mol^{-1} (only $Si^{3+} \to Si^{4+}$ ionization potential is 3890.64 kcal mol^{-1}). In none of the natural or laboratory processes does complete ionization of atoms occur; molecules and complexes, but not free ions, react in solutions, melts and solid states.

The multivalent ions ionization energies are of enormous magnitudes greatly exceeding the heats of melting, dissociation, formation from elements and the bond energies. For instance, the ionization potential $Ti^0 \to Ti^+ \to Ti^{2+} \to Ti^{3+} \to Ti^{4+}$ is 2100 kcal mol^{-1}, while the heat of formation of rutile TiO_2 from Ti and O is equal to 218 kcal mol^{-1} (see Chap. 7.2).

Thus in all calculations for compounds the ionization potentials of atoms have a meaning only as VSIE (see above, Table 28) taken for a given effective charge in a certain compound; in these cases, as noted, the VSIE are only the intermediate in MO calculation values which have not in general the sense of ionization potential.

Ionization potentials of compounds (molecules and complexes) in the MO scheme are determined by the energies of the last occupied bonding or nonbonding molecular orbital from which the least strong bonded electron can be lost. In octahedral and tetrahedral complexes (Fig. 48) that is always the ligand ion nonbonding orbitals (i.e., oxygen, fluorine, sulfur orbital): t_1 in the tetrahedron and t_{1g} in the octahedron. In order to estimate the ionization potential difference for the complexes, the MO self-consistent calculations need to be made with due regard for self-consistency not only for the central but also for the ligand atom.

Table 36. Ionization potentials (IP; in eV) for some simple molecules and radicals

	IP		IP		IP
H_2O	12.67	SiO_2	11.7	H_2S	10.47
HF	16.38	SiF	7.26	S_2	8.3
HCl	12.78	SiF_4	15.4	SF_6	19.3
HBr	11.69	$SiCl_4$	11.6	SO_2	12.34

The complex ionization potential depends in such a case on the oxygen VSIP for self-consistent effective charge. The complex ionization potential varies strongly with coordination: in SiF_4 the t_1 molecular orbital energy is -21.4 eV while in SiF_6 the t_{1g} energy is only -4.5 eV [108].

For the same coordination the complex ionization potential increases with its charge decreasing: the t_{1g} energy in SiF_6^{2-} is 4.5 eV, in PF_6^- 11.8 eV, and in SF_6 20.0 eV [108].

Other things being equal the ionization potential of the compound of the same cation decreases with the changes of the ligand ion: F > O > Cl > Br > S.

The ionization potentials of compounds can be determined experimentally by different methods [105] especially by X-ray absorption spectra and electron spectra giving the energy of the transition from last occupied ligand orbital.

The ionization potentials for some simple molecules and radicals are listed in Table 36.

The experimental and calculated values of ionization potentials can be used for the estimation of acidity–basicity characteristics.

The electron affinity of compounds is determined by the first unoccupied antibonding molecular orbital energy an which on electron can be trapped.

In transition metal complexes the d electron occurs usually in antibonding orbitals that belong mostly to the metal ion, in this case the trapping or loss of the electron leads to the change of the metal oxydation state and the energy of this molecular orbital determines the complex oxydation–reduction potential.

The ionization potentials and electron affinities of complexes and molecules determine also the formation of different types of the electron-hole centers.

3.4 Further Development of Molecular Orbital Methods

3.4.1 About the Methods of the MO Calculations for Isolated Clusters

Discussion of features of the calculations by the different methods is beyond the scope of this book: it needs more mathematical preliminaries. However, it is worthwhile here to review the main distinctions which lie at the basis of the classification of these methods.

One distinguishes three categories of the MO techniques: (1) semi-empirical, (2) semi-quantitative and (3) ab initio calculations [83, 86, 88, 91].

However, basic principles of all these different methods are common: they involve one-electron orbital approximation, include the iterative approach of the

Self-Consistent Field (SCF), use the Linear Combination of Atomic Orbitals (LCAO), and give solutions of the secular equations similar to those considered above for the hydrogen molecule ion; the most essential part of these solutions consists of the determination of the quantum chemistry integrals: Coulomb, exchange, overlap and others.

The ab initio calculations consist of exact evaluation of all integrals in the LCAO–SCF equations (from the "first principles"). This is not only the most rigorous solution, but it is the only one retaining all chemical bond information which is partially lost by the approximations of other methods. Accuracy of the ab initio calculations depends only on the completeness of the basic set of atomic orbitals employed (for example, $3s3p$ or $3s3p3d$ basis) and on the form of the wave functions which are taken usually as Slater-type orbitals (STO).

Of course, this approach is connected with serious computational problems and for large inorganic systems such as minerals is very difficult or virtually impossible.

The semi-quantitative approaches differ from ab initio calculations only by computational simplifications. These simplifications imply mainly the elimination of some types of integrals.

There are various schemes of Neglect of Differential Overlap (NDO) semi-quantitative approaches: Complete Neglect of Differential Overlap (CNDO) approximation (only Coulomb integrals remain to be evaluated), Zero Differential Overlap (ZDO) and others where the number of the integrals is neglected or the methods of simplification of their evaluation are used.

Another variant of the semi-quantitative approach is the SCF X_α SW method (Self Consistent Field X_α Scattered Wave), the only technique of the MO calculations which does not use the LCAO approximation. The solutions of the secular equation are obtained here for spherically or volume-averaged forms of the potential within the three separate regions: (1) spheres centered on the atomic nuclei, (2) an intersphere region and (3) the region outside an outer sphere enclosing all the atomic spheres. The solutions give here the energies of the one-electron molecular orbitals of the cluster, but the distribution of their electron density is related not to atomic orbitals but to the three regions of the cluster.

In the semi-empirical approach, some of the integrals are related to experimentally determined quantities. The adaptation of the extended Huckel MO (EHMO) method, used in organic chemistry, to transition metal complexes made by Wolfsberg and Helmholz [126] and revised by Ballhausen and Gray [78] in the form of self-consistent charge and configuration scheme was considered above (see Chap. 3.2) in the discussion of general features of the MO theory. It was the earliest application of the MO method to inorganic compounds and it enables correlation of some chemical bond features in a series of related complexes.

There now exist results of the different semi-quantitative calculations for the tetrahedral oxyanions SiO_4^{4-}, PO_4^{3-}, ClO_4^-, as well AlO_4^{5-}, MgO_4^{6-}, trigonal oxyanions BO_3^{3-}, CO_3^{2-}, NO_3^-, octahedral complexes of Mg, Al, Si, Ti, Fe^{2+}, Fe^{3+}, transition metal complexes MnO_4^-, CrO_4^{2-}, VO_4^{3-}, hexafluoride and tetrafluoride anions etc. [90, 108, 109, 423, 424, 428, 431, 432, 451, 452, 477–480].

3.4.2 Molecular Orbitals for the Larger Clusters

Thus far the MO were considered for molecules and for isolated clusters, i.e., for isolated groups in crystals consisting from the metal plus the directly coordinated ligands such as the oxyanione SiO_4^{4-} or complexes AlO_6^{9-} etc. More reliable is "shared-cluster" representation especially for polymeric structures such as most silicate minerals.

These calculations are performed by the same methods (for example, CNDO or EHM) and for the same symmetry point group as in the case of the corresponding isolated cluster. Similarly one obtains as a result of the calculations the energies and composition of the molecular orbitals, but taking into account larger size of clusters (for example, $Si_2O_7^{6-}$, $Si_3O_{10}^{8-}$, $Si_4O_{13}^{10-}$ etc. instead of SiO_4^{4-}) leads to a greater number of the molecular orbitals forming the band of the same type as the molecular orbitals of the isolated cluster [426].

Consideration of larger clusters is especially important in the interpretation of the X-ray spectra of crystals. Some recent results of the calculations of the larger clusters in minerals are given later (Chap. 8.2).

3.4.3 Bond Orbital Model

After discussion of isolated cluster separated from crystal structure and larger clusters of various sizes, as a further step to understanding and calculating the electronic structure of crystals one can consider the bond orbital approximation combining the features of a Linear Combination of Atomic Orbitals (LCAO) approach and energy band theory. This model is considered below in Chapter 7.1.

4. Energy Band Theory and Reflectance Spectra of Minerals

Aspects of the Energy Band Theory. The applications of the band theory, its meaning and significance, are determined first of all by the fact that it is this theory which in fact represents general quantum-mechanical descriptions of the electronic structure of solids with all its consequences.

The consideration of the chemical bond in solids can not be limited by the model of the molecular orbitals and needs to be extended to the energy band theory. The chemical bond in crystals is described with due regard for concepts of the band theory.

On this basis one discerns the types of solid: metals and nonmetals; dielectrics – semiconductors – metals; crystals with ionic, covalent and metallic bonds.

The band theory is also the basis of physics of semiconductors representing one of the most important classes of inorganic materials; most sulfides and related minerals belong to this class of materials.

It is also the basis of physics of dielectrics to which most other minerals belong. It represents the basis of the electronic theory of metals.

The band theory laid the foundations for the kinetic theory of luminescence which represents the most important part of the theory of luminophors. Thermoluminescence phenomena are also interpreted within its framework.

Electrical properties (in particular, thermopower, Hall effect and other properties which have found widespread application in the mineralogy of sulfides) and magnetic properties of solids are considered in terms of the band theory.

Understanding of optical properties of ore minerals (and opaque crystals in general) is based on the band theory.

Spectra of solids (from soft and ultrasoft X-ray spectra to those in UV, visible and IR regions) are interpreted from the view point of the band theory.

The band concepts have found application in describing the behavior of the matter at high pressures corresponding to the conditions of the Earth's mantle and core.

Various aspects of the band theory have been presented in textbooks on solid state physics [149, 157], in special monographs [130, 131, 142, 153, 156, 162], in works on group theory [65, 138, 139], on physics of semiconductors, dielectrics and metals, on luminescence theory and on X-ray and electron spectra, as well as in works on behavior of matter at high pressures.

Principal concepts of the band theory and their application to reflectance spectra of minerals are given below.

4.1 Basic Principles and Methods of the Energy Band Theory

The electronic structure of solids is usually considered in one-electron approximation (like the electronic structure of atoms and molecules). In the atom one considers the behavior of each electron in the field of the nucleus and other electrons, in the molecule for each electron occurring in the field of the nuclei and other electrons of the molecule; in crystal one considers the behavior of the same single electron (1s, 2s, 2p,..) in the periodic field created by all the nuclei and electrons of the crystal. The electronic structure of solids is described by the energy band diagrams, the structure of atom by electron energy level scheme (see Chaps. 1.2 and 1.8) and the electronic structure of molecule by the molecular orbital scheme (see Chap. 3.2).

At the end of this section one can realize how the properties of real crystals can be determined by such consideration of the behavior of each electron, and how it is possible to progress from the Bohr model of the atom to the model of electron behavior in solids.

Electron behavior in crystals (as in atoms and molecules) can be described by the Schrödinger equation (see Chap. 1.3) which has principally the same form for electrons in an atom, molecule and crystal:

$$H\psi = E\psi \quad \text{or} \quad \left(-\frac{h^2}{2m}\nabla^2 + V\right)\psi = E\psi,$$

$$\text{or} \quad \nabla^2\psi + \frac{8\pi^2 m}{h^2}(E-V)\psi = 0,$$

where H is the energy operator presenting the sum of kinetic energy operator $(h^2/2m)\nabla^2$ and potential energy operator V; the potential in crystal is a periodic function of coordinates with periods equal to the lattice periods; further consideration is dependent on the choice of the form of the potential V;

ψ denotes an one-electron wave function, in the atom it is an atomic orbital ($\psi_{1s}, \psi_{2p}, \psi_{3d}, \ldots$), in the molecule a molecular orbital (for instance, $\psi_{t_{2g}} = c_1\psi_{3d\text{Mn}} + c_2\psi_{2p\text{O}}$), and in solids it is a crystal orbital describing electron distribution between all crystal atoms; further course of calculation depends on the choice of the form of the wave function.

4.1.1 Wave Vector k in the Free-Electron Case

Consider first the motion of the free electron, i.e., which thus does not interact with the nucleus and other electrons, i.e., at $V=0$. Then the above Schrödinger equation can be written:

$$\nabla^2\psi + \frac{8\pi^2 m}{h^2}E\psi = 0.$$

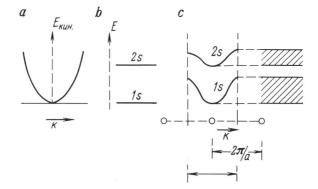

Fig. 59a–c. Electron energy as function of wave vector k. **a** Free electron; **b** electron in a free atom (1s, 2s electron energy levels); **c** electron of an atom in crystal and formation of energy bands

Instead of E representing here the kinetic energy its expression (see Chap. 1.3) $E = \dfrac{h^2}{2m\lambda^2}$ can be put in. Then

$$\nabla^2\psi + \frac{8\pi^2 m}{h^2} \cdot \frac{h^2}{2m\lambda^2}\psi = \nabla^2\psi + \left(\frac{2\pi}{\lambda}\right)^2\psi = \nabla^2\psi + k^2\psi = 0,$$

where $k = \dfrac{2\pi}{\lambda}$ is the wave vector. Hence $k^2 = \dfrac{8\pi^2 m}{h^2} E$ or $E = \dfrac{h^2}{8\pi^2 m} k^2$ i.e., free electron energy depends quadratically on the wave vector k (Fig. 59).

Solution of the Schrödinger equation for the free electron is furnished by the function[6] $\psi = \exp(ikx)$, (i.e., by sinusoidal wave with the wavelength λ and the wave vector k) or for three-dimensional case by

$$\psi(x, y, z) = e^{i(k_x x + k_y y + k_z z)}.$$

The wave vector k is related to the electron impulse p as: $p = h/\lambda = hk/2\pi = \hbar k$. The electron velocity (from $p = mv$) is:

$$v = p/m = \hbar k/m.$$

Thus, the wave vector is of a versatile significance even in the case of the free electron. In addition to its "geometric" meaning as characteristic of the wave motion of the electron, it determines unequivocally the values of electron energy, impulse and velocity.

---

[6] As any periodic function, the wave function of the electron can be represented by the Fourier series

$$f(r) = \sum_k A_k \exp(2\pi i n r/a) = \sum_k A_k \exp(ikr),$$

where $k = h \cdot 2\pi/a$, and $\exp(ikr) = \text{os} kr + i \text{in} kr$. For the representation of the wave processes by complex numbers see (in addition to mathematical reference books): Pole, R.V.: Optiks und Atomphysiks, Berlin-Heidelberg-New York: Springer, 1963 (see §105, p. 149); Porai-Koshitz, M.A.: Practical Course of X-ray Structure Analysis, Vol. II, Moscow Univ. 1960 (see p. 81) (in Russian).

The question now arises: what changes are observed when treating the electron motion in the periodic field of the atoms of lattice?

4.1.2 Two Approximations of the Energy Band Theory: Nearly Free Electrons and Tight Binding Models

Electron behavior in solids can be considered starting from two simpler cases: from the free electron motion or from the electron behavior in atom and molecule. The idea of the free electron moving in the atomic core interstices arose from observations of high electrical conductance of metals at the beginning of this century (Drude, Lorentz), i.e., before the quantum theory was developed.

With the background of this model the nearly free electron approximation was then suggested in the quantum theory of solids. In this approximation the Schrödinger equation is employed to describe the motion of nearly free electrons perturbated by a lattice periodic potential. This is the case when the kinetic energy is greater than the potential energy and the Schrödinger equation approaches the form of the free electron equation. However, the existence of the periodic potential immediately introduces restriction in this continuous motion by definite bands separated by forbidden gaps (Fig. 50).

One can also understand electron transference in insulating crystals as a result of the tunnel effect whose mechanism resides in the fact that the poorly overlapping wave functions of electrons of two atoms overlap by their "tails" and thereby provide a means of electron collectivization.

For nonmetallic crystals however, the tight-binding model (i.e., tight binding of the valence electrons with the atom core) is more adequate. In this case the potential energy (determined by interaction of the electron with the nucleus and other electrons of the atom) is greater than the kinetic one. Here the starting point of the model is not the free electron but the atomic orbitals.

However, as distinct from atomic and molecular orbitals, it is characteristic of electrons in crystal that they move from atom to atom. It results in electron collectivization over all atoms of crystal, the kinetic energy contribution increases, the total energy of the electron depends on its position in the lattice, its long-range interactions with atoms within the large coordination sphere (Madelung interactions) are taken into consideration, the potential becomes the periodic lattice potential, the wave functions become periodic and correspond to atomic orbitals only "within the atom", while outside they approach free electron wave functions.

The energy, for instance of $2s$-, $2p$-electrons of oxygen or $3s$-electrons of magnesium in MgO varies at different points of the lattice. It varies, however, within certain limits, so that instead $2s$, $2p$ oxygen levels and $3s$ magnesium levels, the $2s$-, $2p$- or $3s$-like bands are formed, which are as much related to the atomic orbitals as molecular orbitals, being composed mainly of certain atomic orbitals and related to these atomic orbitals.

The wave functions of the electron in crystal (crystal orbitals) are presented as periodic Bloch functions which are to be considered, starting from the same either nearly free electron or tight binding approximations and with regard to the duality of these functions in crystal, namely, outside the atomic cores, they are similar to the plane waves $\exp(ikr)$ and within the atoms they approach atomic orbitals ψ_{2p}, ψ_{3s} etc.

In the nearly free electron approximation the crystal orbitals $\psi_k(r)$ describe the behavior of collective nearly free electrons, but occurring in the periodic field of lattice atoms:

$$\psi_k(r) = \exp(ikr) U_k(r),$$

where $\exp(ikr)$ is the equation of the plane wave with the wave length $\lambda = 2\pi/k$ describing the behavior of free electrons, $U_k(r)$ is the periodic function (with the periodicity of the crystal lattice) modulating the electron motion and reflecting the perturbating effect of the periodic field of the lattice atoms.

In the tight approximation the crystal orbitals (see also below Chap. 7.1) are obtained as the product of strongly localized atomic orbitals by the periodic factor $\exp(ikR)$:

$$\psi_k(r) = \sum_R \exp(ikR) \Phi_a(r-R),$$

where $\exp(ikR)$ is the periodic structure factor with the wave vector k related to the lattice period (see below), $\Phi_a(r-R)$ are localized wave functions (atomic orbitals) at the point R, for example: Φ_{2p_x}, Φ_{2p_y}, Φ_{2p_z}, Φ_{3s}, ...

4.1.3 Concept of k-Space and Brillouin Zones

Both the wave functions of the free electron ($\exp(ikr)$) and the Bloch functions of the electron in crystal [crystal orbitals $\psi_k(r)$] are characterized by the wave vektor k; the electron energy is dependent on k both in the free electron case ($E = (h^2/8\pi^2 m)k^2$) and in the case of the electron in crystal, where the dependence is much more complicated and represents the main object of energy band calculations. In the one-dimensional case, this dependence is plotted on the diagram, the coordinate axes of which are the energy values E and the wave vector k values (Fig. 59).

In the three-dimensional case (in crystal) the coordinate axes of the diagram are the components of the k-vector along the x, y and z axes of crystal, i.e., k_x, k_y, k_z. Each point on this diagram with the radius vector k describes the state of an electron. These electron states (i.e., the k-points on the diagram k_x, k_y, k_z) correspond to the energy values (as well as to the impulse and electron velocity values). It is this three-dimensional diagram whose axes are the components of the electron wave vector k, and inside of which the points, lines and planes represent the electron energy $E(k)$ varying with k values, called k-space.

However, the wave vector k in crystal and the k-space itself have an additional important significance. This results from the fact that k is related to the crystal lattice.

On reflection, the electron motion in crystal can be considered as diffraction of the valence electrons by the crystal lattice (similar to the diffraction of the electron beam with the wave length λ by the crystal treated in electron diffraction study or to the diffraction of the X-ray beam with the wave length λ considered in X-ray crystallography). All these diffraction phenomena are determined by the same Bragg equation (see textbooks on crystallography)

$$n\lambda = 2d\sin\varphi, \quad \text{or} \quad n\lambda = 2d\cos\chi,$$

where λ is the wave length (of electrons or X-ray photons); n is an integer; d is the interplanar distance of the lattice, φ is the angle between the lattice plane and the beam direction; $\chi = 90 - \varphi$, i.e., the angle between the normal to the plane and the beam direction.

At the normal incidence of the beam

$$n\lambda = 2d, \quad \text{or} \quad \frac{n}{d} = \frac{2}{\lambda}, \quad \text{or} \quad n\frac{\pi}{d} = \frac{2\pi}{\lambda} = k.$$

This condition means that the wave vector k is determined not only as the inverse of the wave length values ($2\pi/\lambda$) but also as the inverse of the interplanar distances in the lattice (π/d), i.e., the k-space is the space of the reciprocal lattice. It will be recalled (see textbooks on structural crystallography), that the reciprocal lattice is constructed from vectors perpendicular to the lattice planes and from points at the vectors with periodicity equal to $1/d$. Exactly this system of points represents the reciprocal lattice, it is rather straightforward to obtain a corresponding diffraction pattern produced by X-ray or electron passing through the crystal.

However, the same Bragg equation means that when $n\lambda = 2d$ (or $k = n\pi/d$) electrons and X-rays are reflected by the lattice planes resulting in reflexes in X-ray patterns. In the case of the moving valence electrons this condition means that such electrons cannot exist as continuous moving waves, i.e., there must be forbidden values of kinetic energy corresponding to the condition $k = n\pi/d$ (with certain $1/d$ values). Therefore, the continuous variations of the electron energy with k are limited by the cases with values corresponding to discontinuity in energy variations when the energy changes by jump (owing to potential energy change) and then again varies continuously till the next gaps at $2\pi/d$, $3\pi/d$ etc.

Thus the k-space (i.e., a diagram with k_x, k_y, k_z axes) appears to be divided into concentric regions, formed by polyhedra with the electron energy within them changing continuously, and at the faces, i.e., at the contact planes of these polyhedra, incorporated into each other, there occur points with values $k = \pi/d$ corresponding to the energy discontinuities and to the forbidden energy values.

The regions within these polyhedra are called Brillouin zones (The notion of "the Brillouin zone" is often restricted to denote the first or "reduced" zone, while all larger zones are referred to as Jones zones). The shape of these polyhedra taken to be Brillouin zone boundaries is determined by the lattice symmetry (by Bravais lattice type) as follows (Fig. 60).

1. The Bravais lattice type (one of 14 crystallographic translation lattice types) is to be found in crystallochemical reference books for crystals with known structure.

Concept of k-Space and Brillouin Zones

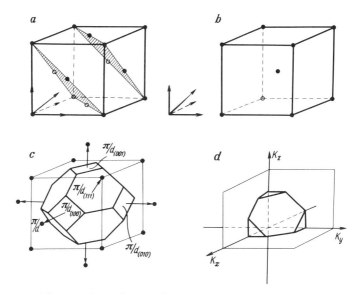

Fig. 60a–d. Brillouin zone and k-space formation for face-centered cubic lattice: **a** face-centered cubic lattice (structure type NaCl–MgO–PbS); normals to the faces (III) (hatched), (100), (010), (001) are used to construct reciprocal lattice; **b** reciprocal lattice for face-centered lattice is body-centered; **c** first Brillouin zone; **d** k-space

2. A reciprocal lattice is constructed from the Bravais lattice by drawing normals to the most important planes and by placing points there at a distance equal to π/d.

3. A unit cell (polyhedron) is set up for this reciprocal lattice, i.e., one sets up the shortest vectors from the origin of reciprocal space to the nearest equivalent points of the reciprocal lattice, and planes perpendicular to these vectors and passing through their midpoints.

This polyhedron forms the boundaries of the first Brillouin zone. Drawing the vectors to the next lattice points and setting up the planes perpendicular to them one obtains the boundaries of the second, third, fourth etc. Brillouin zones. The construction procedure of the first Brillouin zone is similar to that of Wigner-Seitz cells which has been used for separation in the direct lattice of the region attributed to each atom and then approximated by spheres.

They fill the whole space without interstices (like Fedorov parallelohedra). The Brillouin zone may be defined as the Wigner-Seitz unit cell of the reciprocal lattice with periodicity $k=\pi/d$; the points at its plane correspond to k values at which discontinuities in the energy occur; the first Brillouin zones can also fill the k-space without interstices.

For cubic crystals the following first Brillouin zones correspond to the three types of Bravais lattice (Fig. 61):

Bravais lattice → Reciprocal lattice → Brillouin zone
Simple cubic	Simple cubic	Cube
Body-centered	Face-centered	Rombododecahedron
Face-centered	Body-centered	Truncated octahedron.

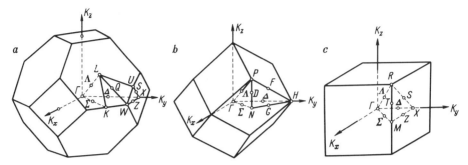

Fig. 61. Brillouin zones for cubic face-centered **a**, body-centered **b**, and primitive **c** lattices. Characteristic points are shown (see text)

The first Brillouin zone is called a reduced zone and k vectors whose values lie within it are called reduced vectors.

Since the wave vector k is related to the electron impulse ($p = \hbar k$ for free electron), instead of k-space one sometimes considers the p-space, also called a space of quasi-impulses ("quasi" because the impulse of electrons in crystal, unlike free electrons, does not remain constant due to the movement of the electron in the alternating (accelerating and decelerating) field of the lattice.

4.1.4 Classification of Orbitals in Crystals with Respect to Symmetry Types

Band structure calculations must yield a numerically continuous variation of the wave function and the energy of the electron within the reduced Brillouin zone. The calculations are performed for certain characteristic points within this zone.

A special designation of these characteristic points for each type of Brillouin zone is used, usually the notation is that of Boucaert et al. [129].

For the Brillouin zones in the forms of cube, rhombic dodecahedron and truncated octahedron, these designations are shown in Figure 61. The Brillouin zones for other types of Bravais lattices are discussed in detail in [62, 65, 146] etc.

These points are characterized first of all by their symmetry: the point Γ at the center of these zones has the cubic symmetry O_h ($m3m$), while the points $\Delta - C_{4v}$, $\Lambda - C_{3v}$, and $\Sigma - C_{2v}$ etc.

Classification of the wave functions according to the symmetry types is of great importance in a description of the behavior of electron of an atom in the ligand field (s-, p- and d-electron states transform according to the symmetry types a_1, t_{1u}, t_{2g} etc; see Chap. 2.2), as well as in considering the molecular orbitals. The molecular orbitals, for instance, with t_{2g} symmetry are formed from atomic orbitals transformed according to the same symmetry type as the molecule (see Chap. 3.4):

$$\psi_{MO}(t_{2g}) = c_1 \psi_{3d}(t_{2g}) + c_2 \psi_{2p_\pi}(t_{2g}).$$

The classification of electron wave functions in crystal with respect to the symmetry types has also the same significance.

Table 37. Symmetry types according to which the s-, p-, d-like wave functions transform at different points of the Brillouin zone for the face-centered cubic lattice (truncated octahedron, see Fig. 61)

	$\Gamma(O_h)$	$\Delta(C_{4v})$	$\Lambda(C_{3v})$	$\Sigma(C_{2v})$	$X(D_{4h})$	$L(D_{3d})$	$K(C_{2v})$
s	Γ_1	Δ_1	Λ_1	Σ_1	X_1	L_1	K_1
p	Γ_{15}	$\Delta_1+\Delta_5$	$\Lambda_1+\Lambda_3$	$\Sigma_1+\Sigma_2+\Sigma_3$	X_3+X_5	L_1+L_3	$K_1+K_3+K_4$
d	$\Gamma_{12}+\Gamma_{25}$						

The points Γ, Δ, Λ and Σ are common to truncated octahedron, rhombic dodecahedron and cube.

Unlike the molecule and atom in the ligand field, the electron wave functions in crystal changing from point to point of the space lattice transform according to the different symmetry types at the different points of the Brillouin zone. To denote the symmetry types one takes the symbol of the point (Γ or Δ or Σ etc.) and numerical subscripts indicating the symmetry type (Γ_1, Γ_2, Γ_{15}, Γ_{25} for the point Γ; or Δ_1, Δ_2, Δ_3, Δ_4 for the point Δ; Σ_1, Σ_2, Σ_3, Σ_4 for the point Σ etc).

It should be noted that these are only different notations of the same transformation under the action of symmetry as in the crystal field theory but these notations are related to the k-space and the Brillouin zones. Let us compare notations of symmetry types, according to which s-, p- and d-like wave functions transform in the crystal field theory (after Mulliken, see Chap. 2.2) for the symmetry of octahedron O_h and in the energy band theory (after Boucaert et al. [129]) at the point Γ of Brillouin zones in the form of cube, truncated octahedron and rhombic dodecahedron with the same O_h symmetry:

$$s \to A_{1g} \to \Gamma_1 \quad (1),$$
$$s \to A_{2g} \to \Gamma_2 \quad (1),$$
$$p \to T_{1u} \to \Gamma_{15} \quad (3),$$
$$d \to E_g + T_{2g} \to \Gamma_{12} + \Gamma_{25} \quad (2+3).$$

(In parentheses are given degeneracies of the states.)

As in the crystal field theory for each point (i.e., for each symmetry group of the k point, that is for the wave vector group) one can construct the character tables which determine the symmetry-type assignments applicable both to the crystal field and to the points of the band structure. For example, the character table for the points Δ and Σ, having C_{2v} symmetry (but different in the wave vector k values) is the same as that for this point group given above in Table 17 (see Chap. 2.2) in which Δ_1 and Σ_1 corresponds to A_1, and Δ_2 and Σ_2 to A_2, and Δ_3, Σ_3 to B_1, and Δ_4 and Σ_4 to B_2.

The notations of the symmetry types, according to which the wave functions in these points transform, are taken to be the notations of the wave functions themselves at these points (Table 37). Just as $3d$ electrons are described in the crystal field as t_{2g} and e_g electrons, so also in crystal these d electrons are denoted at the point Γ of the Brillouin zone as Γ_{12} and Γ_{25}, at the point Σ as $\Sigma_1 + \Sigma_2 + \Sigma_3$

etc, while p electrons at same points are denoted as Γ_{15} (at the point Γ) and as $\Sigma_1 + \Sigma_3 + \Sigma_4$ (at the point Σ) etc.

At the point with cubic symmetry O_h, p levels are degenerate and a single Γ_{15} state corresponds to these levels. At Σ with rhombic symmetry C_{2v}, the degeneracy is completely removed and p levels are split into p_z, p_x, p_y, i.e., into Σ_1, Σ_3, Σ_4 states.

In a number of crystals (e.g., in galena PbS), large spin-orbit splitting of electron levels occurs (see Chap. 2.4). In this case the spin component of the wave function is to be taken into account when specifying the symmetry type of the function. It is important to note that the spin function has special properties as far as symmetry transformations are concerned.

In rotating at the angle of 2π corresponding to identity operation, the spin function changes the sign and only double rotation at 2π leaves it unchanged.

Thus, to each symmetry element considered without taking spin into account, there correspond two sets of symmetry elements when taking the spin component of the function into account. It is for this reason that the representations associated with the spin functions are said to be "double-valued" and the corresponding symmetry groups are called "double groups". The d-type wave functions in cybic crystals can be split by spin-orbit coupling, while s- and p-type wave functions are not split, but transform according to the representations of the double groups. When spin-orbit interactions are taken into account, the symmetry types, for example, at the points Γ and L (of the same cubic Brillouin zones) are designated as follows:

$$\Gamma_1 \to \Gamma_6^+ \quad (\Gamma_1^6) \qquad L_1 \to L_6^+ \quad (L_1^6)$$
$$\Gamma_2 \to \Gamma_7^+ \quad (\Gamma_2^7) \qquad L_2 \to L_6^+ \quad (L_2^6)$$
$$\Gamma_{12} \to \Gamma_8^+ \quad (\Gamma_{12}^8) \qquad L_3 \to L_6^+ + L_{45}^+ (L_3^6 + L_3^{45})$$
$$\Gamma_{15} \to \Gamma_8^+ + \Gamma_6^+ (\Gamma_{15}^8 + \Gamma_{15}^6)$$
$$\Gamma_{25} \to \Gamma_8^+ + \Gamma_7^+ (\Gamma_{25}^8 + \Gamma_{25}^7).$$

Here plus stands for the even function and minus for the odd function (see Chap. 2.2). An alternative way of notation is given in parantheses combining the designation of the spin-orbit state with the designation of the initial state.

4.1.5 Energy Band Structure Schemes

Similar to molecular orbitals in molecules and complexes and as electron energy levels in free atoms or in the atom in the ligand field, energy bands in crystals represent their electronic structure. However, unlike the energy level corresponding to the atomic orbital or to the molecular orbital and characterizing a single state (they are plotted as horizontal lines, but these are, in fact, one-dimensional presentations of the $1s$, $2s$, $2p$, $3s$... or $1a_1$, $1t_2$, $2a_1$, $2t_2$, $1e$... states) the crystal orbitals and their energies are different at the different points of the lattice and hence at the different points of the k space. The problem is thus reduced to representation of the dependence of wave-function type (according to the symmetry types) and their energy on the values of the vector k at different points of the reduced Brillouin zone.

At characteristic points $\Gamma, \Sigma, \Delta, \Lambda, K, X$ etc (in the cubic lattices), for example, the states formed from $2p$ orbitals of oxygen with an admixture of $3s$ orbitals of magnesium in MgO are described now not as $2p$ but as Γ_{15} at the point Γ, as $\Sigma_1 \Sigma_3 \Sigma_4$ at the point Σ (but not $2p_x, 2p_y, 2p_z$) etc. The corresponding energies at these points are different and change from point to point quasicontinuously (quasi because at each point between Γ and Σ and between Σ and K etc. there is a discrete set of levels). This quasicontinuous variation of energy of the levels at the points of the k space (i.e., of the reduced Brillouin zone) and the change of the wave function type at the points of the k space represent the contents of the band structure scheme.

The whole set of states (at all points of the Brillouin zone) arising from, for instance, the $2p$ orbital, forms a band which may be referred to as $2p$-like band; other bands form from $3s$, $3p$ orbitals etc. ($3s$-, $3p$-like bands).

Band-structure schemes are plotted not in the k space itself because a four-dimensional space (k, k_y, k_z, E_k) would have to be used for this purpose; this is why only surfaces of equal energy (e.g., Fermi surfaces, see below) are presented in it.

These schemes are simply diagram $E_k - k$ along the different directions of the k-space (Brillouin zone); one such diagram shows the energy change of different wave functions as a function of the k-value along the direction from the central point $\Gamma \langle 000 \rangle$ to the point $K \langle 110 \rangle$ through the point Σ (in the cubic lattice), the second diagram from the point $K \langle 110 \rangle$ to the point $X \langle 010 \rangle$, the third one from the point $\Gamma \langle 000 \rangle$ to $L \langle 111 \rangle$ through Λ etc. These particular diagrams are usually jointed by the adjacent sides into one energy band diagram.

Consider, for example, the band structure of the periclase crystal MgO (Fig. 62). The valence electrons are here $3s$ electrons of Mg ($1s^2 2s^2 2p^6 3s$) and $2s, 2p$ electrons of O ($1s^2 2s^2 2p^4$). The structure is of NaCl–MgO–PbS type, the Bravais lattice is face-centered cubic, the first Brillouin zone is a truncated octahedron representative points of which are shown in Figure 61. Crystal orbitals formed mainly from $2s$ orbitals of O transform into the states $\Gamma_1 \Sigma_1 K_1 X_1 \Delta_1 \Lambda_1 L_1'$ which form $2s$-like band and the energy of which is changed within narrow ranges.

The $2p$-like band formed mainly from $2p$ orbitals of O (with participation of $3s$ orbitals of Mg) is represented at the central point Γ (the symmetry of which coincides with the point group of the crystal) by the unsplit state Γ_{15}, at the points $\Delta(C_{4v}), X(D_{4h}), \Lambda(C_{3v}), L(D_{3d})$ with axial symmetry by two states (splitting into p_x-, p_y- and p_z-like states) and at the points $\Sigma, K(C_{2v})$ with rhombic and lower symmetry by three p_x-, p_y-, p_z-like states.

An empty $3s$-like band (formed mainly from $3s$ orbitals of Mg) is represented by the same types of state as the $2s$-like band (made from $2s$ orbitals of O). The $3d$-like band composes next in energy unoccupied state (Γ_{25} etc).

The $2s$- and $2p$-like bands are occupied and are called valence bands; the $3s$-, $3d$-like band are unoccupied and are called conduction bands.

These bands are separated by the energy gap corresponding to the forbidden states. The states composing these bands can be compared with molecular orbitals (see Chap. 3.2): valence bands are analogous to bonding molecular orbitals, conduction bands to antibonding ones.

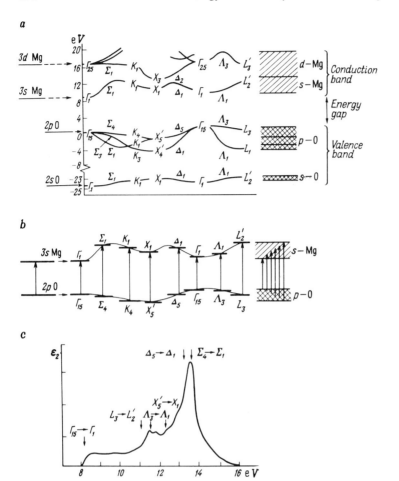

Fig. 62a–c. Energy band scheme and reflectance spectrum (ε_2 spectrum) of periclase, MgO. **a** Energy band structure of MgO. **b** Optical transitions between top levels of valence band and bottom levels of conduction band in characteristic points of Brillouin zone. **c** Assignement of ε_2 spectrum [136]

The same picture of the energy band structure of solids can be obtained in terms of the nearly free electron approximation: a free motion of electrons is limited by conditions similar to those of electron diffraction (see above) thus leading to the same concept of Brillouin zones filling the k-space and concept of energy bands (Fig. 60).

4.1.6 Band Occupation; Densities of States; Fermi Surface

Let us recall that the construction of a molecular orbital scheme includes filling with electrons in consecutive order beginning from the lowest molecular orbital: two electrons in a orbitals, four electrons in e orbitals and six in t orbitals. In a band scheme, however, for example, p orbitals in octahedral coordination transform not into one state t_{2g}, but into the state $\Gamma_{15}(t_{2g})$ at the point Γ of the Brillouin zone, into the states $L_1 + L_3$ at the point L, into the states $\Delta_1 + \Delta_5$ at Δ etc with all intermediate quasicontinuous state sets forming together the band of the p-like states. However, distribution of the densities of states within the band is uneven; it can be obtained experimentally from X-ray spectra where the shape of the emission or absorption band reflects the distribution of the density of states in the filled or empty bands.

Three types of solids are distinguished according to the width of the energy gap: (1) with the width of the gap more than $\sim 3\,\text{eV}$: these are nonmetals; transparent in the visible region; dielectrics (insulators); with mostly ionic bond type; (2) with the band gap less than ~ 3–$5\,\text{eV}$: semiconductors whose absorption edge is in the IR or visible or near UV region, thus determining either their opacity, coloration, or transparence, with mostly covalent bond type; (3) with overlap of the valence and conduction bands: metals; opaque; conductors; with metallic bond type.

Above the filled band there is the Fermi energy level. All states under this energy level are occupied, above it they are unoccupied. In metals this level lies in the conduction band, in nonmetals in the forbidden band.

In the forbidden band levels of impurity ions can occur, then the Fermi level divides them into donor and acceptor levels. Here, also, there are in compounds of transition metals narrow bands of d electrons; if the Fermi level is higher than these bands of d electrons, then the latter are localized; if lower, they are collective (see Chap. 8.3).

4.1.7 The Methods of Band Structure Calculation

For any energy band description of a number of phenomena such as thermoluminescence, luminescence, electrical properties etc. it is often sufficient to use the data on the width of the energy gap which separates the valence and conduction bands. However interpretations of optical reflectance spectra, X-ray and electron spectra, discussions of the chemical bonding etc. need detailed and numerical calculations of the band schemes. The energy bands are the most fundamental descriptions of the electronic of solids comparable in its importance with X-ray diffraction data for crystal structure determinations.

Development of modern computer facilities and inventions of the new methods of the calculations, taking into account the possibilities of these computers, enables to obtain the energy band schemes for many metals, dielectrics and semiconductors (including many main sulfide and related minerals).

General features of these methods are follows: (1) these are mostly not ab initio calculations, satisfactory usually only for obtaining the principle scheme, but semi-empirical approaches utilizing experimental values of parameters compensating inaccuracies of assumed approximations and entering the scheme of claculation in such a way that all other parameters being calculated take on the values giving a good fit to the experimental data; (2) all types of calculations are performed in one-electron approximation and are based on the Schrödinger equation; (3) all the methods of band calculations use the wave functions made from combinations of atomic orbitals and plane waves.

One can distinguish two groups of methods. In the first, the space of crystal is divided into two regions: (1) a set of spheres centered at the atomic sites in the crystal structure and (2) the remaining interstitial region. Inside each sphere the potential is assumed as spherically symmetrical while in the interstitial region it is taken to be constant. There are related to such a "muffin-tin" potential basis wave functions (Augmented Plane Waves – APW) which are obtained inside the spheres by solution of the Schrödinger equation while in the interstitial region they correspond to plane waves.

This approach corresponds to the APW method. Close to it is the orthogonalized plane wave (OPW) method (orthogonalization between valence and core electron): energy bands, analogous to the certain atomic state ($2p$ or $3s$, for example), can be represented by plane wave by adding to them a certain amount of the $2p$ or $3s$ state respectively. The pseudopotential (PP) method and that of empirical pseudopotential (fitting the parameters to experimental data such as energy gap width and certain assigned interband transitions) are of central importance in the description of the electronic structure, cohesive energy and a variety of physical properties of metals and simple covalent crystals.

Special monographs and reviews [131, 134, 138, 140, 142, 146, 156, 157, 161, 162] deal with these methods.

However, the approaches using plane waves are less convenient to correlations of the band structure with the properties of atoms.

Another group of energy band methods is the tight-binding approach using the linear combination of atomic orbitals (LCAO) approximation, among these especially bond orbital method (or equivalent orbital) which will be considered later (see Chap. 7.1) is of great importance in the interpretation of the chemical bond, spectra and physical properties of many ionic-covalent crystals such as, for example, zinc blende and quartz.

4.2 Analysis of the Band Schemes and Reflectance Spectra of Minerals

Optical properties of solids are considered both in crystal optics based on phenomenological Fresnel-Maxwell principles and in spectroscopy, which is actually the most straightforward expression of quantum-mechanical causal relationships between properties of atoms, crystal structure and chemical bond.

Crystal optics of most rock-forming minerals is described by means of the indicatrix of refraction indices; they are transparent in the visible region because their intrinsic absorption is shifted to the UV region; absorption bands in the visible region are usually related to the transition metal ions with d electrons.

The optics of most ore minerals is mainly determined by the spectrum of intrinsic absorption or reflectance spectrum occurring in the visible and near IR or near UV region.

Consideration of reflectivity, the most important optical constant of opaque minerals (just as refraction is the most important constant of rock-forming minerals) was restricted in ore microscopy mainly to classical theories, to its relation to refraction and absorption and to the dielectric constant.

Comparison of the narrow ranges of reflectivity dispersion measurements usual for ore microscopy with the whole reflectance spectrum reveals that these measurements cannot be interpreted because of their incomplete and rather fortuitous information. These are not peaks in reflectance spectrum which can be related to band schemes and compared in different crystals with respect to their position and intensity, but they represent reflectivity values in the general case on the slope of such peaks.

It is possible to gain insight into the nature of reflectivity and the optical properties of ore minerals only on the basis of interpretation of reflectance spectra with the help of the band structure schemes.

The importance of the band theory to ore microscopy has already been noted in general terms. However, it is only in the last years that energy band calculations have been emerging for many structure types, to which a number of the most important sulfides and other minerals belong. On their basis the reflectance spectra have been interpreted, i.e., the peaks therein are assigned to the interband transitions. It is these calculations and interpretations that represent the modern theory of optical properties of ore minerals.

This leads to a new meaning of optics of ore minerals, namely, the reflectivity acquires structural chemical implication, the peaks in reflectance spectra are related to transitions between the valence and conduction bands, and the whole reflectance spectrum is referred to as the most direct manifestation of the band scheme; one can find the peaks arising from the same interband transitions, which can be used for correlations in series of related compounds etc.

This means that the properties of ore minerals are no longer "dumb" but each peak is assigned to the definite transition and can be "read" in the language of crystal structure features (lattice type, interatomic distances and distortions) and of actual descriptions of the chemical bond with atomic and crystal orbitals, their splitting, variations in the contribution of the wave functions of metal and sulfur, covalence etc.

The band schemes of definite minerals become of central importance for minerals, i.e., they provide fundamental characteristics of their electronic structure just as the crystal chemical models represent the atomic structures. On the other hand, the band schemes are of the same significance for crystals as the molecular orbital schemes are for complexes and clusters, or as the atomic-orbital schemes are for atoms and ions.

A number of recent reviews and monographs deal with methods of measurement of reflectance spectra [133, 142, 145, 152, 155, 156, 163, 166, 168] and with the interpretation of optical properties of solids within the framework of the band theory.

The band schemes and reflectance spectra of some ore and other minerals only will be considered below.

4.2.1 Intrinsic Absorption and Reflectance Spectra.
Measured and Calculated Parameters

Intrinsic absorption spectra resulting from transitions from the filled levels of the valence band to the empty levels of the conduction band are characterized by high values of absorption coefficients. Thus, already near by the absorption edge (corresponding to the minimal gap between the top of the valence band and the bottom of the conduction band) the absorption coefficient rises rapidly to $\sim 10^4$ cm^{-1} (compare with absorption coefficients of the crystal field spectra of the order of 0.01–10 cm^{-1}, see Chap. 6.2) and increases further to $\sim 10^5$–10^6 cm^{-1}. The measurements of the absorption coefficient and index of refraction become thus possible only in very thin films of samples (of the order of 10^3 Å). Therefore, instead of absorption and refraction the reflectance spectra (at normal incidence) are usually measured and from this the absorption coefficients and indices of refraction are calculated; for this aim they should be expressed through the dielectric constant.

Between the reflectivity R, extinction coefficient k (related to the linear absorption coefficient α, cm^{-1} as $\alpha = 4\pi k/\lambda$ where λ is the wave length in cm) and the index of refraction n there is following relationship: $R = \dfrac{(n-1)^2 + k^2}{(n+1)^2 + k^2}$ for absorbing crystals, and $R = \dfrac{(n-1)^2}{(n+1)^2}$ for transparent crystals. R may be expressed also in terms of the complex index of refraction $\bar{n} = n - ik$, i.e. $R = \dfrac{(\bar{n}-1)^2}{(\bar{n}+1)^2}$.

Then, n and k may be expressed in terms of real (ε_1) and imaginary (ε_2) parts of the complex dielectric constant $\varepsilon = \varepsilon_1 + \varepsilon_2 = \bar{n}^2$:

$$n^2 - k^2 = \varepsilon_1 \quad \text{and} \quad 2nk = \varepsilon_2.$$

Finally, ε_1 and ε_2 can be determined from each other (on condition that the measurements are made in large spectral ranges) in terms of the Kramers-Kroning relations:

$$\varepsilon_1(\omega_0) = 1 + 1/\mu \int_{-\infty}^{\infty} \frac{\varepsilon_2(\omega)}{\omega - \omega_0} d\omega.$$

Maxima R, k and ε_1, are close to one another and only slightly more than the transition frequency (for R at ~ 0.1 eV) and hence reflectance, absorption and ε_1 spectra alike can be correlated with interband transitions. Of particular importance is that it is possible for reflectance spectrum to be directly measured, though the use of the ε_2 spectrum (the imaginary part of dielectric constant) is more rigorous. On the contrary, ε_2 and n exhibit other spectral dependence: namely, a bend in the ε_2 and n curves corresponds to the reflectance as well as absorption and ε_1 maxima (Fig. 63). This explains the behavior of the refraction index of ore minerals as quite different from that of transparent minerals (the change of n values from very high to very low ones).

The values of n and k are derived from reflectivity measurements in air R_a and in immersion R_i. In reflectance spectra band crests of the order of 0.1 eV are more

Fig. 63. Relation between reflectance R and complex dielectric constant: real (ε_1) and imaginary (ε_2)

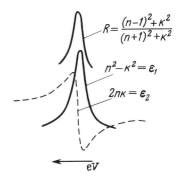

$$R = \frac{(n-1)^2 + \kappa^2}{(n+1)^2 + \kappa^2}$$

$$n^2 - \kappa^2 = \varepsilon_1$$

$$2n\kappa = \varepsilon_2$$

often observed than peaks. When analyzing spectra, one specifies at first the types of critical point related to the types of analytical feature of functions of the interband density of states [156]. Then the oscillator strengths of some transitions responsible for their higher or lower intensity are considered.

The oscillator strength (see Chap. 2.5) is determined by the overlap of the wave functions of the valence band and conduction bands state and transition momentum, for example [156]:

$$f_{jj'}(k) = \frac{\langle \varphi_{3sNa} | P | \varphi_{3pCl} \rangle}{m(E_j - E_{j'})}.$$

Since in the cases of crystals with high absorption coefficients the light being reflected interacts with the very thin layer of sample surface (skin effect), in order to obtain exact absolute values of R, one must treat the surface chemically or electrolytically, or use freshly cleaved surfaces.

4.2.2 Structure Type of NaCl–MgO–PbS

The systematics of band structure schemes is based first of all on the Bravais lattice type determining the type of the Brillouin zone. Binary compounds with a NaCl–MgO–PbS-type of structure have face-centered cubic lattice and one and the same space symmetry group. Hence, they are characterized by the same Brillouin zone type, namely by the truncated octahedron, with the same characteristic points in this zone (see Fig. 61). Further, the systematics of the band schemes within the structure type can be based on the electronic configuration of atoms in crystal, which are similar for NaCl (Na $3s$, Cl $3s^2 3p^5$) and MgO (Mg $3s^3$, O $2s^2 2p^4$), i.e., the same s and p orbitals of anion and s orbitals of cation, but are different for PbS (Pb $5d^{10} 6s^2 6p^2$, S $3s^3 3p^4$).

Finally, the systematics within the compounds of the same type, for example, PbS–PbSe–PbTe, is based on interatomic distances.

Periclase MgO [136]. The band structure features (Fig. 62) determines the form of the intrinsic absorption (reflectance) spectrum of MgO. The large energy gap (determined here by the distance between the Γ_{15} state at the top of the

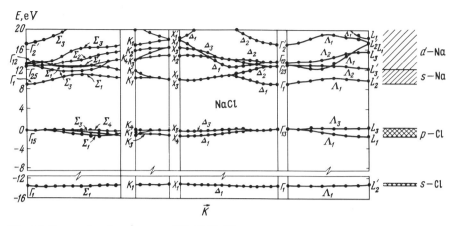

Fig. 64. Energy band structure of NaCl [145]

valence band and Γ_1 at the bottom of the conduction band) explains the transparence of MgO in the visible and near UV. A group of bands crests in the reflectance spectrum in the region of 8–16 eV are assigned to the transitions from the fourth valence band $(\Gamma_{15}-\Sigma_4-K_4...)$ to the fifth conduction band $(\Gamma_1-\Sigma_1-K_1...)$. In the case of the molecule of MgO there would be merely a single transition from the 2p orbital of O to the unoccupied 3s orbital of Mg ("charge transfer transition").

However, in crystals due to the energy variation of these orbitals (crystal orbitals similar to them) a number of peaks appear in the reflectance spectrum with different oscillator strength owing to the different overlap of s-wave functions oxygen 2p and magnesium at different points of the Brillouin zone (see Fig. 62).

NaCl and Other Alkali Halides. The band scheme of NaCl (Fig. 64) is quite analogous to that of MgO (as well as of other compounds with the face centered cubic lattice and the electron configurations of the same type). The characteristic features of the chemical bond in terms of the energy band theory can be considered not only with the help of such schemes but also using the semiempirical tight-binding methods [153].

In the absorption (reflectance) spectra of alkali halides one observes the same transitions from the valence p band to the s conduction band, from p band to d band etc. However, other distances between the levels at different points of the Brillouin zone and different overlaps of the wave functions result in both the shift of the whole spectrum to the UV region and the redistribution of position and intensity of the peaks. The large energy gap determines the transparence of alkali halides up to 200–100 nm. The peculiarity of alkali halides spectra is the existence of sharp peaks related to exciton absorption which are superimposed on the bands arising from the interband transitions. As distinct from interband transitions the exciton transitions are characterized by electron excitation to discrete levels below the conduction band; the hole and electron are bound and can be considered as an electron hole pair. Such systems are referred to as hydrogen-like and transitions therein as transitions in hydrogen atom between $1s, 2s...$ orbitals [156, 157].

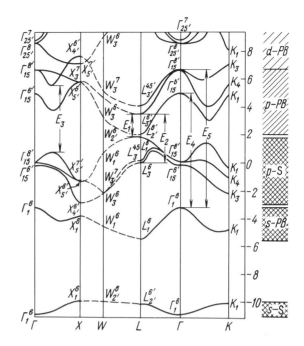

Fig. 65. Energy band structure of altaite PbTe [127]

Interband transitions at different points of the Brillouin zone are accompanied by their own systems of exciton peaks: two sharp peaks of Γ excitons are observed from the long-wave side of the transition $\Gamma_{15} \rightarrow \Gamma_1$ and L exciton peaks from the $L_3 \rightarrow L_2$ transition. Strong exciton absorption overlaps the absorption edge in ionic crystals nearly completely. On the contrary, in semiconductors, very weak exciton lines can hardly be discerned from strong interband transitions. The nature of these exciton transitions is of importance for the consideration of the mechanism of processes occurring in crystals on irradiation, formation of electron-hole pairs, photoconductivity luminescence etc.

PbS (Galena)–PbSe–PbTe (Altaite). Band structure and reflectance spectra of these semiconductor compounds have been studied in great detail [127, 128].

In ore microscopy galena is one of the standards in reflectivity estimation [151]. Band structure of these compounds (Fig. 65) is characterized by the same Brillouin zone as NaCl and MgO (see Fig. 61), hence the same points $\Gamma - \Sigma - K - X - \Delta - \Lambda - L$ with their symmetry are considered for this scheme. However, a different electron configuration of Pb ($5d^{10}6s^26p^2$) with different ratios of atomic orbital energies results in other order of bands, namely, the valence bands are s band of S (or Se, Te), s band of Pb and p band of S (Se, Te), the conduction bands are p band of Pb and d band of Pb. The types of symmetry transformation and splitting of these bands are the same as in the schemes of NaCl–MgO.

The top of the valence band and the bottom of the conduction band fall on the point L, and not on the point Γ as in NaCl–MgO. The transition $L_2^6 \rightarrow L_1^6$ corresponds to the absorption edge and is responsible for the width of the energy gap. Semiconductor properties of these crystals and their opacity up to the near IR

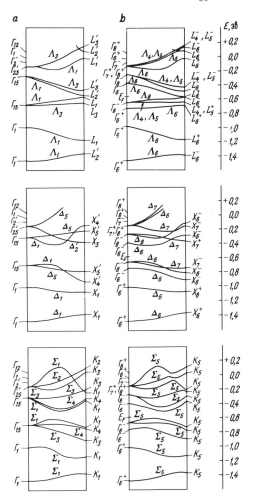

Fig. 66a and b. Energy band formation in PbTe [127] **a** without spin-orbit interaction; **b** with spin-orbit interaction

region (0.37 eV PbS ≈ 3000 cm^{-1} ≈ 3.3 microns) are dependent on a very small energy gap (0.37 eV in PbS, 0.26 eV in PbSe, and 0.29 eV in PbTe).

The magnitude of spin-orbit splitting is very high for compounds of heavy elements. Therefore, in the band scheme of PbS–PbSe–PbTe (Fig. 66) splitting of states $\Gamma_{15} \to \Gamma_{15}^6 + \Gamma_{15}^8$; $X_5 \to X_4 + X_6$ etc., or their symmetry transformation $\Gamma_1 \to \Gamma_1^6, X_1 \to X_1^6$ etc., must be taken into account. Very often these states split by spin-orbit interactions are denoted simply as Γ_5, Γ_8 etc. (see p. 163).

Thus, interband transitions in the reflectance spectrum of PbS–PbSe–PbTe are the transitions between spin-orbit sublevels; a peak in the reflectance spectrum usually presents a superposition of two or three transitions close to each other in energy.

There are in the spectra of PbS–PbSe–PbTe besides the absorption edge six main band crests forming two groups of three crests (Figs. 67–69).

The assignment of the peaks to the interband transitions is shown in Figure 70. In the tight-binding terms the crests $E_2 - E_5$ are determined by transitions from

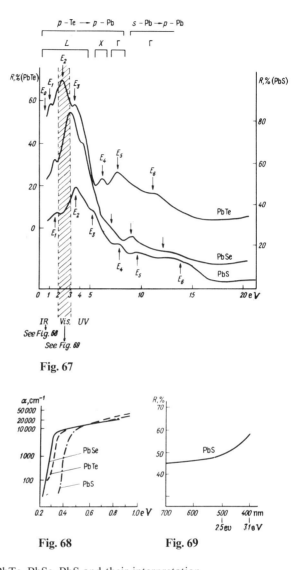

Fig. 67. Reflectance spectra of PbTe, PbSe, PbS and their interpretation

Fig. 68. A portion of reflectance spectra of PbTe, PbSe, PbS (absorption edge in the infrared)

Fig. 69. A portion of reflectance spectrum of PbS (dispersion of reflectivity in the visible)

the valence p band of S (Se, Te) to the conduction p band of Pb, the crest E_6 from s-band of Pb to p band of Pb.

Comparison of PbS–PbSe–PbTe spectra shows that energies of the interband transitions decrease with increasing interatomic distances (lattice constant). On the contrary, when the cations are changed (PbTe–SnTe–GeTe) the energies of one and the same transitions increase with the increasing lattice constant [156, 468].

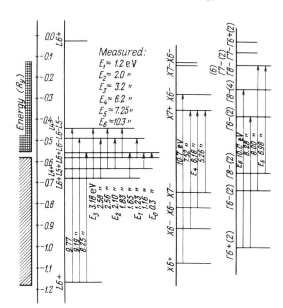

Fig. 70. Interpretation of several peaks in reflectance spectrum of PbTe [127]

4.2.3 Structure Type of Sphalerite (Cubic ZnS) [161, 165]

Sphalerite has the same face-centered cubic Bravais lattice as NaCl–MgO–PbS and hence the same Brillouin zone (truncated octahedron), but it has a quite different symmetry T_d (tetrahedral) and hence the same points $\Gamma - \Delta - X - K - \Sigma - \Lambda - L$ have here a different symmetry, while the wave functions transform at these points in other representations.

Valence bands s–S and p–S and conduction bands s–Zn and p–Zn are distinguished in the energy band scheme of ZnS (Fig. 71). The width of the energy gap (between Γ_{15} and Γ_1) determines the position of the absorption edge in the near UV region.

In the reflectance spectrum (Fig. 72) the peaks are assigned to the transition from the p–S conduction band to s–Zn conduction band ($\Gamma_{15} \to \Gamma_1$, $\Lambda_3 \to \Lambda_1$, $X_5 \to X_1$, $L_3 \to L_1$) and from p–S valence band to the p–Zn conduction band ($\Gamma_{15} \to \Gamma_{15}$, $L_3 \to L_3$ etc.).

4.2.4 Structure Type of Wurtzite (ZnS Hexagonal) [161]

The hexagonal symmetry of wurtzite determines the type of the Brillouin zone (hexagonal prism) as quite different from that of sphalerite and the more complex band structure (Fig. 72). But coordination polyhedra Zn–S$_4$ in both structures are very similar and are only slightly distorted in wurtzite. Therefore, the reflectance spectra of sphalerite and wurtzite are similar; there are the same transitions between the valence bands s–S and p–S and the conduction bands s–Zn and p–Zn, but are characterized by quite different symmetry. There are differences, though small, in wurtzite spectra parallel and perpendicular to the c axis of the crystal.

Fig. 71. Energy band structure of ZnS [161]

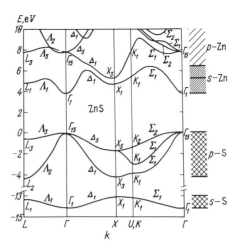

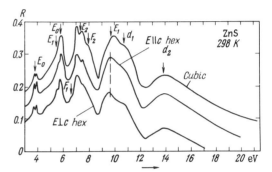

Fig. 72. Reflectance spectra of sphalerite and wurtzite in the UV and the interpretation for sphalerite [161]

4.2.5 Data for Other Minerals

In addition to the band schemes of alkali halides, of systems of ZnS- and PbS-type, considered above, as well as of metals (gold, silver and copper), diamond and graphite (see [156]), the energy band schemes have been calculated for molybdenite, MoS_2, stannisulite, SnS_2, stibnite, Sb_2S_3 chalcopyrite, $CuFeS_2$, gallite, $CuGaS_2$, rockesite, $CuInS_2$ cinnabar, HgS, fluorite, CaF_2, quartz, monooxides of transition metals etc. [127, 138, 141, 143, 150, 154–160, 165].

For a number of sulfide and related minerals the reflectance spectra obtained for the region from near UV to near IR and hence suited for energy band interpretation are now available [151, 163, 164].

5. Spectroscopy and the Chemical Bond

Spectra of solids have the same significance for the description of the chemical bond as atomic spectra for the understanding of the electron structure of atoms: these are the spectra of atoms only in the state of chemical bonding, they are spectra of individual atoms, but occurring in the concrete crystals or molecules in a definite structural position with all the pecularities of this position: its local symmetry, overlapping of atomic orbitals and other features of the binding. Solid state spectroscopy is the most direct and complete method for obtaining information about the properties of atoms in compounds.

The development of all branches of spectroscopy (from the 1950s) is closely connected with the development of theories of electronic structure of solids (crystal field, molecular orbitals, energy band theories) and the development of solid state spectroscopy provided theories of solids with the only parameters which have direct quantum-mechanical meaning and which arise from the schemes of these theories. Understanding of chemical bonding and the estimation of its features in different classes of compounds and in the actual compound are more and more related to spectroscopic parameters described in the framework of quantum-mechanical theories of chemical bonding.

A spectroscopically i.e., experimentally obtained picture of bonding is completely compatible with its quantum-mechanical presentation (as distinct from macroscopic or termodynamic bond characteristics revealing summarized effects of bonding). Different methods of spectroscopy of solids and their mineralogical applications are considered in detail in the next book of the author [1]. Hence only a most general outline of the spectroscopic parameters related to general concepts of the chemical bond theories and used in descriptions of chemical bonds in individual classes and groups of minerals will be given here.

5.1 General Outline and Parameters of Solid State Spectroscopy

As distinct from spectrochemical analysis and atomic absorption spectrometry dealing only with elemental composition of matter being dissociated into free atoms, solid state spectroscopy studies only solids with all the peculiarities of the states of atoms in crystals, as well as the electronic structure of the crystals themselves.

Spectroscopy of solids arose as a consequence of: (1) discovery of new phenomena and (2) development of crystal field, molecular orbital and energy band theories, without which the branches of spectroscopy could not have been developed.

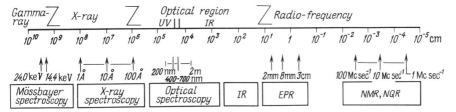

Fig. 73. Branches of spectroscopy of solids

The phenomenon of electron paramagnetic resonance was discovered in 1944 by Zavoisky, nuclear magnetic resonance was discovered in 1945 by Bloch and Pursell; nuclear quadrupole resonance was discovered in 1949 by Demelt and Krüger; Mössbauer effect (nuclear gamma resonance) by Mössbauer in 1958. From the 1950s optical absorption spectra began to be interpreted on the basis of the crystal field theory; from the 1960s investigation of reflectance spectra started on the basis of the energy band theory and X-ray spectra on the basis of molecular orbital theory. In the middle of the 1950s method of electron spectroscopy (ESCA) was developed with commercial production of spectrometers from the end of the 1960s.

In 1950–1960 all these led to the intense development of different branches of spectroscopy, which together formed the complete system of spectroscopy of solids.

Branches of spectroscopy of solids can be classified by wavelength (Fig. 73) and by the types of level between which transitions occur (Fig. 74). In the order of increase of wavelength one can classify: Mössbauer (nuclear gamma resonance) spectroscopy (NGR) covering the spectral region of gamma radiation; X-ray and electron spectroscopy, covering the region of X-ray radiation, optical spectroscopy i.e., UV, visible and near IR; IR spectroscopy; electron paramagnetic resonance (EPR), region of centimeter—millimeter wavelengths; nuclear magnetic and nuclear quadrupole resonance (NMR and NQR), regions of meter–kilometer wavelengths.

X-ray, optical and EPR-spectroscopies compose electron spectroscopy; X-ray spectroscopy is the spectroscopy of inner electrons, optical spectroscopy that of outer (valence) electrons; EPR is spectroscopy of the transitions between spin states of the electron.

Mössbauer, NMR and NQR spectroscopies compose nuclear spectroscopy related to transitions between different nucleus states: Mössbauer spectroscopy deals with isomeric transitions, NMR deals with transitions between spin levels of the nucleus, NQR deals with transitions between the quadrupole levels of the nucleus.

However, application of these branches of nuclear spectroscopy for the investigation of solids is determined by the electron structure of the solids, which influences nuclear levels. Infrared and Raman spectroscopy deals with vibrational levels; they are not discussed in this book.

Here we consider the characteristics of the branches of spectroscopy arising from the level schemes, which describe electronic and nuclear transitions. This is

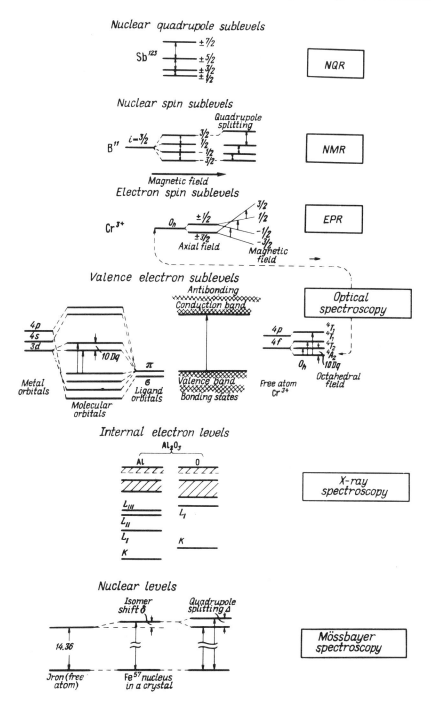

Fig. 74. Energy levels and related branches of spectroscopy of solids

the essence of the application of spectroscopy to the investigation of solids, especially the chemical bond in solids.

Mössbauer (nuclear gamma resonance) spectroscopy [178] is based on the transitions between excited and ground levels of a nucleus (as in atomic spectroscopy, between excited and ground electron levels). For example: excited (called isomeric) state of nucleus Fe^{57} arises from radioactive isotope Co^{57} by the capture of an electron; transition into the ground state is accompanied by gamma radiation.

For investigation of spectrum of iron in crystal one takes the source Fe^{57}, for Sn the source Sn^{119} etc. Radiation of a nucleus of source, entering solids (for example, Fe^{57} in stainless steel or in chromium, platinum and others), occurs without changes of energy, that is the radiation is recoilless.

Obtaining the Mössbauer spectrum consists essentially of comparison of the energy of the source nucleus radiation with the energy of the same nucleus radiation in a sample; when nuclei in the sample and in the source are in a similar environment, then resonance absorption of gamma radiation by the sample takes place.

If the nucleus in the sample occurs in a different electronic environment, then under its influence the change of nuclear levels positions takes place and the condition of resonance is violated. For restoration of this condition of resonance one uses the Doppler change of energy during the movement of the sample relative to the source with different speed. This speed is the measure of the energy difference and determines the dimensions of the Mössbauer spectra parameters expressed in $mm\,s^{-1}$ or $cm\,s^{-1}$.

Three principle parameters describe the displacement and splitting of nucleus levels in a sample: (1) chemical or isomeric shift is related to the displacement of levels due to the redistribution of electron density of a given atom in the condition of the chemical bond in a sample, which influences nuclear levels through the atom's s electrons; (2) quadrupole splitting arises from the splitting of the nuclei levels, having quadrupole moment in a ground or excited state interacting with the crystal field gradient at the nucleus; (3) magnetic superfine structure is related to the splitting of nuclear levels in ferromagnetic or antiferromagnetic crystals under the influence of the local magnetic field.

The most suitable Mössbauer nuclei are: Fe^{57}, Sn^{119}, I^{127}, I^{129}, Ir^{191}, Ir^{193}, Te^{125}, Au^{197} and others [178].

In X-ray spectroscopy of solids there are two directions of study [170, 176, 178].

From spectra related to transitions between inner electron levels, it is possible to measure the change of energy differences of these levels. For example, for Si ($1s^2 2s^2 2p^6 3s^2 3p^2$) transition into 1s-level (K-shell) from 2p-level (K_α-line in spectrum) is the transition between both inner levels i.e., between inner 2p- and 1s-atomic orbitals of Si, not overlapping with oxygen orbitals and not forming molecular orbitals. The difference of these transition energies ΔK_α permits the determination of effective charges of atoms in a given compound in comparison with other compounds containing this atom.

In spectra resulting from transitions between outer electron levels and inner levels there occur bands, but not lines. Outer levels are represented by molecular orbitals formed due to the splitting and overlapping of atomic orbitals. For example, transitions into inner $1s$ level for Si occur from outer electron levels indicated not as $3s$, $3p$, $3d$ (K_β spectrum), but as a_1, t_2, e, t_1 molecular orbitals. From these spectra one determines the positions of the molecular orbitals and, using selection rules and relative band intensities, the type of the molecular orbital and contribution of different atomic orbitals.

In electron spectroscopy [192, 193], or electron spectroscopy for chemical analysis (ESCA), one does not measure the energy difference of the levels between which the transitions occur, but the kinetic energy of electrons knocked out from different inner and outer levels. It is thus possible to determine immediately the energies of all electronic levels. From these energies one estimates the effective charges (but within the framework of a theoretical scheme other than in X-ray spectroscopy) as well as positions of the molecular orbitals (all MO independently of selection rules and thus without their assignments).

In optical spectroscopy one determines the energies of systems of absorption or reflection bands. Very intense interband transitions (occurring in UV for transparent crystals and in visible and near IR for opaque crystals) are assigned to transitions between the states described by crystal orbitals at different points of Brillouin zone using calculated energy band schemes.

Spectra of d and f electrons are described by means of crystal field parameters Dq, B, C while charge transfer spectra are described in terms of the molecular orbitals scheme (see Chaps. 2 and 6).

In electron paramagnetic resonance (EPR), spectra are obtained as the result of transitions between spin sublevels of lower orbital level [171, 172, 174, 177, 180, 183, 184, 186, 190], and splitting of these sublevels occurs only under the action of the external magnetic field. In the case of electron spin $S=1/2$, two sublevels arise in the magnetic field: with $m_S = +1/2$ (spins of electrons are directed along the field) and with $m_S = -1/2$ (antiparallel to the magnetic field spins). Transitions between them are accompanied by absorption of radiofrequency radiation, corresponding to cm or mm wavelengths (frequency 9000–38,000 Mc s^{-1}).

Spectra EPR can be observed from substances which contain ions with unpaired electron spins (transition metal and rare earth ions in trace quantities to 1%), as well aw from electron-hole centers and free radicals.

The principal EPR parameters are:

The g factor, which determines the position of a line in the spectrum dependent upon the frequency of the used radiation hv and the external magnetic field H value in accordance with resonance condition $hv = g\beta H$, where β is the electron magnetic moment;

B_l^m are parameters of initial splitting, describing the splitting of spin sublevels by crystal field; values are connected with lattice sums A_l^m (see Chap. 7.3);

A is the parameter of super-fine structure, describing the splitting of spin sublevels due to the interaction with nucleus magnetic moment of paramagnetic ion (in case this nucleus possesses magnetic moment);

A' is the parameter of super-hyperfine (ligand) structure, arising due to the interaction with magnetic nuclei of surrounding (ligand) ions (in the event of their possesssing magnetic moments).

All these interactions in EPR are described by means of the spin Hamiltonian, taking the particular forms for the ion or center with a given spin and in the field with a given local symmetry.

In nuclear magnetic resonance (NMR) [173, 182, 187] as in EPR, transitions occur between spin states, however not electron but of the nucleus. Nuclear g_N-factor and nuclear magneton β_N enter resonance condition $hv = g_N \beta_N H$. Because nuclear magneton is 1836 times smaller than Bohr's magneton β (since the mass of proton is 1836 times larger than the mass of electron) the resonance in NMR is observed at frequencies approximately 1800 times less than in EPR, that is near 5–20 Mc s^{-1} (i.e., with wavelength 20–50 m).

The most suitable nuclei for NMR are H^1, Li^7, Be^9, B^{11}, F^{19}, Na^{13}, Al^{27}, P^{31} and others entering diamagnetic substances.

There are two types of NMR study: (1) high-resolution NMR in liquids (2) NMR of broad lines in solids.

Many studies have been carried out by proton resonance: proton position in a structure, interproton distances, water types in crystals etc. For this, orientation dependence of distances between resonance lines from different protons or the second moment of the absorption band are measured.

The principal parameter in NMR characterizing the state of electron environment of the nucleus in a crystal is crystal field gradient and its orientation in crystal. Chemical shift, owing to the large width of a line is not usually observed in crystals.

In nuclear quadrupole resonance (NQR) [181, 191] transitions occur between quadrupole levels of a nucleus, split by noncubic crystal fields. The frequency, with which the NQR signal is observed, is determined by the value of the quadrupole moment of the nucleus eQ (which is the measure of deviation of the nucleus charge distribution from the spherical) and by crystal field gradient q.

The principal NQR parameters are: quadrupole interaction constant eQq and the parameter of asymmetry, showing the deviation crystal field symmetry from axial.

Value eQq is the same quadrupole splitting as in NMR and NGR, but it is observed for another nuclei: As^{75}, Sb^{121}, Sb^{123}, Bi^{209}, and others.

Consider now the relation between these branches of solid state spectroscopy. They can be classified on the following three bases: by related chemical bond concepts, by the spectral regions and by the dependence of spectroscopical parameters upon the chemical bond parameters.

5.2 Principal Concepts and Parameters of the Chemical Bond from the Standpoint of Spectroscopy

1. Spectroscopic observations confirm the general model of chemical bonding described by the molecular orbital scheme and determine the choice between valence bond and molecular orbital methods in favor of the MO. Theoretically obtained MO schemes for different types of complexes (number, order and approximate energy values of the MO levels) correspond completely to the lines

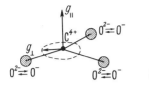 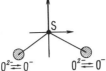

Fig. 75. Delocalization of electron in free radicals CO_3^- and SO_2^- according to EPR data

and bands in X-ray and optical spectra. MO and energy band schemes are direct expressions of the experimental data of absorption and reflectance spectra.

The characteristic feature of the MO model is electron delocalization, i.e., distribution of electron density all over the molecule. It is revealed definitely in EPR spectra: (1) in symmetry of the spectra and from orientation of g-factor axis, (2) in superhyperfine structure arising due to interaction of the unpaired electron with the magnetic nuclei of ligand ions. Thus in tetrahedral complexes of transition metals, instead of four spectra with g-factor axes directed along the M–L bonds, there occurs only one spectrum with the axes coinciding with symmetry axes of the tetrahedron (in the case of rhombic symmetry).

In free radicals, for example CO_3^-, SO_2^- (Fig. 75), a hole (that is one missed electron: $O^{2-} 2p^6 \rightarrow O^- 2p^5$) is trapped not by one oxygen in these radical groups, but is distributed all over the radical; it is revealed in the symmetry and orientation of g-factor axes: in CO_3^- one spectrum with axial symmetry is observed (with $g_\perp$ the in the plane of the radical and $g_\parallel$ normal to the plane) but not three spectra with axis along C–O directions; in SO_2^- also one spectrum is observed with symmetry C_{2v} with g_z-axis coinsiding with C_{2z} and with bisectrix O–S–O angle but not two spectra with axes along S–O bonds.

Distribution of an electron over the whole complex is revealed also by the superhyperfine structure of EPR spectra: the number of lines in it shows the number of ligand ions nuclei with which the electron interacts.

Super-hyperfine structure of EPR is the direct confirmation of the existence of "tunneling" which explains the collectivization of electrons in the energy band theory: in EPR spectra of $ZnS:Mn^{2+}$ one observes the interaction of electrons of impurity Mn^{2+} with the nuclei of Zn atoms separated from Mn^{2+} by sulfur atoms. Double resonance of F centers shows the interaction of the unpaired electron of the F center with nuclei of Na and Cl ions in 3-d and 4th coordination spheres.

2. Integer valencies and fractional charges differ unequivocally by optical and EPR spectra (see Chap. 3.3). Sharp principal differences in these spectra are caused by the change of valence, as well as by the change of high or low spin state and coordination, while the variation of the effective charge is accompanied by gradual displacement of the line or bands within a given type of spectrum. These changes and variations differ also in Mössbauer spectra.

3. Effective charges determined by K_α-transitions in X-ray emission spectra, together with data obtained from dielectric properties (for simple compounds) are considered to be the most real and can be obtained for any element in any combination. For compounds of Fe and Sn, the most detailed characteristic of effective charge and effective electron configuration is obtained by a combination of X-ray and Mössbauer data.

4. Ionicity-covalency of bonding is revealed in all types of spectra and some of their parameters show the most convincing dependence upon it. The most detailed characteristics can be obtained from EPR spectra. Observation of superfine structure is of principal importance, as it shows the existence of electron density transfer, that is the presence of covalency even in most ionic compounds. Values of the superfine structure parameter allowes experimental estimation of relative ionicity in actual compounds. From the measured values of g factors, it is possible to estimate coefficients c_i in MO. However, all these EPR data are obtained only for the impurity of transition metal ions or for electron hole centers in crystals. Moreover, this information on ionicity (c_A/c_B) is obtained only for one of molecular orbitals which in the case of d ions is the antibonding orbital.

While optical spectra are interpreted usually within the framework of the crystal field theory, though empirically their parameters Dq, B, C are connected with relative covalency of bonding, this makes it possible to order compounds in spectrochemical and nephelouxetic series. Calculations of crystal field parameters taking into account of the overlapping of wave functions and change to the MO scheme correspond to taking covalency into account.

Sulfides, sulfosalts and related compounds are characterized essentially by the covalent type of bonding. However, Mössbauer spectra of iron show a considerable variation in their covalency. Interpretations of their reflectance spectra are of especial importance owing to the determination of ionicity-covalency from energy band calculations (see Chap. 7.1). NQR spectra provide important information for sulfides and sulfosalts of As, Sb, Bi. Moreover, reflectance spectra and X-ray spectra measurements are of central importance, as they produce experimental values which enter the crucial stages of the theoretical calculations of the MO schemes and energy band structures; these empirical parameters take upon themselves approximations of the theories. The more accurate schemes thus obtained include ionicity-covalency as inherent characteristics.

5. Ionization potentials of compounds are obtained immediately from X-ray spectra or from optical spectra in the UV region as the energies corresponding to a last occupied nonbonding molecular orbital.

6. Hybridization of sp electrons is determined quite evidently from the ratio of isotropic and anisotropic components of the EPR superfine structure, but it is usually possible for free radicals only because ions with s and p electrons usually have filled electron shells and are not paramagnetic in crystals. The participation of different atomic orbitals (for example, the $3d$ orbital of Si) in bonding can be determined by analysis of X-ray spectra from the point of view of the molecular orbital sheme and in particular cases of elements with Mössbauer nuclei by using both Mössbauer and X-ray parameters.

7. Contribution of metallic components of bonding, determined by the position of the Fermi surface relative to the levels of the valent d electrons, can be estimated from reflectance spectra (and from electric and magnetic properties of compounds).

8. Cohesive energy is not determined directly from spectroscopy data. However, the most reliable calculations of cohesive energy by energy band methods use empirical parameters, obtained from reflectance or X-ray spectra. Crystal field stabilization energies are determined from optical absorption spectra

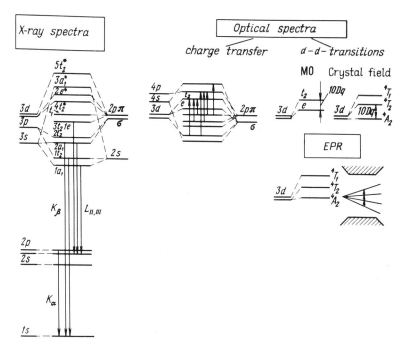

Fig. 76. Relation between electron levels describing X-ray, optical, and EPR spectra and molecular orbital scheme

(from terms including A_l^m with $l \neq 0$; term A_0^0 in the expression of Madelung energy displace equally all electron levels, and thus A_0^0 is not reflected on spectra).

9. Charge distribution in lattice determining crystal field gradient is reflected in spectroscopic parameters corresponding to the term A_2^0 in quadrupole splitting in NQR, NMR and Mössbauer spectra or to B_2^0 in EPR spectra and to the splitting by trigonal and tetragonal crystalline fields in optical spectra.

Estimation of the Chemical Bonding in Different Branches of Spectroscopy. The relation of different branches of spectroscopy (see Fig. 73) to the chemical bond is first of all the relation of corresponding energy levels schemes to molecular orbital scheme (Fig. 76).

In X-ray spectra one can distinguish from the standpoint of chemical bonding spectra arising as the result of transitions between: (1) the levels of inner (core) electrons not entering the MO scheme (K_α lines in spectra of the third group elements Al, Si, P, S, Cl); (2) the valent electron levels, presented not by atomic orbitals, but by molecular orbitals, and inner electron levels (K_β- and L-spectra in spectra of the third group elements). The K_α-transitions are used to determine effective charges, the K_β- and L spectra to determine energies of MO levels.

In optical absorption spectra one distinguishes charge transfer bands occurring usually in UV, and $d-d$ transitions spectra (crystal field spectra) i.e., transitions between split levels of d electrons of transition metals in crystal,

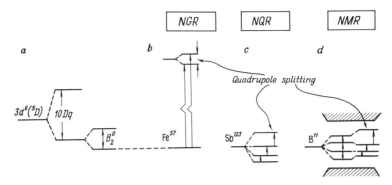

Fig. 77a–d. Relation between splitting by crystal field of nuclear levels describing Mössbauer spectra (NGR), NQR and NMR. Scales of electron and nuclear levels are different. **a** Splitting by crystal field (B_2^0 is crystal field gradient acting upon quadrupole moment of the nucleus); **b** nuclear levels of Fe^{57}; **c** quadrupole sublevels of the lower nuclear level; **d** spin sublevels of magnetic nucleus

occurring in visible and adjacent regions of the spectrum. Charge transfer bands are related to transitions from filled bonding and nonbonding molecular orbitals formed mainly of atomic orbitals of ligand ions to empty or partially occupied molecular orbitals made mainly from central ion atomic orbitals. Levels of d electrons represent in molecular orbital schemes antibonding molecular orbitals (two MO, t_{2g}^* and e_g^*, in the octahedron and two, e^* and t_2^*, in the tetrahedron), which are occupied by different numbers of d electrons (d^0–d^{10}); one of them is the σ orbital (e_g in the octahedron; t_2 is partially σ and partially π in the tetrahedron) while the other is the π orbital (t_{2g} in the octahedron; e in the tetrahedron). Only these two MO (which can be split up to five in the case of lowering the symmetry) determine all crystal field spectra, however, transitions occur not directly between these MO, but between states, as described by terms arising as the result of spin-orbit interaction (different numbers of terms arise from these two MO for different numbers of electrons and hence different numbers of transitions are obtained; for example, for $t_{2g}^1 e_g^0$ there are only two terms, for $t_{2g}^3 e_g^0$ there are three terms etc., see Chap. 3.2).

The same antibonding d-molecular orbitals can be characterized by g-factor values and superfine structure in the EPR spectra of transition metals ions.

In EPR spectra of free radicals, paramagnetic electron may occur in bonding, nonbonding and antibonding MO; since free radicals are formed by capture of one electron or one hole, the arising terms are usually singlets.

In all nuclear spectra: Mössbauer (NGR), NQR, NMR, there are mainly two types of parameter: chemical shift and quadrupole splitting. In NGR quadrupole splitting arises as a result of transitions to the excited level of the nucleus quadrupole sublevels; in NQR it arises from transition between quadrupole sublevels of the lower nuclear level; in NMR it arises from transitions between spin sublevels of the lower nuclear level (split by an external magnetic field) which can be displaced by quadrupole interactions (Fig. 77). It is evident then that NGR on the one hand and NQR and NMR on the other correspond to contrary regions of the spectrum (Fig. 73). However, quadrupole splitting is explained in all these types of spectra by the same causative action of the crystal field gradient arising due to noncubic local symmetries.

Mechanisms Determining the Relation of Spectroscopic Parameters to the State of the Chemical Bond. Each spectroscopic parameter presents a value summarizing an intricate complex of elementary physical interactions, part of which only is compared with one of the aspects of another complicated concept presented by one of the chemical bonding parameters.

It is useful to consider the interactions as revealing different spectroscopic parameters which are related to chemical bonding characteristics.

1. Formation of MO from AO. Type of the MO scheme and the MO energies.	Energies of transitions in X-ray. K_β and L spectra (absolute value of MO energies); energies of transition in optical charge transfer spectra (relative MO energies).
Formation of orbital terms. Spin-orbit interaction. Contribution of atomic orbitals in MO.	Optical $d-d$ transitions. EPR spectra: g factor, B_l^m. Experimental energies and intensities of X-ray and optical bands; g factor and EPR superhyperfine structure.
2. Interaction of inner parts of wave functions of outer (valent) electrons with wave functions of core electrons.	Shift ΔK_α of energies of X-ray transitions between inner electron levels.
3. Variation electron density at nucleus.	
A. Displacement of nuclear levels due to the change of density in s electrons at a nucleus as the result of electron redistribution in the course of formation of chemical bonding.	Chemical shift in NGR and NMR.
B. Splitting of spin sublevels of paramagnetic electron due to interaction:	Superfine structure in EPR of transition metal ions.
a) with magnetic nucleus of paramagnetic ion (d electrons – s electrons – nucleus),	
b) with magnetic nuclei of ligand ions.	Superhyperfine structure in EPR of transition metal ions and electron-hole centers; ratio of isotropic-anisotropic components (A_{iso} and B).

Principal Concepts and Parameters of the Chemical Bond   189

4. Interaction of crystalline fields:
a) with nuclear quadrupole moment,   Quadrupole splitting in NGR, NGR, NMR; orientation of crystal field gradient axes.

b) with atomic orbitals,   Splitting in optical spectra (crystal field components with B_l^m where $l \neq 0$).

c) with electron spin.   Splitting of spin sublevels (B_l^m). Orientation of g-factor axes.

5. Band formation; interaction of atomic orbitals with periodic potentials of lattice.   Valence band width; energy gaps; amounts of atomic orbitals and plane waves in valence and conduction bands.

6. Optical Absorption Spectra and Nature of Colors of Minerals

6.1 Parameters of Optical Absorption Spectra

When the wavelength of monochromatic light passing through crystal corresponds to the difference of the energy levels of an ion in crystal (or in solution), light energy absorption occurs, resulting in the appearance of an absorption band in the optical spectrum of the crystal (Fig. 78).

These energy levels are ion levels split by the crystal field (see Chap. 2). The energy difference is the separation between ground state level, taken as zero, and one of the excited state levels. The absorption condition is the equality of this difference to the monochromatic light energy:

$$hv = E_{exc} - E_{gr}.$$

Transition from the ground to the excited state corresponds to the transfer of an electron to one of the excited configurations or to the excited states described by the ion terms in a crystal field. For example, in the case of the d^1 configuration, the electron is transferred from the t_{2g}^1-level (d_{xy}, d_{xz}, d_{yz}) to the e_g^1-level ($d_{x^2-y^2}, d_{z^2}$) separated by the crystal field. In the case of d^3 configuration, the transition occurs from the $^4A_2$ ground level to the excited $^2E, {}^4T_2, {}^4T_1$ levels (Fig. 78).

Absorption bands correspond to energies of excited levels of ions in crystal, and optical absorption spectra are basic experimental data for ion level energies in the crystals.

The energy differences between levels for ions with complete shells unsplit in crystals (S terms are transformed to A_1 – see Table 7 and Table 14) are in most cases too great and correspond to transitions occurring in the far UV or soft X-ray region.

Levels of the ions with incomplete d and p subshells only split by the crystal field and the strength of the crystal field for these ions has such a value that the energy separation between the split levels corresponds to energies of visible, near infrared and near UV regions of the spectrum.

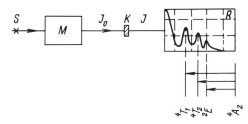

Fig. 78. Experimental arrangement for obtaining of optical absorption spectra. I, source of light; M, monochromator; K, crystal; P, registration of spectrum; I_0 and I, incident and transmitted intensities. The energy levels and transitions are shown corresponding to the absorption bands (for Cr^{3+})

6.1.1 Units of Measurement of Optical Transition Energies

So far as the energies of levels of ions in crystal are determined by absorption band positions in optical spectra, these energies are often expressed through the frequencies of the absorption bands. That is, a duality exists in the determination of level energies by frequencies of the corresponding absorption bands and of the absorption band position not only in wavelength and frequency units but also by the energy of the corresponding transitions (Fig. 79).

Transition energy is measured in erg. However since the optical region covers energy ranges of 10^{-11}–10^{-12} erg, optical transition energies are expressed usually in electron volt (1 eV = $1.602 \cdot 10^{-12}$ erg). The visible region of spectra corresponds to the energies of the order 1.5–3.0 eV.

However electron volt units in their turn are only suitable for the determination of an approximate value of transition energy; for more exact definition of the position of levels, frequency units are used proportional to energy. Frequency v can be expressed in s^{-1} as in radiospectroscopy (s^{-1} = Hz), in the optical region they correspond to a frequency of the order 10^{14}–$10^{15}\,s^{-1}$ and it is more convenient to express frequency in wave numbers: $\bar{v}$ (cm^{-1}). Frequency $\bar{v}$ is reciprocal to wavelength: $\bar{v}(cm^{-1}) = 1/\lambda(cm)$ or $\bar{v} \cdot \lambda = 1$. Thus:

$$E = h \cdot v (erg = erg \cdot s \cdot s^{-1}),$$
$$v = \bar{v} \cdot c (s^{-1} = cm^{-1} \cdot cm \cdot s^{-1}),$$
$$\bar{v} = 1/\lambda (cm^{-1} = 1/cm),$$
$$\bar{v} = v/c (cm^{-1} = s^{-1}/cm \cdot s^{-1}),$$
$$v \cdot \lambda = c (s^{-1} \cdot cm = cm \cdot s^{-1}).$$

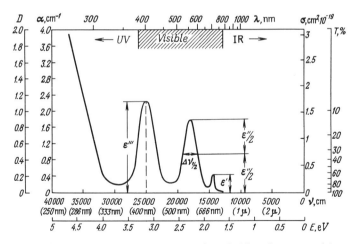

Fig. 79. Relation between optical spectra parameters: wavelength (λ) – frequency (v) – transition energy (E) and transmittance (T) – optical density (D) extinction coefficient (α) – absorption cross-section σ; ε is molar absorption coefficient; $\Delta v_{1/2}$ is the half intensity band width

It is worthwhile to remember the relations between the wavelength in *nm*, frequency in cm^{-1} and energy in eV, because all three units of measurement are used in optical spectroscopy:

$\lambda(nm) \cdot \bar{v}(cm^{-1}) = 10^7$, because $\lambda(cm) \cdot v(cm^{-1}) = 1$
and $1\,cm = 10^7\,nm$
$1\,eV = 8066\,cm^{-1} \approx 1240\,nm$
$E(eV) \cdot \lambda(nm) \approx 1240$
$E(eV)/\bar{v}(cm^{-1}) = 1240 \cdot 10^{-7} = 1.24 \cdot 10^{-4}$
$1\,cm^{-1} = 10^{-7}\,nm$; $10{,}000\,cm^{-1} = 1{,}000\,nm = 1\,\mu$
$1\,cm^{-1} = 10^8\,\text{Å} = 10^7\,nm = 10^4\,\mu$; $1\,nm = 10\,\text{Å}$.

It should be noted that representation of spectrum in the wavelength scale and not in cm^{-1} (or in eV) distorts the form of the lines; the spectrum is broadened, which is especially large when passing from the visible to the near infrared.

6.1.2 Intensity of Absorption

While the position of absorption bands is related to energy level differences, their intensities are determined by the integral of the product of the ground and excited states wave functions and transition moment. For example, for Cr^{3+}:

$$I = \int \psi(^4A_2) \cdot M \cdot \psi(^4T_1),$$

where $\psi_{^4A_2}$ is a wave function of the $^4A_2$ state; $\psi_{^4T_1}$ is a wave function of $^4T_1$ state, M is the electric dipole transition moment for the transition between these states (magnetic dipole and electric quadrupole transitions have intensities less by some order of magnitude).

Intensity is equal to zero, when the product of these states and the transition moment is equal to zero; in this case transition is forbidden. Transition is allowed when the product is not equal to zero. Equality or inequality to zero follows from a simple consideration of symmetry, leading to selection rules (see Chap. 2.3).

For the allowed transitions, intensity is determined by oscillator strength f, which can be found from the following expression:

$$f = \frac{mc^2}{\pi e^2} \int \psi_1 \cdot M \cdot \psi_2 \cdot d\tau,$$

where the term before the integral is represented by the atomic constants (m and e are the mass and charge of electron, c is the velocity of light), and the integral is a term determined by the peculiarities of the ground and excited states. The name "oscillator strength" comes from the classical model for the transition between states, considered as oscillating systems.

The oscillator strength is usually estimated from absorption spectra:

$$f = \frac{mc^2}{\pi e^2} \int \varepsilon \, dv,$$

where integral $\int \varepsilon \, dv$, equal to $\int \psi_1 \cdot M \cdot \psi_2 \cdot d\tau$, is an experimentally determined integrated absorption; ε is the molar extinction coefficient; v is frequency.

In this case, this expression for the oscillator strength, f, is an equation describing absorption as a function of frequency, and the integral itself corresponds to the area under the absorption curve.

Absorption intensity (as follows from its physical meaning) is the property of individual transitions between given pairs of levels of absorption ion in crystal (or in solution). Hence with equal concentration of an ion, absorption intensity can be different in various crystals in dependence on detail of local symmetry (especially upon the presence or absence of the center of symmetry, see Chap. 2.5), because in different point groups of symmetry there will be different transitions. For example, for Cr^{3+} instead of $^4A_{2g} \to {}^4T_{2g}$ in periclase (cubic symmetry O_h), the transition $^4A_{2g} \to {}^4E_g$ arises in spinel (local symmetry of Cr^{3+} D_{3d}), the transition $^4A_2 \to {}^4E$ in beryl (D_3), and the transition $^4A \to {}^4E$ in corundum (C_3).

To reduce the experimentally estimated absorption intensity to comparable values, it is necessary to take into account its dependence upon the crystal thickness and concentration of the absorption ion.

Dependence upon crystal thickness is expressed by Buger-Lambert's law, which means simply that every next layer of a substance absorbs equally; this corresponds to the exponential dependence of absorption intensity upon thickness. This dependence is obtained as follows.

Let us mark the initial intensity of light entering crystal as I_0, and the intensity of light passed through the first layer of a substance as I_1. Then $I_1/I_0 = T$ or $I_1 = I_0 T_1$, where T is transmittance, i.e., positive fraction, showing by how much I_1 is smaller than I_0. When an already weakened flow of light enters the next layer, the situation is repeated: $I_2 = I_1 T$ or $I_2 = I_0 T^2$, which leads to the general law: $I = I_0 T^t$ or $\ln I/I_0 = t \ln T$ where t is the thickness of crystal, $\ln T$ the natural logarithm of the fraction, which is why $T = \exp(-\alpha)$ where α is the extinction coefficient. Hence one obtains Buger-Lambert's law:

$$I = I_0 \exp(-\alpha t) \quad \text{or} \quad \ln I/I_0 = \ln T = -\alpha t \quad \text{or} \quad \ln I_0/I = \ln 1/T = \alpha t.$$

The same can be expressed in common logarithms (the ground of natural logarithms $e = 2.718...$, $\lg e = 0.4343$; $\ln 10 = 1/\lg e = 2.303$): $I = I_0 \, 10^{-0.4343 \alpha t} = I_0 \, 10^{-kt}$; $\alpha(\ln) = 2.303 k(\lg)$; $k(\lg) = 0.4343 \alpha(\ln)$; $\lg I_0/I = \lg 1/T = D$ is called optical density. It is worth noticing that the absorption law $I = I_0 \exp(-\alpha t)$ is common to all kinds of electromagnetic radiations, including X-rays and gamma-rays.

The extiction coefficient is the characteristic of the individual absorption band in a given crystal with a given concentration of absorption ion. To obtain the characteristic of an absorption band (per unit of concentration) it is necessary to take into account dependence upon concentration.

Dependence upon concentration is expressed by Beer's law and can be considered in four aspects.

1. As the absorption characteristic of substance per concentration unit. When concentration is expressed in mol l^{-1} (Beer's law was obtained initially for solutions), then the molar coefficient of absorption ε is used, which is equal to:

$$\varepsilon = \frac{\alpha \, \text{cm}^{-1}}{C \, \text{mol} \cdot \text{l}^{-1}} \quad \text{(that is in mol} \cdot \text{cm}^{-1} \cdot \text{l)}, \quad \text{or} \quad \alpha = \varepsilon \cdot C, \quad \text{or} \quad D = \varepsilon \cdot C \cdot t.$$

2. As Buger-Lambert's law is reduced to the dependence of absorption upon thickness, so Beer's law is reduced to the dependence of absorption upon concentration. This dependence is the base of colorimetry, which is the analytical method for determining the concentration in solutions by intensity of absorption. Then Buger-Lambert-Beer's law reads as follows:

$$I = I_0 \cdot e^{-\varepsilon Ct} \quad \text{or} \quad I = I_0 \cdot 10^{-kCt}.$$

3. In solutions complexes like $M(H_2O)_6^{n+}$ usually develop, where M is the transition metal ion; thus with the increase of concentration, only the number of these complexes is increased, which leads to linear dependence between absorption intensity and concentration for diluted solutions.

In solid solutions of minerals and inorganic compounds, the linear dependence upon concentration ceases to be a rule; it can be observed for some series of solid solutions while for others, deviations from linear dependence exist.

4. When considering, not series of solid solutions with retaining local symmetry of a crystal field acting upon the absorption ion, but different structural types, it is necessary to take into account changes of local symmetry and crystal field strength and especially the splitting of an absorption band into two or three bands, as well as the presence or absence of the center of symmetry. These are the additional factors influencing the intensity and positions of absorption bands.

Absorption intensity can be characterized by means of different parameters distinguished by: use of natural and common logarithms ($\ln 1/T = E$ or $\lg 1/T = D$); expression of absorption per total thickness of a crystal (optical density, extinction) or unit of thickness (linear extinction coefficient); expression per unit of concentration, in this case the parameters differ by the way of expression of concentration.

Parameters of absorption intensity are:

I_0: intensity of light enetering the crystal;
I: intensity of light leaving the crystal.

Initial experimental data:

$T = I/I_0$: transmittance (in %);
$A = I_0 - I/I_0 = 1 - T$: absorbance (in %).

Taking Into Account the Exponential Weakening of Light Passing Through Successive Layers of a Substance:

$\ln 1/T = \ln I_0/I = E$: extinction;
$\lg 1/T = \lg I_0/I = D$: optical density, extinction.

Reduction to a unit of thickness of crystal:

$\alpha = \ln I_0/I \cdot 1/t = E/t$: linear absorption coefficient (natural) (in cm^{-1} or mm^{-1});
$K = \lg I_0/I \cdot 1/t = D/t$: linear coefficient of extinction (decimal) (in cm^{-1} or mm^{-1}); $K = 0.4343\alpha$; $\alpha = 2.303 K$.

Reduction to a unit of concentration:

$$\varepsilon = \frac{\alpha}{C} = \frac{\ln I_0/I}{t \cdot C}$$: molar (natural) absorption coefficient (C in mol l^{-1}, t in cm, then α in cm$^{-1}\cdot$l$\cdot$mol^{-1});

$$k = \frac{K}{C} = \frac{\lg I_0/I}{t \cdot C}$$: molar (decimal) coefficient of extinction (in cm$^{-1}\cdot$l$\cdot$mol^{-1});

$$\sigma = \frac{\alpha}{n_0} = \frac{\ln I_0/I}{t \cdot n_0}$$: absorption cross-section (n_0 is density, number of atoms per 1 cm^3, that is n_0 in cm^{-3}, α in cm^{-1}, then σ in cm^2).

Expression by integral absorption: $f = 4.6 \cdot 10^{-9} \cdot \varepsilon \cdot \Delta v_{1/2}$: oscillator strength (ε is natural molar absorption coefficient, $\Delta v_{1/2}$ half width of absorption band, see Fig. 79); $f = 1.88 \cdot 10^{-9} \cdot k \Delta v_{1/2}$: oscillator strength (k is decimal molar absorption coefficient; $k = 0.4343\varepsilon$).

In comparing transmittance $T = I/I_0$ with optical density $D = \lg 1/T$, it is necessary to pay attention to the fact that D is the logarithm of the inverse value T: the scale of optical density D in spectrophotometers is calibrated from 0 to 2 because value of $D = 2$ corresponds only to 1% of transmittance, i.e., to weakening of the light by 100 times ($T = 1/100$; $1/T = 100$; $\lg 100 = 2$). Values of D from 2 to ∞ correspond to transmittance from 1% to 0 (these are practically opaque crystals); that from $D = 1$ to $D = 2$ (that is 10% to 1% of transmittance) correspond to dark-colored, strongly absorbing crystals. Optimal values of D are from 0.1 to 0.8–1.0 (that is from 80% to ~10–17% of transmittance), outside of these values, relative mistakes in the determination of optical density are strongly increased. The relation between optical density and the linear extinction coefficient $K = Dt$, where t is the thickness of crystal in cm, is used for determination of such thickness of a crystal which correspond to values with in the limits 0.1–0.8 (for example when $K = 30$ cm^{-1} and the thickness $t = 0.003$ cm, then $D = K \cdot t = t = 30 \cdot 0.003 = 0.09$).

The molar absorption coefficient is the most common way of expressing absorption intensity per unit of concentration, because the expression of concentration in mol l^{-1} is usual for solutions and it is for a solution that the quantitative data are usually given. However, it is an artificial way for crystals corresponding to the number of gram-atoms of absorbing ion per 1000 cm³ ("liter") of crystal. It is more natural to use the expression of concentration in number of absorbing ions (n_0) per 1 cm³ of crystal and to express the absorption intensity by the absorption cross-section: $\sigma = \alpha/n_0$.

The number of absorbing ions in 1 cm³ can be determined in different ways.

1. Proceeding from the crystal-chemical formula of a mineral (a) the number of "molecules" of mineral in 1 cm³ is: $x = N_{Av} =$ specific gravity/M, where N_{Av} is the Avogadro number ($6.02 \cdot 10^{23}$) M is molecular weight ($M_A \cdot \% A + M_B \cdot \% B$); (b) the number of atoms of absorbent ion per 1 cm³ is equal to x multiplied by the ratio of the number of absorbing atoms to the total number of atoms in the molecule formula unit.

2. Proceeding from a number of formula units per unit cell, volume of unit cell and crystal-chemical formula. For example, the volume of unit cell of $(Al_{99.98}Cr_{0.02})_2O_3$, approximately equal to the volume of Al_2O_3, is equal to 20.3 Å³ ($20.3 \cdot 10^{-24}$ cm³); in this volume there are two formula units of Al_2O_3, that is, four atoms of Al. Hence in 1 cm³ there are $0.2 \cdot 10^{24}$ atoms of Al. In $Al_{99.98}Cr_{0.02}$, the number of atoms of Cr in 1 cm³ is equal to $4 \cdot 10^{19}$.

The relation between the molar absorption coefficient ε(cm⁻¹·l·mol⁻¹) and the absorption cross-section σ(cm²) is: 1 cm⁻¹·l·mol⁻¹ ≈ $0.4 \cdot 10^{-20}$ cm². The oscillator strength:

$$f = \frac{mc^2}{\pi e^2}\int \varepsilon dv = 4.315 \cdot 10^{-9} \int \varepsilon dv.$$

However, assuming the Gaussian form of absorption curve and expressing the area under a curve by $\varepsilon_{max}\Delta v_{1/2}$ where $\Delta_{1/2}$ is the half-intensity band width, i.e., width at $1/2\varepsilon_{max}$, we can write: $f = 4.6 \cdot 10^{-9} \cdot \varepsilon_{max} \cdot \Delta v_{1/2}$ (ε is the natural molar absorption coefficient), i.e., the oscillator strength value differs from molar absorption coefficient not only by the coefficient corresponding to atomic units, but also by the width of the absorption band. (Oscillator strength enters Smakula's formula for the determination of the number of F centers from the absorption spectrum, but in this formula the half width of the absorption band is expressed in eV, extinction coefficient in cm⁻¹, concentration in atoms per cm³.)

Practical determination of the absorption coefficient values is complicated by the necessity of taking into account the reflection from the crystal's surface and scattering by cracks, inclusions and other defects.

Correction of reflection is taken into account in the expression

$$I = I_0(1-R)^2 e^{-\alpha t},$$

where R is reflectance. Because of the difficulties of taking reflection and scattering into account, absorption intensities are often expressed by the optical density or in arbitrary units.

6.1.3 Diffuse Reflectance Spectra

A method of practical importance for obtaining optical spectra is that of registration of diffuse reflectance spectra, i.e., reflectance spectra from powder crystals and not from the polished surfaces or faces of crystals [168, 240, 244, 245, 262]. These must be used in the case of dark-colored minerals, powder samples, and monocrystals with inside defects, where this method gives more expressive spectra than absorption spectra.

The similarity of diffuse reflectance spectra to absorption and their distinction from specular reflectance spectra is connected with the mechanism of the light reflection from powder crystals. There are two components in reflected light.

1. The specular component arising from the reflection of light by a surface without transmission through the crystal is described by Fresnel's formula (in the case of normal incidence of light)

$$R_{spec} = \frac{(n-1)^2 + n^2[K]^2}{(n+1)^2 + n^2[K]^2},$$

where $[K] = K \cdot \lambda/4\pi n$ is the absorption index; K is the linear extinction coefficient, n is the refractive index, λ is the wavelength.

2. The diffuse component arising from absorption of light by the specimen and its reappearance at the surface of the particles of the specimen after multiple scattering is described by:

$$R_\infty = \frac{1 - \sqrt{\frac{K}{K+2S}}}{1 + \sqrt{\frac{K}{K+2S}}},$$

where R_∞ is the reflectance relative an nonabsorbing standard, K the linear extinction coefficient, and S the scattering coefficient, practically independent of the wavelength and connected mainly with the size of the grain.

The value $\lg I_0/I$, experimentally obtained in spectrophotometers, consists in the case of reflectance spectra of these two components. The specular component prevails in crystals with strong absorption (with metallic luster), while the diffuse component prevails in comparatively weakly absorbing crystals. In obtaining diffuse reflectance spectra, measures are taken to reduce the specular component (for example with the help of crossed polarizers [168]).

The ordinate in diffuse reflectance spectra is the so-called Kubelko-Munk function [168, 244, 245]:

$$F(R_\infty) = \frac{(1-R_\infty)^2}{2R_\infty} = \frac{K}{S}.$$

Since the scattering coefficient S is independent of the wavelength, the positions of the minima in diffuse reflectance spectra coincide with the positions of maxima of optical density in absorption spectra.

Since the relative diffuse reflectance R_∞ is connected with the extinction coefficient multiplied by the average grain diameter, it is possible to estimate the K value but, because $K = \varepsilon \cdot C$, it is possible to estimate the dependence upon concentration in actual systems [168].

6.2 Types of Optical Absorption Spectra and Selection Rules

One distinguishes mainly three types of optical absorption in minerals and inorganic materials which are described respectively within the framework of the three theories: crystal field spectra (crystal field theory), charge transfer spectra (molecular orbital theory) and absorption edge (energy band theory).

1. Crystal field spectra are related to the transitions between d electron levels (of transition metal ions) split by the crystal fields. These transitions give rise to absorption bands occurring in the visible region and described by the crystal field theory considered above in Chapter 2. Here only these crystal field spectra are principally discussed. Spectra of lanthanide and acitinide ions with f electrons are considered in the other book of the present author [1].

It should be emphasized that these are transitions between the terms (Figs. 32, 35, and see below), not between the d-orbital levels, i.e., for example, for Cr^{3+} between the ground term $^4A_2$ (originating from the free ion term ^4F) and excited $^4T_2$, $^4T_1$, ^2E etc. terms, but not between the split d orbitals. It is not sufficient to use schemes of the splitting of d-orbitals in the octahedral, tetrahedral and other crystal fields (or types of Figs. 26, 27) but it is necessary to employ the diagrams of crystal field terms (Figs. 32, 33, and see below the Tanabe-Sagano diagrams). If the energy levels of a complex are represented in the form of a molecular orbital diagram, one has to obtain the set of the terms (or irreducible representations, or symmetry types, see Chaps. 2.1 and 2.3), corresponding to a given scheme of the molecular orbitals (i.e., to electron configuration of the complex described by the types and occupations of all the molecular orbitals).

2. Charge transfer spectra are described in the framework of molecular orbital diagram as the transitions from the filled nonbonding oxygen orbital to the antibonding partially occupied molecular orbitals composed mainly of metal atomic orbitals. For example, in Fig. 48a and b, these are the transitions of the types $t_{1g} \to t_{2g}$ or $t_1 \to e_2$ and others.

These transitions are usually pictorially represented as the transfer of the electron from the oxygen (or other ligand) to the metal, but it should be noted that reduction of the metal does not occur here and the very name "charge transfer" is descriptive rather than completely adequate: it is only the transition to an excited state of the complex.

In the energy band scheme it corresponds to the transition between the valence band and d-electron levels in the forbidden band.

For the most transition metal ions in the complexes with oxygen ligands the charge transfer band occurs in UV, except that of Fe^{3+} ion, a tail of which moves with increasing iron concentration into the visible and strongly influence Fe^{3+}

spectra. In practice the charge transfer spectra are mainly the spectra of the single band of the single Fe^{3+} ion.

3. Absorption edge represents a beginning of the inter-band transition spectra and corresponds to the minimum separation between the top of the valence band and the bottom of the conduction band, i.e., corresponds to the band gap (see Chap. 4.2). For essentially ionic crystals these transitions occur in UV and even in far UV.

Spectra of pairs of exchange-coupled transition metal ions (for example, $Mn^{2+} - Ni^{2+}$ in $KZnF_3$ or $Cr^{3+} - Cr^{3+}$ in Al_2O_3) reveal absorption bands characteristic of each of the ions, as well as those associated with the pairs [226, 241]. However, the spectrum of the pair represents in fact the bands of each of the ions arising from transitions between the levels of each of these ions split by the exchange coupling.

Consider, for example, the pair Mn^{2+} ($3d^5$, $^6S \to {}^6A_1$; spin $S_{Mn^{2+}} = 5/2$) and Ni^{2+} ($3d^8$; $^3F \to {}^3A_2$, $^3T_2^3$, T_1; spin $S_{Ni^{2+}} = 1$) in $KZnF_3$ (226). The total spins S of the ground state of Mn^{2+} and Ni^{2+}: $(S_{Mn} + S_{Ni}) = 5/2 + 1 = 7/2$; $5/2$ and $(S_{Mn^{2+}} - S_{Ni}) = 3/2$; i.e., the ground level $^6A_1$ of Mn^{2+} is split into the three sublevels $^8A_1$, $^6A_1$ and $^4A_1$ (with $S = 7/2$, $5/2$, $3/2$) and the ground level $^3A_2$ of Ni^{2+} is split into the sublevels $^8A_2$, $^6A_2$, $^4A_2$.

Consider the transition $^6A_1 \to {}^4A_1$ of Mn^{2+} (see below Figs. 92 and 94); it is spin-forbidden. However, the excited level $^4A_1$ (with $S_{Mn^{2+}} = 3/2$) of Mn^{2+} is also split by the exchange coupling into the three sublevels: $^6A_1$, $^4A_1$, $^2A_1$ (with $S = 5/2$, $3/2$, $1/2$). Thus the transitions $^4A_1$ (ground state) $\to {}^4A_1$ (excited state) and $^6A_1 \to {}^6A_1$ become spin-allowed.

In another part of the spectrum, one observes absorption bands of the Ni^{2+} arising from the transitions between the sublevels obtained in the same way from the splitting of the ground level $^3A_2$ of Ni^{2+} and excited level ^1E of Ni^{2+} (see below Fig. 109).

The separation of the exchange-coupled sublevels is of the order of some tenths cm^{-1}. Thus the interpretations of these spectra can be quite unequivocal for the cases of narrow bands (as in $KZnF_3$: Mn, Ni or Al_2O_3: Cr) and with observations at low temperatures.

Extensive literature exists concerning so-called intervalence-transfer absorption, or metal–metal charge transfer spectra [197, 198, 199, 221, 233, 234, 255, 289 etc.]. Some broad bands in spectra of mineral and inorganic crystal structure where neighboring cations ($Fe^{2+} - Fe^{3+}$, $Ti^{3+} - Fe^{2+}$ etc.) occurring in edge-sharing polyhedra are stronger than the usual spin-forbidden bands. They were assigned to these metal–metal charge transfer transitions.

More rigorous theoretical and experimental treatment is necessary to explain these absorption bands. It should be noted now that there is here no transfer of an electron from one metal atom to other (as there is no transfer of an electron in ligand–metal charge transfer spectra) but transitions occur between ground and excited levels of the cation interacting with neighboring cations, probably in the same manner as the Mn–Ni pair discussed above.

Absorption bands in optical spectra of crystals may be very different in intensity and width. Thus, the molar extinction coefficient acquires values from 10^5 to $10^{-2} cm^{-1} \cdot l \cdot mol^{-1}$.

Optical transitions of different types have a different order of magnitude of the extinction coefficient value. This is controlled by the selection rules, determining between which electronic states (energy levels) transitions are allowed. Transitions forbidden by the selection rules nevertheless can give rise to bands, but with one or more orders of magnitude less than with corresponding allowed transitions.

Different selection rules compare electronic state characteristics by the orbital quantum number of electrons (s, p, d, f), by spin S (multiplicity $2S+1$), by the orbit state of ion in a crystal (i.e., by the symmetry type, or irreducible representation, see Chap. 2.1). The weakening of these selection rules is connected with the mixing of different states (for example d and p states), with spin orbit interaction, with the interaction of electronic and vibrational states. Let us consider the different selection rules and possibilities of their weakening.

Parity Selection Rule (Laporte Rule). Recall that d and s orbitals are called even orbitals where positive and negative charges are distributed center-symmetrically, while p and f orbitals are odd orbitals with noncentersymmetric charge distribution.

Transitions between states with the same parity are "parity-forbidden" or forbidden by the Laporte selection rule: $g \not\rightleftarrows g$, $u \not\rightleftarrows u$ (see Chap. 2.1). Transitions between "even–odd" states ($g \rightleftarrows u$) are parity-allowed because the integral of the transition moment in this case turns into zero (see Chap. 2.1). It means that transitions d–d, p–p, s–s are forbidden, but s–p, p–d, d–f are allowed. This corresponds to the selection rule by orbital quantum number l in atomic spectra: transitions are allowed if $\Delta l = \pm 1$ (see Fig. 3).

This rule is common for atomic spectra and for spectra of atoms in crystals. In atomic spectra the d–d, p–p, s–s transitions are in fact completely absent and s, p, d, f series are connected with this selection rule (see Fig. 4). However in crystals the d–d transitions (and f–f transitions) occur, and it is these transitions that give rise to crystal field spectra, i.e., transition metal spectra, which are most common in the visible and determine the color of most minerals. The fact that the crystal field spectra (d–d transitions) are partially forbidden causes their lower intensity in comparison with parity-allowed charge-transfer transitions (between p ligand orbitals and d orbitals of transition metals).

Weakening of the parity selection rule is connected mainly with the following three factors: (a) absence of the center of symmetry; (b) mixing of d and p states; (c) interaction of electronic d states with odd vibrational states.

Absence of the center of symmetry (in local symmetry of a complex) relaxes this selection rule, which compares the orbitals by their centersymmetric or non-centersymmetric characteristics. This is why spectra of tetrahedral complexes have essentially greater intensity than spectra of similar octahedral complexes. Thus cis-complexes and states without the center of symmetry also have greater intensity than trans-complexes and other centersymmetric structural positions, and distortions of coordination polyhedra can be accompanied by a relative increase in intensity (though in general, intensities in all octahedral complexes are similar, independent of the distortion).

By mixing transition metal d orbitals with ligands p orbitals (with formation of the molecular orbital), the even character of d orbital is lost. That is why the increase of the bond covalency causes an increase in the intensity of d–d

transitions. Mixing of the states takes place also in the case of proximity of the forbidden transition to the intense allowed transition; it causes the absorption bands "with borrowed intensity" (for example, increase of intensity of the forbidden transitions superimposed on charge transfer band). The weakening of the parity selection rule in center-symmetric complexes is related to the vibrational-electronic (vibronic) coupling, making for temporary destruction of the center of symmetry owing to odd vibration.

Transition moment M_x, M_y, M_z in the octahedral symmetry group transforms as an odd irreducible representation T_{1u}, and so transitions between all even d states are forbidden in this group (see Chap. 2.1). However, as far as vibrations transform in 0_h as $A_{1g}+E_g+T_{2g}+2T_{1u}+T_{2u}$ (that is T_{1u} is contained here); T_{1u} (vibr)$\cdot T_{1u}$ (transition moment)$=A_{1g}$ and the transition becomes allowed (see [68], pp. 128–131).

Spin (Multiplicity) Selection Rule. The electron state of atoms is determined by orbital and spin components. Spin S remains in the atom in crystal from the free atom (so does corresponding multiplicity $2S+1$). Thus, the spin selection rule is common for spectra of the free atom and of the atom in crystal: transitions are allowed between states with the same spin ($\Delta S=0$). For example quartet–quartet transitions $^4A_{2g}\to{}^4T_{2g}$ of Cr^{3+} ion between states with the same multiplicity 4 (spin 3/2) are spin-allowed, but sextet–quartet transitions of Mn^{2+} and Fe^{3+} $^6A_{1g}\to{}^4T_{1g}$ between states with different multiplicity 6 and 4 (spin 5/2 and 3/2) are spin-forbidden. Weakening of this selection rule occurs by spin-orbit coupling.

Selection Rule of the Number of Electrons Involved in Transition: one-electron transitions are allowed. For example, $t_{2g}^3 \to t_{2g}^2 e_g^1$ one-electron transition is allowed, while $t_{2g}^3 \to t_{2g}^1 e_g^2$ two-electron transition is forbidden and its intensity is essentially lower (see below, description of the Cr^{3+} spectra).

Symmetry-Related Selection Rules connected with electronic state transformations in crystal according to the symmetry types (irreducible representations) are discussed earlier in Chapter 2.1. They are common for transitions allowed or forbidden by other selection rules and are related to the peculiarities of the ion state in the crystal field of a point group of symmetry and cause the dichroism of absorption spectra. All absorption bands are dichroic in fields of noncubic symmetry: both charge transfer and crystal field bands.

For example, transitions $A \to E$ are allowed in perpendicular with respect to the three-fold axis orientation while transitions $A \to A$ are allowed in the parallel one.

Selection rules are different for the three types of transitions: electric dipole, magnetic dipole, electric quadrupole. However, for the ions with d configurations, only electric dipole transitions are usually realized and only for them can the molar absorption coefficient take the value of more than one. However, in some cases magnetic dipole transitions are observed (when electric dipole transitions are forbidden) with an intensity below $1-0.1\,\text{sm}^{-1}\cdot 1\cdot\text{mol}^{-1}$. Magnetic dipole transitions occur often in lanthanide and actinide compounds (with f configurations).

Intensities and Width of the Different Types of Absorption Band. The role of the different selection rules is revealed in different aspects of absorption spectra characteristics. Some selection rules are related to the type of transition: parity-allowed are charge transfer and interband transitions with strong intensity

Table 38. Selection rules and intensity of different type absorption bands

Transition characteristics	f	ε	Spectra characteristics
I. Parity forbidden			Crystal field spectra
1. Spin-forbidden	10^{-7}–10^{-8}	0.1–0.01	Linear spectra
2. Spin-allowed	10^{-5}–10^{-4}	10	Band spectra
a) With d–p mixing		10–100	
b) With borrowed intensity	10^{-3}	100	Band spectra
II. Parity-allowed and spin-allowed	10^{-2}–10	10^{-3}–10^{-6}	Charge transfer bonds: very broad with maximum in the UV far, interband transition bands

occurring usually in the UV for most ionic crystals; parity-forbidden are transitions between d electron levels split by crystalline fields.

Other selection rules determine less strong changes of the absorption intensity: spin-allowed transitions in crystal field spectra give rise to a rather intensive wide absorption band (approximately $1000\,\text{cm}^{-1}$), while spin-forbidden transitions lead to narrow, weak bands.

The next selection rules determine the dichroism of absorption bands: both crystal field and charge-transfer bands, both spin-allowed and spin-forbidden bands are dichroic in noncubic crystal fields. Transitions between terms (states of different symmetry type) may be allowed or forbidden, dependent upon the orientation of the symmetry axis with respect to the plane of polarization of incident light.

These types of dependence of absorption intensity on the selection rules are compared in Table 38 (where intensity is expressed by the molar absorption coefficient ε corresponding to the maximum of the absorption band and by the oscillator strength describing the integral absorption taking into account the absorption width).

6.3 Analysis and Experimental Survey of Transition Metal Ions Spectra

Consider the general features of the spectra of ions with $3d^n$ configurations in their usual valence states.

The analysis of spectra given below is connected with the previous description and if necessary one can return to the corresponding sections; some excellent reviews of transition metal optical spectra are also available [52, 198, 204, 225, 235, and see especially 276].

The common order of the spectra analysis: (1) electronic configuration (see Table 6 and Fig. 33), (2) free ion terms (see Table 8), (3) term splitting by crystal fields of different symmetry (see Table 14), (4) selection rules (see Tables 23–27), (5) general features of spectra and colors, (6) spectra of typical compounds interpreted

Fig. 80. Absorption spectra of TiO_2 rutile and anatase (charge transfer spectra)

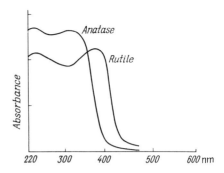

in detail, (7) references to the works on spectra of the given ion in minerals and some inorganic compounds.

Designations of octahedral and tetrahedral coordination will be used for complexes with six and four ligands at the corners of the octahedron and tetrahedron, but it may be they do not correspond to the O_h or T_d symmetry. Local symmetry of an ion is determined by the point group of the ion position in the crystal structure.

Titanium. $3d^0:Ti^{4+}$. In this stable valence state, most usual in minerals, titanium has no d electron and because of this does not give crystal field spectra (i.e., there are no absorption bands in the visible). The charge transfer band for oxygen compounds of Ti^{4+} occurs in near UV, so Ti^{4+} compounds without impurities and color centers are colorless.

In the pure stoichiometric TiO_2, maxima of charge transfer bands are (Fig. 80): in anatase (TiO_2) 3.7 eV (29,850 cm^{-1}, or 335 mm), in rutile (TiO_2) 3.3 eV (26,600 cm^{-1}, or 376 nm) and 3.67 eV (the 3.3 eV band is related to inner transition in Ti^{4+} with the change of configuration: $3d^0 4s^m \rightarrow 3d^1 4s^{m-1}$; the 3.67 eV band corresponds to charge transfer from oxygen to titanium).

In nonstoichiometric TiO_2 (with oxygen vacancies) one observes partial reduction of Ti^{4+} to Ti^{3+} with the formation of electron center ($Ti^{4+} + e^-$). When impurities, especially the more usual impurities of other ions of iron group (Fe^{3+}, Cr^{3+} and others), enter TiO_2, complex centers with Ti^{3+}, Fe^{4+}, Cr^{4+} form. Reversible darkening of rutile exposed to light is connected with these centers. However, the Ti^{3+} band near 500 nm already in the presence of 0.1% of Fe^{3+}, Cr^{3+} in rutile is rapidly overlapped by more intense charge transfer bands or (with the occurrence of impurities) the shifted absorption edge. Additional complications can be connected with possible intervalence transitions ($Ti^{4+}-Ti^{3+}$, $Ti^{3+}-Fe^3$ and others).

$3d^1:Ti^{3+}$. Presence of a single electron over completely filled shells causes the difference of Ti^{3+} (and of other ions with d^1 configuration) from ions with other d^n configurations and its similarity with Cu^{2+} ion (with d^9 configuration, i.e., with the lack of a single electron in completed d^{10} shell):

1. Free ion with $3d^1$ configuration has the single term ^2D since there is here a single electron, then the orbital moment of the ion L is equal to the orbital moment of electron l and since it is the d electron, than $L=l=d=2=D$; spin of the ion S-spin of the electron: $S=s=1/2$; multiplicity $2S+1=2$, term ^2D).

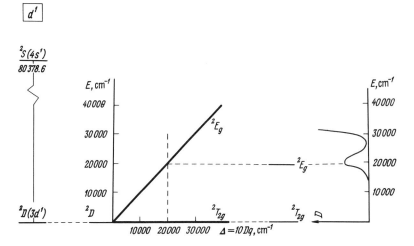

Fig. 81. Energy level diagram for ions with d^1 configuration in octahedral field (Ti^{3+}, V^{4+} and others). The only term ^2D is formed from d^1 configuration, which split up in cubic field into two levels $^2T_{2g}$ and ^2E_g; next excited configuration occurs only at 80378.6 cm^{-1}

2. All levels of an ion with a single nonpaired electron have the same spin, i.e., the same multiplicity, hence this configuration (and only d^1 and d^9) does not give spin-forbidden transitions (between levels with different multiplicity) as in the d^2–d^8 systems, so there are no narrow bands in the spectra of ions with d^1 and d^9 configurations.

3. Because there is in this case only one electron, parameters of the interelectron interaction B and C (the Racah parameters) are absent here; splitting of the ^2D term in crystal fields is determined only by crystal field strength (and also by spin orbital interaction, Jahn-Teller effect, lowering of symmetry).

4. The ^2D term splits into two levels in the cubic crystal field (Fig. 81): $^2T_{2g}$ (lower in octahedral coordination; upper in cube and tetrahedron) and ^2E_g; transition between them has to give only one absorption band, but in tetragonal and lower fields the ^2E_g level splits into two, owing to the lower symmetry; in the cubic and trigonal fields, ^2E_g splits due to the Jahn-Teller effect, which is why a very broad two-humped band is always observed in $3d^1$ ion spectra (sometimes the second transition is not resolved and the splitting is revealed in band broadening or more often in the appearance of a shoulder on the longwave side of the broad absorption band).

5. Since $3d^1$ ion states are one-electron states, their levels in crystal fields are designated both as T_{2g}, E_g, A_g, B_g etc., and as t_{2g}, e_g, a_{1g}, b_{1g} etc., as well as the orbitals corresponding to them: $a_{1g} - d_{x^2-y^2}$, $b_{1g} - d_{z^2}$, $t_{2g} - d_{xy}$, d_{yz}, d_{xz} etc.

Schemes of d^1 ion levels splitting in different crystal fields are given in Figure 82.

In considering the absorption spectra of Ti^{3+} ion (see Figs. 83 and 84), one can distinguish two groups of spectra: (1) spectra of solutions, glasses, chemical compounds, reductive conditions obtained with Ti^3; these spectra can be con-

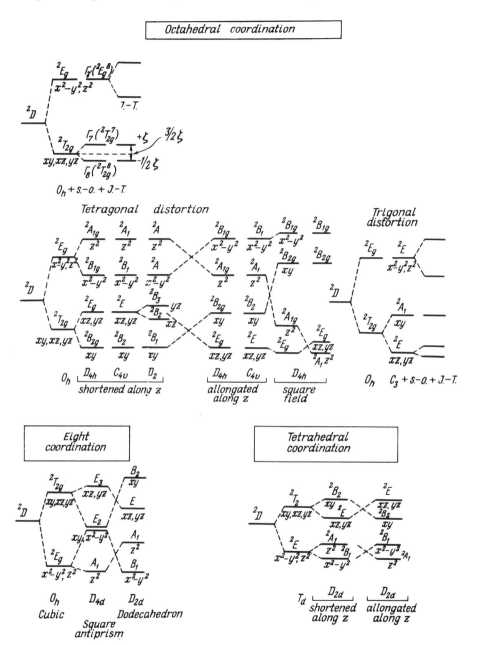

Fig. 82. Splitting of the only ^2D term of d^1 configuration in different coordinations and crystal fields of different symmetry

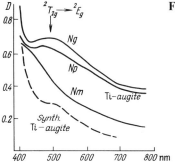

Fig. 83. Absorption spectra of Ti^{3+} in Ti-augite [206]

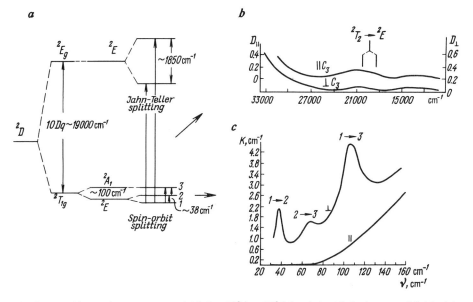

Fig. 84a–c. Absorption spectrum of $Al_2O_3:Ti^{3+}$. **a** Ti^{3+} levels in Al_2O_3 in crystal field with C_3 symmetry and with account of spin-orbit and Jahn-Teller splitting; **b** spectrum of $Al_2O_3:Ti^{3+}$ in the visible and near [256]; **c** spectrum of $Al_2O_3:Ti^{3+}$ in the far IR [256]

sidered as the standard ones revealing the general features of the Ti^{3+} spectrum [263, 295]; (2) spectra of minerals and inorganic compounds containing Ti^{3+} as an impurity (as electron center). Two subgroups of Ti^{3+} spectra can be distinguished among minerals:

a) Spectra of minerals containing iron: Ti-bearing varieties of Fe minerals (Ti garnet, Ti augite etc.) and as Ti minerals nearly always contain iron, it is difficult in these cases to resolve the Ti^{3+} band overlapped by the broad intense charge transfer band as well as by the Fe^{3+} crystal field bands. Spectra of Ti^{3+} have been observed in Ti augite, violet rutile, pink sphene, and benitoite.

In Ti minerals such as ramsayite, astrophyllite, lamprophyllite, neptunite and others, only Fe^{3+} absorption bands are discerned [276].

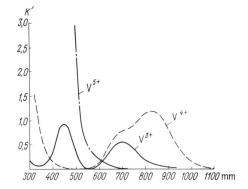

Fig. 85. Absorption spectra of glass 20 K$_2$O. 30 Zn 0.50 P$_2$O$_5$ with V^{5+}, V^{4+}, V^{3+} [296]. k is extinction coefficient per 1% of vanadium for specimen thickness 1 mm

b) Ti^{3+} absorption spectra in minerals where the presence of Ti^{3+} was determined by electron paramagnetic resonance (EPR). By this method the wide occurrence of Ti^3 impurities in minerals was detected.

Like the other d^1-ions (V^{4+}, Y^{2+}, Zr^{3+}, Mo^{5+}, W^{5+}), Ti^{3+} ions occur often in minerals as electron center Ti^{3+} = Ti^{4+} + e^-. Although, owing to the small concentrations (0.n–0.00n%), EPR. is the main method of observation, optical absorption spectra are important for obtaining detailed characteristic of these centers. Only in the octahedral and cubal coordinations must Ti^{3+} absorption be observed in the visible region, while in the tetrahedral coordination it must occur in the near IR region (for Ti^{3+} Dq_{oct} = 17,000–20,000 cm^{-1}, and since Dq_{tetr} = 4/9 Dq_{oct} then Dq_{tetr} ≈ 7500–9000 cm^{-1}).

Vanadium. Vanadium is observed in crystals and glasses [276, 296] in the form of V^{5+}(3d^0), V^{4+}(3d^1), V^{3+}(3d^2), V^{2+}(3d^3) (Fig. 85). Vanadium minerals are represented mainly by vanadates with V^{5+} in the form of tetrahedral complexes VO$_4^{3-}$ and of tetragonal pyramidal or trigonal dipyramidal complexes VO$_5$, as well as distorted octahedra VO$_6$.

Minerals of V^{3+} and V^{4+} are comparatively rare: karelianite V$_2$O$_3$, montroseite VOOH, roscoelite KV$_2$ AlSi$_3$O$_8$ (OH)$_2$, paramontroseite VO$_2$, haradaite SrV(Si$_2$O$_7$), and others.

However, the major part of vanadium in the earth's crust is contained in the form of V^{3+} isomorphously substituting impurity Fe^{3+} and Al^{3+} (in magnetites, pyroxenes, micas, tourmalines, beryls etc). In a number of minerals (beryl, diopside-lavrovite, spodumene-hiddenite, zoisite-tansanite, apophyllite, amblygonite, wavellite and others [276]), it is possible to observe V^{3+} absorption bands; but in most cases of practical importance (V^{3+} in magnetites and others), the simultaneous presence of V^{3+} with Fe^{3+} causes the overlapping of vanadium spectrum by intense absorption bands of Fe. Impurities of V^{4+} (like Ti^{3+}) usually represent an electron center: V^{4+} = V^{5+} + e^- (3d^1).

3d^0, V^{5+}, like Ti^{4+}, has no d electrons and thus does not produce crystal field bands. Spectra of V^{5+} in the case of isolated VO$_4^{3-}$ complexes are interpreted by the molecular orbital theory as charge transfer spectra [276]. However, spectra of most mineral vanadates with V–O$_5$ and V–O$_6$ clusters involved in the complex groups can be interpreted by using the energy band theory (as transitions between the valence band made mainly from 2p oxygen orbitals and conduction band consisted mainly of 3d vanadium orbitals).

$3d^1:V^{4+}$. Diagrams of V^{4+} levels splitting in crystal fields of different symmetry are identical with those for isoelectronic Ti^{3+} ion (see Fig. 81). However, as V^{4+} usually occurs in minerals in the form of vanadile VO^{2+}, its spectra (including charge-transfer bands) are described by molecular orbital diagrams.

$3d^2:V^{3+}$. The only common ion with $3d^2$ configuration is V^{3+}; its spectra are studied in detail both theoretically and experimentally.

$3d^2$ configuration has been considered earlier in this book on two occasions: in illustrating the term derivation of free-ion (see Fig. 25), and as an example in the comparison of strong, medium and weak crystal fields (see Fig. 32), where term splitting in the cubic crystal field is shown.

The Racah parameters of the V^{3+} free ion are $B=862\,cm^{-1}$, $C=3815\,cm^{-1}$, $C/B=4.1$. The energy level diagram of $3d^2$ ion in the octahedral field is shown in Fig. 86. The free ion ground term ^3F splits into $^3T_{1g}$ (ground state), $^3T_{2g}$ and $^3A_{2g}$. Besides, there is one more triplet state $^3T_{1g}$ derived from ^3P. Transitions from the ground state $^3T_{1g}(^3F)$ to $^3T_{2g}(^3F)$ and $^3T_{1g}(^3P)$ produce two intense and broad spin-allowed absorption bands. Transitions to $^3A_{2g}(^3F)$ and higher levels are often overlapped by a charge transfer band. The remaining states of $3d^2$ ions are singlets; transitions to these are spin-forbidden, but are observed due to electron-vibrational interaction: weak linear spectrum arise as a result of transitions to $^1T_2$, $^1E(^1D)$ and $^1A_1(^1G)$ levels which are independent of crystal field strength Dq. Transitions to $^1T_2$ and $^1T_1(^1G)$ levels depending on Dq produce not only weak but also broad bands which are often obscured by neighboring absorption bands.

The energy level diagram and absorption spectrum of $V^{3+}(3d^2)$ are similar to those of $Cr^{3+}(3d^3)$, though these ions are not isoelectronic; having the different number of electrons and hence different multiplicity these ions have the same orbital ground state (^3F and ^4F) and one more state with the same multiplicity (^3P and ^4P) that leads to the three intense (spin-allowed) absorption bands with some weak narrow lines related to absorption due to spin-forbidden transitions.

Similar Dq, B, C values lead not only to the same general features of V^{3+} and Cr^{3+} spectra but also to similar positions of absorption bands, and to the same color of V^{3+} and Cr^{3+} compounds: red (with lower Dq value) or green (with greater Dq). Hence the "alexandrite" color of corundum arises due to the presence of both of Cr^{3+} and V^{3+} impurities (determination of Dq, B, C for V^{3+} see p. 84).

The spectrum of $Al_2O_3:V^{3+}$ (see Fig. 86) has been studied experimentally in polarized light and theoretically the absorption bands [256] are: $17,510\|$ ($f=0.279\cdot10^{-4}$) and $17,420\perp$ ($f=0.360\cdot10^{-4}$) ($^3T_2$), $24,930\|$ ($f=5.61\cdot10^{-4}$) and $25,310\perp$ ($f=1.60\cdot10^{-4}$) ($^3T_1$) and $31,240\,cm^{-1}$ ($^3A_2$). Parameters (249): $10\,Dq=18,000\,cm^{-1}$; $B=610\,cm^{-1}$; $C=2500\,cm^{-1}$; $v=800\,cm^{-1}$; $v'=200\,cm^{-1}$; $\zeta=1500\,cm^{-1}$.

The absorption bands of $V(H_2O)_6^{3+}$ in $NH_4V(SO_4)_2\,12H_2O$: $17,800\,cm^{-1}$ ($^3T_{2g}$) $\varepsilon=3.5\,cm^{-1}\cdot l\cdot mol^{-1}$; oscillator strength $f=0.6\cdot10^{-4}$; band width $\Delta v_{1/2}=3200\,cm^{-1}$; $25,700(^3T_{1g})$; $\varepsilon=6.6$; $f=1.1\cdot10^{-4}$; $\Delta v_{1/2}=3300\,cm^{-1}$; $10Dq=18,600$; $B=665\,cm^{-1}$ [225].

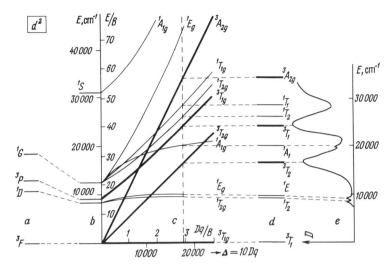

Fig. 86a–e. Energy level diagram for ions with d^2 configuration (V^{3+}) in octahedral field. **a** Free ion terms for V^{3+} ($3d^2$) with $B=862\,\text{cm}^{-1}$, $C=3815\,\text{cm}^{-1}$; **b** the same for $B=610$, $C=2500$, $C/B=4.1$ (in crystal of $Al_2O_3:V^{3+}$) but $Dq=0$; **c** term splitting in octahedral field; **d** V^{3+} levels in Al_2O_3 in cubic approximation ($Dq=1800\,\text{cm}^{-1}$); **e** nonpolarized spectrum of $Al_2O_3:V^{3+}$ [256]

V^{3+} in tetrahedral coordination was observed in synthetic ZnS, CdS [225]; the d^2 energy level diagram in the tetrahedron corresponds to d^8_{oct}, that is to the diagram of Ni^{2+} levels in the octahedron but with $Dq=550\,\text{cm}^{-1}$, $B=456\,\text{cm}^{-1}$ (for ZnS) or $Dq=480\,\text{cm}^{-1}$ (for CdS).

$3d^3:V^{2+}$. The V^{2+} ion is isoelectronic with Cr^{3+} ($3d^3$), hence all energy levels and spectra characteristics described below for Cr^{3+} are completely suited for V^{2+}.

The difference between them lies in the lower crystal field strength Dq for V^{2+} in comparison with Cr^{3+} (Dq is always lower for M^{2+}, than for M^{3+}), hence V^{2+} bands are shifted to the longwave side, in comparison with identical bands of Cr^{3+}. The V^{2+} ion attracted attention in connection with linear fluorescence from the same ^2E levels as Cr^{3+}, but shifted in the IR close to red (11,679.15 cm^{-1} and 11,691.5 cm^{-1}, i.e., 856 nm and 855 nm in $Al_2O_3:V^{2+}$; 11,498.6 cm^{-1} in MgO:V^{2+}) [225]. Absorption bands of V^{2+} are usually overlapped by the bands of V^{3+} presented in greater concentrations (V^{2+} ion is unstable and easily transfers into V^{3+}). In the course of irradiation, the compounds with V^{3+} capture electrons with the formation of $V^{2+}(V^{3+}+e^-)$, which in these cases can be considered as an electron center. Absorption spectra of $V(H_2O)_6^{2+}:A_{2g}\to T_{2g}$ 12,350 cm^{-1} ($\varepsilon=4.1\,\text{cm}^{-1}\cdot l\cdot\text{mol}^{-1}$), $^4T_{1g}$ 18,500 cm^{-1} ($\varepsilon=6$), $^4T_{1g}(^4P)$ 27,900 cm^{-1} ($\varepsilon=3$), $^2T_{1g}$, $^3E_g=13,100\,\text{cm}^{-1}$; $Dq=1235\,\text{cm}^{-1}$, $B=680\,\text{cm}^{-1}$ (225). Absorption spectra and fluorescence of V^{2+} have been studied in detail in MgO and Al_2O_3.

Chromium. $3d^3:Cr^{3+}$. The free ion terms derived from electronic configuration $3d^3$ are: ground term ^4F, excited terms ^4P, ^2P, ^2G, ^2D, ^2H, ^2H, ^2F, i.e., the quartet

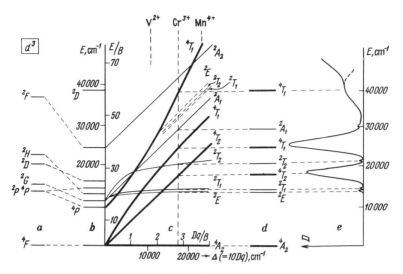

Fig. 87a–e. Energy level diagram for d^3 configuration (Cr^{3+}) in octahedral field. **a** Free ion terms Cr^{3+} ($3d^3$) with $B = 918\,cm^{-1}$, $C = 3600\,cm^{-1}$; **b** the same with $B = 650$, $C = 3210\,cm^{-1}$, $Dq = 0$; **c** splitting of lowest terms ^4F, ^4P, ^2G in octahedral fields (O_h) plotted against crystal field strength $\Delta = 10\,Dq$; **d** Cr^{3+} levels in Al_2O_3; **e** Cr^{3+} spectrum (nonpolarized) in Al_2O_3 (D is optical density)

and doublet terms. Positions of these terms determined from atomic spectra [38] give the best fit with the Racah parameters of the free ion: $B = 918\,cm^{-1}$, $C = 3600\,cm^{-1}$, $C/B = 4$. In the crystal field the terms are compressed (that is expressed in reduction of B and C) and split into levels, corresponding to their transformations according to symmetry types (Fig. 87).

The Tanabe-Sugano diagram (see Fig. 87) shows the aplitting of lower terms ^4F, ^4P, ^2G (and some other doublet terms) in the cubic (octahedral) field as a function of crystal field strength Dq. The other term levels move into the UV region, where the crystal field transitions are overlapped by the charge transfer band. The $^4A_2$, $^4T_2$, $^2A_1$, $^2A_2$ levels exhibit linear dependence on the crystal field strength because each of them is the only level of a given type, while two similar states: $^4T_1$ (^4F) arising from the splitting of ^4F, and $^4T_1$ (^4P) arising from ^4P repel one another and corresponding levels diverge (for the same reason the ^2E, $^2T_1$, $^2T_2$ levels deviate from linear). Only ^2E, $^2T_1$, $^2T_2$ levels (from ^2G) are nearly field-independent.

The ground state in the cubic field is $^4A_{2g}$ (which transforms to $^4A_2$ and ^4A with the lowering of symmetry). Other quartet terms are formed from the ^4F ground term splitting ($^4F \rightarrow {}^4A_{2g}$, $^4T_{2g}$, $^4T_{1g}$) and from ^4P transformation into $^4T_{1g}$. Transitions between $^4A_{2g}$ and these terms (quartet–quartet transitions) are spin-allowed, while transitions between $^4A_{2g}$ and other Cr^{3+} states (quartet–doublet) are spin-forbidden. For determination of Dq, B, C see p. 84) Dq is determined from the position of the first spin-allowed absorption band $^4A_2{}^4T_2$, and B is determined from the position of the second, $^4A_2{}^4T_1$. The obtained values are used for the determination of the remaining absorption line and band positions from the

Tanabe-Sugano diagram (see Fig. 87), or Dq, B and C values can be obtained by fitting the best agreement of all the calculated and experimental energy levels by trial and error using computers.

The general feature of Cr^{3+} spectra is the presence of two (as for all ions with ground F state) intense broad bands in the visible region (arising from the transitions between split ^4F term levels: $^4A_2 \rightarrow {}^4T_2$ and $^4A_2 \rightarrow {}^4T_1$) and the third broad band in the UV region $^4A_2 \rightarrow {}^4T_1(^4P)$, less intensive because it corresponds to a two-electron transition: $^4A_2$ has t_{2g}^3 configuration while $^4T_1(^4P)$ has $t_{2g}^1 e_g^2$, that is two electrons transfer from the state with t_{2g}^3 configuration into the state with $t_{2g}^1 e_g^2$ configuration. These transitions between the states with the same multiplicity (quartet–quartet) are spin-allowed and hence of strong intensity. This band spectrum is accompanied by the linear spectrum representing a series of weak narrow lines occurring from the red part of the visible region to the UV; these are transitions between the state with different multiplicity (quartet–doublet, i.e., from $^4A_2$ to different doublet states ^2E, $^2T_1$, $^2T_2$, $^2A_1$) and hence are spin-forbidden and have small intensity.

The broad intense bands in the visible region cause the color of the Cr-bearing compounds; the Dq and B parameters can be determined from their positions. The first narrow lines in the red region are of great importance because they are related to laser emission and fluorescence of Cr^{3+} ions.

Charge transfer band occurs in the UV region (beginning from approximately 38,000 cm^{-1}, i.e., from 260 nm). With the concentrations of Cr^{3+} near to 1 atom % and more, the spectrum of Cr^{3+} ion pairs is observed in the form of weak narrow lines adjacent to the lines of quartet–doublet transitions.

Cr^{3+} occurs in minerals and inorganic crystals usually in octahedral coordination (but see Cr^{3+} in the tetrahedron [276]). However, the octahedral local symmetry of Cr^{3+} is seldom retained (for example in MgO). The most common for oxides is local trigonal symmetry of Cr^{3+} (in corundum, beryl, spinel and others), for silicates, lower symmetries. The lowering of the local symmetry leads to the further energy level splitting which causes, however, only small displacement of the levels with respect to their position in the cubic field: the $^4A_{2g} \rightarrow {}^4T_{2g}$ transition energy is at 16,000–18,000 cm^{-1}, while the distance between $^4E_2$ and $^4A_2$ states formed from $^4T_{2g}$ in the trigonal field makes the value only 400–600 cm^{-1}. This splitting is revealed in the dichroism of absorption bands according to the selection rules: in one orientation (parallel to the crystal field axis coinciding with C_3 in trigonal symmetry and with C_4 in tetragonal symmetry) a transition can be allowed to one of the split levels, while in an other orientation (perpendicular to the axis), this transition is forbidden, while the other becomes allowed. For example, in trigonal symmetry, the transitions from $^4A_2$ to $^4A_2$ and ^2A levels are allowed in parallel orientation, while the transition to ^4E and ^2E levels is allowed in perpendicular orientation.

Parameters of trigonal splitting (Fig. 88) are designated as v and v' [257] or K and K' [242] or D_σ and D_τ [271].

The relations between them are:

$$v = 3K = 1/3(9D_\sigma + 20D_\tau),$$
$$v' = K' = 2/3(3D_\sigma - 5D_\tau).$$

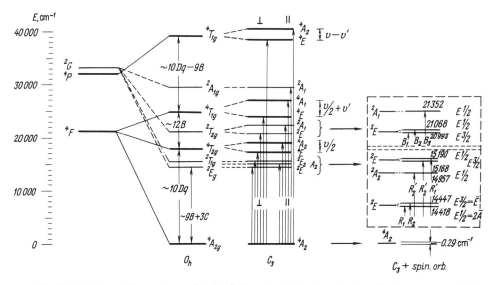

Fig. 88. Splitting of lower levels ^4F, ^4P, ^2G in the fields of octahedral (O_h) and trigonal (C_3) symmetry. Approximate values of splitting are shown expressed in parameters Dq, B, C for cubic field and v, v^1 for trigonal field. *Right*: Action of spin-orbit splitting on the linear spectrum of Cr^{3+} in Al_2O_3

Spin-orbit interaction in the case of Cr^{3+} is small and hardly influences the broad absorption bands, but is shown in further small splitting of narrow lines. The one-electron constant of the spin-orbit interaction for Cr^{3+} free ion is $\xi = 270\,\mathrm{cm}^{-1}$. Hence, the spin-orbit interaction $\lambda = \xi/2S$ (where S is spin of the ion) for quartet terms ^4F ($2S+1=4$, that is $S=3/2$) is equal to $90\,\mathrm{cm}^{-1}$ and for doublet ^2G, ^2P and others ($2S+1=2$, that is $S=1/2$) $\lambda = \xi = 270\,\mathrm{cm}^{-1}$. However, in crystals considerable reduction of this value is observed (similar to reduction of B and C parameters).

Calculations of energy levels of Cr^{3+}-ion are being carried out for the octahedral trigonal and tetragonal fields [292, 242, 271]. Color by trivalent chromium is completely related to the two absorption band $^4A_2 \rightarrow {}^4T_2$ and $^4A_2 \rightarrow {}^4T_1$ positions determined by the Dq and B values. Pink, red and violet colors are observed with greater Dq values, and green with the smaller values. Since Dq is inversely proportional to interatomic distance, the pink, red and violet colors are observed in the cases of Cr^{3+} impurity entering Al octahedra having smaller sizes than Mg, Ca, Cr octahedra with which the green color of Cr-containing compounds is connected.

The Cr^{3+} ion is convenient both for observation by optical absorption spectra, EPR spectra and fluorescence and for calculations of corresponding parameters. It was studied quite extensively for these reasons and as the cause of fluorescence and colors in crystals including the colors of gems: ruby, emerald, spinel, alexandrite, and others. Cr^{3+} ion is characteristic for a number of diamond-accompanying minerals (pyrope, Cr-diopside).

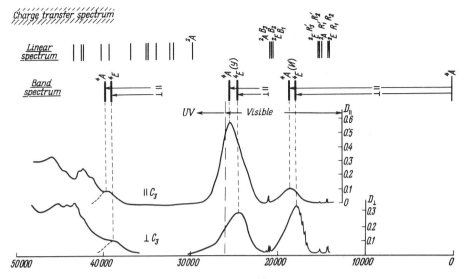

Fig. 89. Absorption spectrum of ruby, $Al_2O_3:Cr^{3+}$ [256]

Hence Cr^{3+} absorption spectra have been studied in a great number of compounds [203, 204, 220, 264, 276, 277, 301, 302, 310, and others]. The absorption spectra studied in most detail are those of ruby $Al_2O_3:Cr^{3+}$ (Fig. 89), emerald and spinel [257, 306, 309].

Manganese. The usual valence state of manganese in minerals is Mn^{2+} ($3d^5$). We consider also Mn^{3+} ($3d^4$) as the representative of $3d^4$ configuration. Mn^{5+} has been observed in dark blue apatite in the form of MnO_4^{3-}; Mn^{4+} is formed after irradiation of apophyllite and kunzite with Mn^{3+} [276].

$3d^4: Mn^{3+}$. The only stable $3d^4$ ions are: Cr^{2+} observed in synthetic ZnS and Mn^{3+} whose spectra have been described in a number of compounds [198, 225, 234, 297]. Mn^{3+} is seldom observed in minerals: in epidote-piemontite, lepidolite, tourmaline, rubellite, mourmanite, eudialyte, purpurite, thulite, red beryl [225, 276].

The ground term ^5D is the only quintet term, the other terms are triplet and singlet. In the octahedral field ^5D splits to the lower ^5E_g and upper $^5T_{2g}$ terms. Splitting of some other terms is shown in Figure 90.

The spectra of Mn^{3+} consist of a broad, often two-humped band due to $^5E_g \rightarrow {^5T_{2g}}$ transition with Jahn-Teller splitting of the $^5T_{2g}$ level or of two dichroic bands in the lower symmetry fields and of seldom-observed, narrow, weak lines due to spin-forbidden quintet–triplet transitions (that is from the ^5E_g ground state to $^3T_{1g}$, ^3E_g and others), while quintet–singlet transitions, still more forbidden because of the change of spin on 2 (in ^5E_g $2S+1=5$, $S=2$, while in $^1T_{2g}$ $2S+1=1$, $S=0$) are not observed in spectra.

As far as $^5E_g \rightarrow {^5T_{2g}}$ intense spin-allowed transition is concerned (or other transitions formed from it in the fields of lower symmetry), features of Mn^{3+}-spectra are quite similar to those of Cu^{2+} spectra ($3d^9$; $^2E_g \rightarrow {^2T_{2g}}$ but without

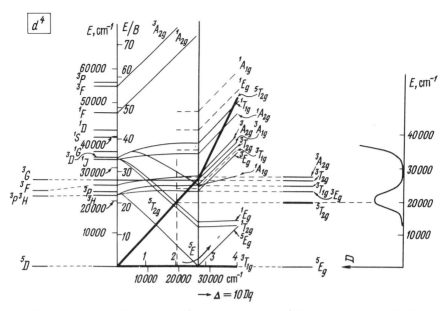

Fig. 90. Energy level diagram for d^4 configuration (Mn^{3+}) in octahedral field; $C/B = 4,6$. *Right:* Absorption spectrum of $Al_2O_3 : Mn^{3+}$ ($Dq = 1950\, cm^{-1}$, $B = 965\, cm^{-1}$, $C/B = 4.66$)

linear spectra, such as Ti^{3+}), but the bands are shifted to the higher energies because the Dq value of Mn^{3+} is greater than Dq of Cu^{2+} (as Dq of all $M^{3+} > M^{2+}$). The energy level diagram of d^4 ions (see Fig. 90) is divided (by vertical line with $Dq/B = 2.7$) into two diagrams: (a) the left is for the weak fields and correspondingly for high-spin state, i.e., with $t_{2g}^3 e_g^1$ configuration with spin $S = 2$ and ground state ^5E_g; (b) the right is for the strong fields and low-spin state, with t_{2g}^4 configuration (with a pair of paired electrons) with spin $S = 1$ and ground state $^3T_{1g}$.

Unlike the energy level diagrams of d^1, d^2 and d^3 configurations where all levels have a positive slope, here some triplet and singlet levels have negative slopes and with further increase of the crystal field strength, the $^3T_{1g}$ level becomes the ground state (see also description of Fe^{2+} spectra). The low-spin state of d^4 ions (the right side of the diagrams of Fig. 90) correspond to absorption spectrum quite different from that of high-spin state (the left side of the diagram): five partially overlapped triplet–triplet transitions arise in the comparatively narrow violet – UV region.

Examples of Mn^{3+} spectra in the high-spin state may be the following: (1) $CsMn(SO_4)_2\, 12H_2O$ [297] with a broad band $^5E_g \to {}^5T_{2g}\, 21{,}000\, cm^{-1}$ (width $\Delta\nu_{1/2} = 5000\, cm^{-1}$, $\varepsilon = 5\, cm^{-1} \cdot 1 \cdot mol^{-1}$; $f = 1.1 \cdot 10^{-4}$); (2) $Al_2O_3 : Mn^{3+}$ [256] with $^5T_{2g}$ level split in trigonal field to lower $^5A_1$ and upper ^5E, the $^5E \to {}^5A_1$ transition is allowed in perpendicular orientation ($18{,}750\, cm^{-1}$, $f = 2.67 \cdot 10^{-4}$) while the $^5E \to {}^5E$ transition is allowed both in perpendicular and parallel orientations ($20{,}600\, cm^{-1}$, $f = 1.77 \cdot 10^{-4}$), $Dq = 1947\, cm^{-1}$ (Fig. 91); (3) Mn^{3+} spectra have also been observed in many solutions and glasses.

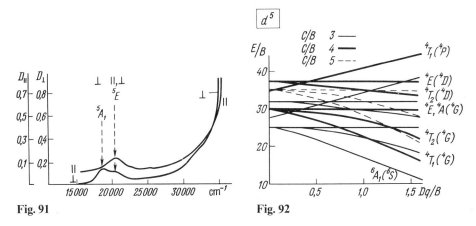

Fig. 91. Polarized absorption spectra of Mn^{3+} in Al_2O_3 [256]
Fig. 92. Lower quartet levels in d^5 configuration for $C/B = 3, 4, 5$

The splitting of $d^3 =$ ions levels in the fields of different symmetries is similar to that of d^1 but with an inverse order of all levels.

$3d^5$: Mn^{2+}. Electronic configuration $3d^5$ corresponding to a half-filled d shell is particularly stable, and the most stable ions, such as Mn^{2+} and Fe^{3+} belong to it.

In crystalline fields the usual high-spin state is common i.e., $t_{2g}^3 e_g^2$ configuration with one unpaired electron in each of the orbitals and with the spin S equal to 5/2 (multiplicity $2S+1=6$). The low-spin state, which was not observed for Mn^{2+} and is less common for Fe^{3+}, has the t_{2g}^5 configuration with two pairs of paired and one unpaired electron and with spin 1/2.

The behavior of d^5 ion energy levels in the crystal field is characterized by the following features.

1. The ground state of the d^5 ions is ^6S; this term transforms into $^6A_{1g}(^6A_1)$ orbit singlet state in the fields of any symmetry and is the only sextet term. Unlike all other d^1–d^4 and d^6–d^9 configurations, where the most intense (spin-allowed) transitions occur between the levels of the ground state split by the crystal field, the ^6S term does not split and is the only one of a given multiplicity. Hence all transitions in d^5 ions are spin-forbidden; their intensities are very small and Mn^{2+} compounds have pale colors.

2. Excited states of d^5 ions are quartet (^4G, ^4F, ^4D, ^4P) and doublet (^2I, ^2H, ^2G, ^2F, ^2D, ^2P, ^2S) terms. Transitions from the ground sextet $^6S(A_{1g})$ into doublet terms are still more forbidden because the spin number is changed with it by two, hence they are practically never detected in Fe^3 spectra and rarely in Mn^{2+} spectra. Thus practically only sextet–quartet transitions are observed.

They can be divided into: transitions to levels dependent on crystal field strength Dq (represented by sloping lines in Fig. 92) giving broad bands (usually $^6A_{1g} \rightarrow {}^4T_{1g}$ and $^6A_{1g} \rightarrow {}^4T_{2g}$) and transitions to Dq-independent levels (plotted as horizontal or almost horizontal straight lines in Fig. 92) which give rather sharp bands [$^6A_1 \rightarrow {}^4A_1 + {}^4E$; $^6A_{1g} \rightarrow {}^4E(^4D)$ and others].

Energy level diagrams of d^5 ions are available in two variants: for quartet levels only, and for quartet and doublet levels showing a much more complicated scheme of the energy levels [237]. However, in most cases, it is sufficient to use the variant with quartet levels only.

3. The behavior of the $^6S(^6A_{1g})$ level unsplit in crystal fields is similar in octahedral and tetrahedral coordinations. Unlike all other electron configurations with tetrahedral level schemes which are the inverse of the octahedral for the same ion, i.e., $d^4_{tetr} = d^{10-n}_{oct}$ (see Chap. 2.3), the energy level diagrams here are the same for octahedral, tetrahedral and cubic coordination ($d^5_{oct} = d^5_{tetr}$), being distinguished only by the order of magnitude of crystal field strength ($Dq_{tetr} = 4/9 Dq_{oct}$; $Dq_{cub} = 8/9 Dq_{oct}$).

Moreover, the energy level diagrams of d^5 ions have similar values of Dq, B, C for Mn^{2+} and Fe^{3+} ions; the Racah parameters for Mn^{2+} in crystals are: $B = 570-790\,cm^{-1}$, $C = 3200-3770\,cm^{-1}$, $C/B = 4-6.5$ and B-, C-values for Fe^{3+} are similar to these ranges. These ranges show, however, that a set of diagrams with different C/B ratios both for Mn^{3+} and Fe^{3+} is necessary in different crystals.

Hence in Fig. 92 the energy levels are plotted for the three C/B ratios equal to 3, 4 and 5; by interpolation, it is possible to obtain from them the level positions for intermediate C/B values. Only parallel displacement of all levels occurs, and for the determination of their position, only the diagram calibration is necessary, which can be made conveniently by using the transitions: $^6A_{1g} \rightarrow {}^4A_{1g} + {}^4E_g$ and $^6A_{1g} \rightarrow {}^4E_g(^4D)$ which are Dq-independent and give sharp absorption bands easily observed in the spectra. For estimation of Dq, B and C values see p. 84.

It should be noted that B and C values are assumed differently in various calculations already for the free ion: besides the values $B = 860$ and $C = 3853\,cm^{-1}$ by Tanabe-Sugano, the values 786 and $3790\,cm^{-1}$, 950 and $3280\,cm^{-1}$, 910 and $3270\,cm^{-1}$ can also be taken [74]. Probably it is expedient to take the Tanabe-Sugano values (because their values correspond to that determined from the ^4G and ^4D terms from which $^4A_{1g}$, $^4E_g(^4G)$ and $^4E_g(^4D)$ are derived in the crystal field and used for the determination of B and C in crystals.

Thus the free ion B and C values determined from the energies of free ion levels are [38]:

$$\left. \begin{array}{l} {}^4G\,10B + 5C = 26{,}325\,cm^{-1} \\ {}^4D\,17B + 5C = 32{,}308\,cm^{-1} \end{array} \right\} \quad B = 855\,cm^{-1}$$

$$\left. \begin{array}{l} 4{}^4P\,7B + 7C = 29{,}168\,cm^{-1} \\ {}^4F\,22B + 7C = 43{,}574\,cm^{-1} \end{array} \right\} \quad B = 960\,cm^{-1}.$$

The same discrepancies occur in the estimation of B and C in crystals. The determination from the $^4A_{1g}$, $^4E_g(^4G)$ and $^4E_g(^4D)$ level positions lead to results different than the determination of B from the $^4T_{1g}$ and $^4T_{2g}(^4G)$ level separations or determination by computer fitting of all parameters to best agreement between all observed and calculated transitions. However, the Dq value is strongly dependent on B and C values, both if determined from equations for $^4T_{1g}$, $^4T_{2g}(^4G)$, where each gives different Dq values, and from the energy level diagram.

Spectra of Mn^{2+} (Fig. 93) have been interpreted for many minerals [201, 228, 239, 258, 276, 291, especially 237].

Fig. 93. Absorption spectra of Mn^{2+} in minerals [237]

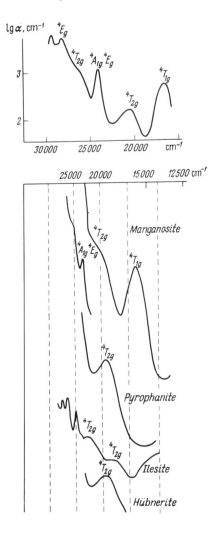

Iron. The significance of optical absorption spectra of iron is determined first of all by the fact that its ions are the most important chromophores in minerals. However, besides this usual presence of iron, which isomorphously substitutes other transition metal ions in minerals leads to superimposition of the Ti^{3+}, V^{3+}, Cr^{3+}, Mn^{3+}, Mn^{2+}, Ni^{2+} spectra on the iron spectra; hence it is often necessary in investigating the spectra of these ions to separate them first from the spectrum of iron.

Analysis of optical iron spectra is complicated by a number of difficulties in comparison with other transition metal ions spectra: (1) all crystal field bands of Fe^{3+} and Fe^{2+}, except one Fe^{2+} transition, are spin-forbidden and hence are of small intensity; (2) these bands are superimposed on very intensive (especially in case of Fe^{3+}) charge transfer band; (3) often Fe^{3+} and Fe^{2+} ions are presented simultaneously and their similar forbidden bands occur in the same region of spectrum and are overlapped; (4) in a number of cases, ferrous–ferric interactions, as well as the occurrence of these cations with other transition metal ions (Ti^{4+},

Cr^{3+} and others) lead to so-called intervalence transfer absorptions (see Chap. 6.2).

In silicate minerals local symmetry deviations from the octahedral lead to large splittings and make it impossible to restrict interpretation by cubic crystal field approximation. The occurrence of the cations in several sites in a structure, each of them giving distinct spectrum, can lead to further complications.

It is useful in analyzing the optical absorption spectra to correlate them with Mössbauer spectra to distinguish Fe^{3+} and Fe^{2+}, to determine the existence of each of these ions in nonequivalent positions, and to estimate by quadrupole splitting value the deviation of the crystal field from cubic symmetry.

$3d^5 : Fe^{3+}$. The Fe^{3+} ion is isoelectronic with Mn^{2+}, hence descriptions of general features of their behavior in crystal fields are completely common for both ions.

However, Fe^{3+} spectra differ greatly from spectra of Mn^{2+}; colors of Fe^{3+} compounds are intensely brown already for small concentrations of ferric iron, while colors of Mn compounds are pink, even with great concentration. This difference is connected with the fact that weak crystal field bands in the case of Fe^{3+} are superimposed on the tail of the intense charge transfer band, while in the case of Mn^{2+} the charge transfer band occurs in the UV region. The influence of the charge transfer band is of two kinds: on the one hand interaction with it increases the intensity of the superimposed forbidden transition, on the other hand the charge transfer band intensity is so great that already in the violet and all the more in the near UV, it overlaps in most cases weak crystal field bands. Hence, only two or three transitions are observed usually in spectra of Fe^{3+} compounds: first two broad bands $^6A_{1g} \to {}^4T_{1g}$ and $^6A_{1g} \to {}^4T_{2g}$ and the sharp band $^6A_{1g} \to {}^4E_g + {}^4E_{1g}({}^4G)$. Band $^6A_{1g} \to E_g({}^4D)$, observed in most Mn^{2+} compounds, and of importance for the determination of B and C parameters, occurs rather seldom in Fe^{3+} spectra.

The charge transfer band of Fe^{3+} differs strongly from crystal field bands: its maximum occurs in the far UV region (about 200 nm), its intensity is two or three orders greater than of crystal field bands (absorption coefficients 6000–7000 cm^{-1} in comparison with 10 cm^{-1} for spin-allowed and 0.1–0.5 cm^{-1} for spin-forbidden crystal field transitions).

Only the tail of the charge transfer band occurs in the visible region. Due to its great intensity, the charge transfer band tail occurs already at the Fe^{3+} concentration of about 0.5% in the violet part of the visible region. With the increase of the Fe^{3+} concentration, this band rapidly covers all the visible region and reaches the near IR.

In rock-forming silicates, the charge-transfer band represents the main factor determining (together with the Fe^{2+}-allowed transition) the color and pleochroism of these minerals. The mechanism of the formation of this band differs from that of the crystal field band. It can be considered as the transfer of the electron from the oxygen to the iron cation. Its interpretation can be obtained by use of the molecular orbital diagrams (see Chap. 3.2 and Fig. 48). The quartet energy levels diagram of d^5 configuration computed for different B and C values typical for Fe^{3+} [260] is shown in Fig. 94. For the complete diagram both with quartet and doublet levels see [215, 237].

Consideration of Fe^{3+} spectra in minerals shows that the number of absorption bands and general features of the spectrum, as well as the intensity of the same

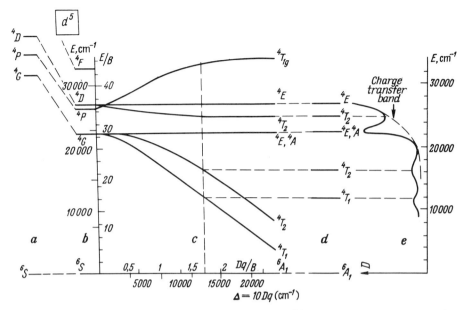

Fig. 94a–e. Energy level diagram for d^5 configuration (Fe^{3+}); quartet levels only. **a** Free ion terms for Fe^{3+} ($B=992\,cm^{-1}$, $C=4430\,cm^{-1}$; $C/B=4.46$); **b** the same for $B=760\,cm^{-1}$, $C=3220\,cm^{-1}$, $C/B=3.24$ but $Dq=0$; **c** energy level diagram for d^5 configuration; for $C/B=4.24$; **d** position of levels for $Dq=1300\,cm^{-1}$; **e** spectrum of garnet-demantoid $Ca_3Fe_2(SiO_4)_3$

transition in the spectra of different crystals are primarily dependent on the Fe^{3+} concentration (Fig. 95).

The moving of the charge transfer band tail into the visible region leads to a sharp change in the crystal field spectra: absorption intensity increases in that transition which is reached by the tail of the charge-transfer band. However, further increase of the charge-transfer band leads to overlapping one by one (beginning from the UV-violet) of the crystal field bands.

Hence, with high Fe^{3+} concentrations and in specimens that are not too thin, only one $^4T_{1g}$ band in the near infrared can be observed with much greater intensity and superimposed on the charge-transfer band moving rapidly in the UV (see Fig. 95). In very thin specimens (about 10μ) it is possible to observe two bands: $^4T_{1g}$ and $^4T_{2g}$ even in ferric oxides and hydroxides.

In most silicate minerals with Fe^{3+}, in addition to the $^4T_{1g}$ and $^4T_{2g}$ bands, the narrow band $^4E_g+^4A_{1g}$ is observed in the form of a shoulder on the more or less steep charge transfer band (Fig. 96).

With a small concentration of Fe^{3+}, the charge transfer band does not reach frequencies corresponding to $^4T_{1g}$ and $^4T_{2g}$ transitions, hence the latter are exhibited weakly or even not observed at all. However, then there is a series of narrow bands $^4E_g+^4A_{1g}$, $^4T_{2g}(D)$, $^4E_g(^4D)$, but with still lower concentrations (for example, 0.005% of Fe^{3+} in Al_2O_3 in Fig. 96) even $^4A_{2g}(F)$ 38,500 cm^{-1} band is observed, while all other less intense bands are not observed. This band can be used for the determination of small Fe^{3+} concentrations equal (in wt-%) to $0.02K$ (cm^{-1}) at $\lambda=260\,nm$ [276, 229].

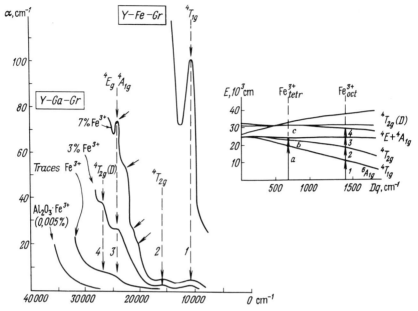

Fig. 95. Variation of Fe^{3+} spectra with the Fe^{3+} concentration in Y–Fe garnet ($Y_3Fe_2Fe_3O_{12}$), Y–Ga garnet ($Y_3Ga_2Ga_3O_{12}$) with Ga^{3+} substitution in octahedral site by Fe^{3+} (7 at % Fe^{3+}, 3 at % Fe^{3+} and traces of Fe^{3+}) and in $Al_2O_3:Fe^{3+}$ with 0.005 at % Fe^{3+} [303]. Absorption bands of Fe^{3+} in tetrahedral coordination are indicated by slanting pointers (in Y–Ga–garnet Fe^{3+} enters in a grater amounts in octahedral sites than in tetrahedral ones, hence these bands are weak here). Fe^{3}_{tet} ($Dq = 690\, cm^{-1}$, a, 24,000; b, 23,700; c, 22,500, 21,200, 20,200 cm^{-1}); Fe^{3+}_{oct}: $Dq = 1450\, cm^{-1}$; 1, 2700; 2, 24,000; 3, 16,000; 4, 11,000 cm^{-1}

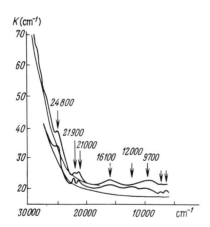

Fig. 96. Absorption spectrum of Fe^{3+} in epidote

Three types of transition in Fe^{3+} spectrum can be distinguished:

1. Narrow, relatively intense bands corresponding to the Dq-independent transitions (horizontal lines in the energy level diagram, see Fig. 94): one observes usually the $^4E_g + {}^4A_{1g}$ band while the $^4E_g(D)$ and $^4A_2(F)$ are seldom observed (for example, in corundum).

2. Two weak, broad bands: T_{1g} 11,000–12,000 cm^{-1} (900–800 nm) and $^4T_{2g}$ 16,000–17,000 cm^{-1} (625–580 nm); their positions are dependent on crystal field strength.

3. Weak bands of doublet transitions: $^2T_2(I)$ and $^2T_1 + ^2A_1(I)$; these transitions are twice forbidden: spin-forbidden and as two-electron transitions (i.e., with the excitation of two electrons from $t_{2g}^3 e_g^2$ configuration of the ground state $^6A_1$ to t_{2g}^5 configuration of $^2T_2$, $^2T_1$, $^2A_1$ and other states), therefore these transitions are weak and are not usually exhibited in spectra of Fe^{3+}. However, they are clearly visible in Mn^{2+} spectra. In Fe^{3+} spectra of some compounds there are absorption bands 14,000–15,000 cm^{-1} (710–660 nm) and 20,000 cm^{-1} (500 nm) which cannot be assigned to $^4T_1$ and $^4T_2$ and coincide with the positions of $^2T_2(I)$ and $^2T_1{}^2A_1(I)$ levels in the energy level diagram. Probably the mixing with transitions close to them and the interaction with the charge-transfer bands lead to the fact that these transitions are exhibited in the absorption spectra of some compounds. The assignment of the Fe^{3+} absorption bands in different compounds can be correlated by taking into account the following relations:

1. Crystal field strength Dq inversely proportional to interatomic distances increases in the series (Fe^{3+}, Mg)O_6 (2.10–2.20 Å)–$Fe^{3+}O_6$ (2.00 Å)–(Fe^{3+}, Al)O_6 (1.91–1.95 Å).

2. In the same series, covalency is increased and hence the B-value (parameter of interelectronic repulsion) is reduced;

3. These relations are revealed in spectra in different ways: with increasing Dq (see Fig. 94), the energies of Dq-dependent levels decrease (in contrast to levels of most other ions), that is $^4T_{1g}$ and $^4T_{2g}$ bands are shifted to the smaller energies with increasing Dq (because of shorter interatomic distances); reduction of B leads to an increase of C/B ratio (because the C values vary relatively little), that is, lead to the shift of all levels up the energy scale (see Fig. 92): energies of one and the same level are increased with it since the E/B ratio increases more rapidly than E decreases, due to B reduction.

4. The energy difference of some transitions can be determined approximately from following expressions [68]:

$$t_2^3 e^2 : {}^6A_1 = 0,$$
$$^4A_1{}^4E = 10B + 5C,$$
$$^4T_2(D) = 13B + 5C + \ldots,$$
$$^4E(D) = 17B + 5C,$$
$$^4T_1 = 18B + 7C - \ldots,$$
$$^4A_2(F) = 22B + 7C;$$
$$t_2^4 e^1 \; ^4T_1 = -10Dq + 10B + 6C - 26B^2/10Dq,$$
$$^4T_2 = 10Dq + 18B + 6C - 38B^2/10Dq,$$
$$t_2^5 \; ^2T_2 = -20Dq + 15B + 10C - 140B^2/10Dq$$

(the levels energy including $B^2/10Dq$ are given for $C/B = 4$).

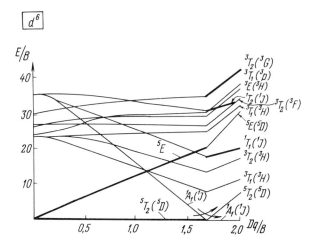

Fig. 97. Energy level diagram for d^6 configuration in the octahedral field for $C/B=4$

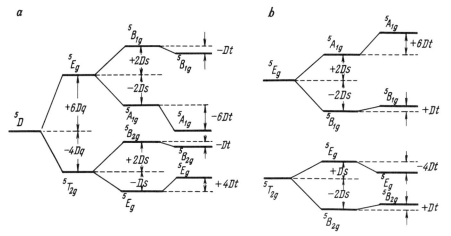

Fig. 98a and b. Splitting of spin-allowed transition $^5T_{2g} \rightarrow {}^5E_g$ (for Fe^{2+} in tetragonal distorted octahedra). **a** Elongated octahedron ($E_g = 4Dq - Ds + 4Dt$, $B_{2g} = -4Dq + 2Ds - Dt$, $A_{1g} = 6Dq - 2Ds - 6Dt$, $B_{1g} = 6Dq + 2Ds - Dt$). **b** Shortened octahedron ($B_{2g} = -4Dq - 2Ds + Dt$, $E_g = 4Dq + Ds - 4Dt$, $B_{1g} = 6Dq - 2Ds + Dt$, $A_{1g} = 6Dq + 2Ds + 6Dt$)

One can see from these equations that the distance between ^4A, ^4E and $^4E(D)$ is equal to $7B$ and that between $^4T_1$ and $^4T_2$ is approximately $8B$. Absorption spectra of Fe^{3+} in tetrahedral coordination have been observed in ferriferous orthoclase and in silicate glasses. As mentioned above (see Mn^{2+}), the energy level diagram for the d^5 configuration is the same for the octahedral and tetrahedral coordination. Even the position of the $^4E + {}^4A_1$ transition in Fe-orthoclase is close to the position of the corresponding transition in Fe^{3+} spectra in octahedral coordination. However, the parameters of the spectra are essentially different (reduced B, increased C/B and $Dq_{tetr} = 4/9 Dq_{oct}$).

$3d^6$: Fe^{2+}. The free ion ground term is ^5D (the only quintet term), excited terms are triplet ^3H, ^3P, ^3F, ^3G, ^3D and singlet ^1I, ^1D.

In the octahedral field the ^5D term splits in the ground $^5T_{2g}$ level and excited ^5E_g (Fig. 97). The $^5T_{2g} \to {}^5E_g$ transition is the only allowed transition in the octahedral field, the related broad intense absorption band ($\varepsilon = 1.6\,\text{cm}^{-1} \cdot \text{l} \cdot \text{mol}^{-1}$, oscillator strength $f = 0.4 \cdot 10^{-4}$ for $FeSO_4 7H_2O$) at the border of the red and IR is the only intense band in spectra of Fe^{2+} compounds.

The ^5E_g state is subject to the strong influence of the Jahn-Teller effect, that causes the splitting of ^5E_g due to the inner configuration instability of this state. However, the presence of the Jahn-Teller effect can be claimed only in the case of cubic local symmetry, since its action with the lowering of symmetry is inseparable from splitting due to noncubic components of the crystal field.

According to the local symmetry, the following cases of ^5E state behavior ($^5T_{2g} \to {}^5E_g$ transition) can be distinguished.

1. Cubic local symmetry (O_h): the splitting of an absorption band is not observed; the Jahn-Teller effect is shown only as moderate broadening of the absorption band (1800–2200 cm^{-1}).

2. Trigonal local symmetry: (a) ^5E_g level is not split in the trigonal field; (b) the trigonal field reduces the action of the Jahn-Teller effect upon other levels; as a result the $^5T_{2g} \to {}^5E_g$ band is not split in the trigonal field and its width is reduced to 1400–1800 cm^{-1}.

3. Tetragonal local symmetry (see Fig. 98) and cases of lower symmetry with predominance of the tetragonal (and not trigonal) component of the crystal field: (a) ^5E_g level split in two levels in the tetragonal field: $^5B_{2g}$ and $^5B_{1g}$ (or others in dependence on a symmetry point group), (b) the tetragonal field and the Jahn-Teller effect mutually increase their action, (c) further lowering of symmetry does not lead to further splitting, but gives rise only to the transformation of sublevels according to different types of symmetry; this results in the change of the selection rules, but the maximum number of absorption bands related to allowed transitions to the split ^5E_g level is no more than two.

With small tetragonal and lower distortions this is shown as a shoulder of the longwave (IR) side of the absorption band. With large distortions, larger or smaller splitting is observed: to ~ 4000 or even to $6000\,\text{cm}^{-1}$ between maxima of the bands.

With the usual values of crystal field strength Dq (about 1000 cm^{-1}), and when splitting is absent, the Fe^{2+} absorption band occurs in the near IR region. Because of this, oxygen and haloid compounds of Fe^{2+} with cubic and trigonal symmetry are colorless (for example siderite $FeCO_3$). In tetragonal and lower local symmetry, one of the split bands shifts into the visible (red) region that causes the green color of the compounds.

Transitions between the levels derived from the ground term ^5D (i.e., $^5T_{2g}$ and ^5E_g in cubic symmetry and quintet terms formed from them in the fields of lower symmetry), are determined only by crystal field strength and are independent of B and C parameters of interelectronic repulsion, because these are transitions between states of the same multiplicity. Hence, positions of $^5T_{2g}$ and ^5E_g states in the energy level diagrams remain identical in all variants of these diagrams with different C/B ratios.

In addition to the broad intense $^5T_{2g} \to {}^5E_g$ band (or to two intense bands formed when the symmetry is reduced), many more weak spin-forbidden tran-

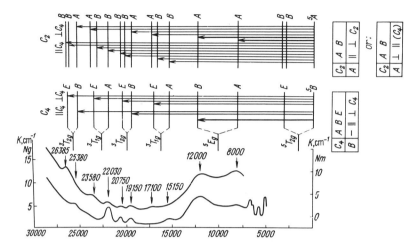

Fig. 99. Absorption spectra of colorless vivianite $Fe_3(PO_4)_2 \cdot 8H_2O$ in platelet (1.17 mm thickness) parallel to (010) from Cameroun. Two pseudotetragonal sites of $Fe^{2+}:Fe_1^{2+}-O_2(H_2O)_4$ – shortened along the C_4 octahedron (Fe–O = 1.97, Fe–H_2O = 2.00 Å), local symmetry C_2; $Fe_2^{2+}-O_4(H_2O)_2$ shortened along the C_4 octahedron (Fe–O = 0.95, Fe–H_2O = 2.00 Å), local symmetry C_2. The C_2 axes of both sites coincide with $\perp(010)$; the C_4 axes in plane (010) are approximately perpendicular to each other; $Fe_1:Fe_2 = 1:2$. Splitting of the $^5T_{2g} \to {}^5E_g$ absorption band into two bands $^5A(^5T_{2g}) \to {}^5T(^5E_g)$ with the separation 4000 cm^{-1} can be distinctly seen. Splitting of forbidden transition bands is also distinctly observed $Dq = 970$ cm^{-1}; $B = 1110$ cm^{-1}; $C = 440$ cm^{-1}

sitions from the ground quintet $^5T_{2g}$ level to triplet levels occur in Fe^{2+} spectra. It seems that the transitions to singlet levels have not been observed, as being twice forbidden (by spin and as two-electron transitions). The positions of these triplet and singlet levels depend not only on Dq but also on B and C parameters and vary with different C/B ratios.

The Tanabe-Sugano diagram for d^6 configuration has been computed for Co^{3+} ion isoelectronic with Fe^{2+} [74]. The system of levels for both ions remain the same, but in the case of Co^{3+}, the crystal field strength is greater than in the case of Fe^{2+} (Dq for $M^{3+} > Dq$ for M^{2+}) and it leads to the fact that the singlet level $^1A_{1g}(^1D)$ becomes the ground state corresponding to the strong field electronic configuration.

This is why in the diagram for Co^{3+}, mainly singlet levels are shown, which give rise in this case to the allowed transitions, while some triplet levels important for Fe^{2+} are not shown (in order not to complicate the diagram).

Hence the diagram shown in Fig. 97 is constructed specially for Fe^{2+} [260]. The strong field case in Fe^{2+} compounds is observed only in some sulfides (by Mössbauer spectra of pyrite and marcasite) but is not realized in transparent minerals.

Trigonal (and pseudotrigonal) distortion of Fe^{2+} octahedra causes the appearance of a shoulder on the absorption band with the splitting about 1600–1800 cm^{-1}. In the case of tetragonal distortion (with possible superimposed rhombic or lower field) the splitting is increased to 4000 cm^{-1} in viviante and to 3670 cm^{-1} in FeF$_2$. Compare as an example FeCl$_2$ spectra (trigonal crystal field) with unsplit $^5T_{2g} \to {}^5E_g$ band and FeF$_2$ (rutile structure; mainly tetragonal field with rhombic distortion) with two bands. $^5B_{1g} \to {}^5B_{1g}$ (6990 cm^{-1}) and $^5B_{1g} \to {}^5A_{1g}$ (10,660 cm^{-1}).

For tetragonal distortion one distinguishes the case of the elongated octahedron when in splitting mainly the infrared component is shifted, and the case of the shortened octahedron, when splitting causes the shift of more a high-energy component to the visible region (see Fig. 98).

Forbidden quintet–triplet transitions are observed in Fe^{2+} as very weak bands and not in all compounds. These transitions have been assigned in the spectra of siderite and vivianite (Fig. 99).

Energy levels values in cubic field approximation are [68]:

$$t_2^4 e^2 \;{}^5T_2 = -4Dq$$

$$t_2^3 e^3 \;{}^5E = +6Dq$$

$$t_2^5 e^1 \;{}^3T_1 = -14Dq + 5B + 5C - 70B^2/10Dq$$

$$ {}^3T_2 = -14Dq + 13B + 5C - 106B^2/10Dq$$

$$t_2^5 e^1 \;{}^1T_1 = -14Dq + 5B + 7C - 34B^2/10Dq$$

$$ {}^1T_2 = -14Dq + 21B + 7C - 118B^2/10Dq$$

$$t_2^6 \;{}^1A = -24Dq + 5B + 8C - 130B^2/10Dq.$$

Fe^{2+} in Tetrahedral Coordination. Since in the tetrahedral field the order of levels is inverse to the octahedral one, the ground state of Fe$_4^{2+}$ is ^5E while $^5T_2$ is an excited state. In Fe$_4^{2+}$ spectra, as in Fe$_6^{2+}$ spectra, one allowed transition (with the absence of distortion) and some weak forbidden quintet–triplet transitions occur. Since $Dq_{\text{tetr}} = 4/9 Dq_{\text{oct}}$, then this transition (independent of B and C and only Dq-dependent as in Fe$_6^{2+}$) occurs in the IR region ($10Dq_{\text{tetr}}$ for Fe$_4^{2+}$ is about 3500–4500 cm^{-1}, i.e., corresponds to the position of the $^5E \to {}^5T_2$ band near $3 \div 2\,\mu$). The ^5E level in tetrahedral coordination, as distinct from ^2E_g in case of Cu^{2+} in octahedral coordination, is not subject to the Jahn-Teller effect.

Spin-orbit interaction is not great (λ is less than 100 cm^{-1} in crystals) and splits in the first order only T states and T and E states in second order with total splitting at about 50–60 cm^{-1}.

Intensity of $^5E \to {}^5T_2$ transition in Fe$_4^{2+}$ spectra is essentially stronger than intensity of $^5T_{2g} \to {}^5E_g$ in Fe$_6^{2+}$ spectra, due to the partial removal of parity forbiddenness; more intensive are also the spin-forbidden quintet–triplet transitions.

One of the most extensively studied absorption spectra is ZnS:Fe^{2+}-spectrum where Fe^{2+} occurs in the field with T_d symmetry. In spectra of different samples of natural sphalerites, Fe^{2+}, Co^{2+}, Ni^{2+} bands have been observed (Fig. 100). The complete Fe^{2+} spectrum in sphaletite, including both $^5E \to {}^5T_2$ transition in the infrared and forbidden transitions observed in natural crystals of good quality

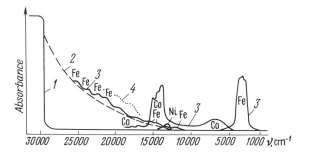

Fig. 100. General scheme of absorption spectra of natural sphalerites and their interpretation: *1*, interband transition; *2*, charge transfer band; *3*, crystal field bands; *4*, donor–acceptor transition bands

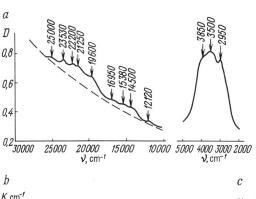

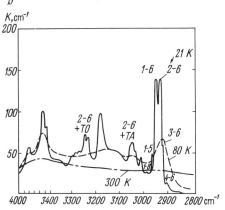

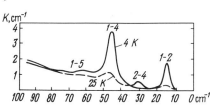

Fig. 101a–c. Absorption spectra of Fe^{2+} in ZnS in different region of spectrum **a** [276], **b** [286], **c** [287]

with about 7% of Fe^{2+}, has been interpreted [276] in accordance with the diagram shown in Figure 101. The $^5E \rightarrow {}^5T_2$ transition has been studied in detail, both experimentally and theoretically [232, 286, 287]. At the temperatures of liquid hydrogen and liquid helium the $^5E \rightarrow {}^5T_2$ band is resolved into series of narrow intense lines (see Fig. 101b) related to: (1) pure electronic transition (zero phonon) from five sublevels derived (Fig. 102) from splitting of the ground ^5E state by spin-orbit interaction of the second order into spin-orbit sublevels of $^5T_2$ states [232, 286] and (2) phonon-assisted ZnS lattice vibrational sidebands.

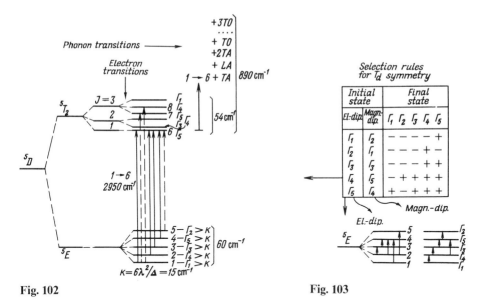

Fig. 102. Energy levels describing the structure of band $^5E \rightarrow {}^5T_2$ in ZnS:Fe^{2+} spectrum [286]

Fig. 103. Spin-orbit sublevels of ^5E state, between which electric dipole and magnetic dipole transitions are allowed corresponding to the absorption bands in the far IR region (after [287])

Transitions between spin-orbit sublevels of the ^5E ground state (Fig. 103) give rise to the absorption bands [287] in far infrared region (see Fig. 101b). Fe$_4^{2+}$ spectra have been studied extensively also in natural MgAl$_2$O$_4$ spinel from Burma and Ceylon and in synthetic CdTe [286].

Cobalt. $3d^7$:Co^{2+}. The $3d^7$ electron configuration can be considered as a configuration with three holes in a filled d shell ($3d^{10-3} = 3d^7$). Hence, free ion terms for d^7 configuration are the same as that for d^3-ions (Cr^{3+}): ground ^4F term and excited ^4P, ^2G, ^2H, ^2D, ^2F terms. Thus the diagram of these terms splitting in the octahedral field (Fig. 104) is similar to the diagram of d^3 ions (see Fig. 87) but the ordering levels, arising from the terms of free d ions in the crystal field is inverse to that of d^3 ions. Hence, the ground state of d^7 ions in octahedral coordination is $^4T_{1g}$ (for d^3 it is $^4A_{2g}$). More complete diagrams including the splitting of all doublet terms and taking into account spin orbit splitting are given in [271] and [205].

Two main absorption regions are observed in Co^{2+} spectra in octahedral coordination that corresponds to quartet–quartet transitions: (1) 7000–10,000 cm^{-1}, i.e., in the near IR: $^4T_{1g} \rightarrow {}^4T_{2g}$ transition ($\varepsilon = 1$–10 cm$^{-1} \cdot$l$\cdot$mol^{-1}); and (2) 17,000–20,000 cm^{-1} most intense $^4T_{1g}(^4F) \rightarrow {}^4T_{1g}(^4P)$ transition ($\varepsilon = 5 \div 40$ cm$^{-1} \cdot$l$\cdot$mol^{-1}). There is one more spin-allowed transition $^4T_{1g} \rightarrow {}^4A_{2g}$, but this is a two-electron transition: $^4T_{1g}(t_{2g}^5 e_g^2) \rightarrow {}^4A_{2g}(t_{2g}^3 e_g^4)$, hence it is either not observed in absorption spectra or gives rise to a very weak band.

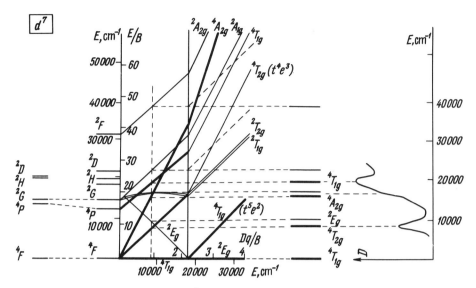

Fig. 104. Energy level diagram for d^7 configuration in octahedral field

The $^4T_{1g}(^4F) \rightarrow {}^4T_{1g}(^4P)$ transition, with intensity one order greater than the other two allowed transitions, determines the color of the Co^{2+} compounds. Among the spin-forbidden the $^4T_{1g} \rightarrow {}^2T_{1g}, {}^2A_{1g}$ quartet-doublet transition should be mentioned. This is mixed with quartet $^4T_{1g}(^4P)$ transition and superimposed on it in the form of a shoulder on the shortwave side of the absorption band; the rather strong intensity of this forbidden transition is the consequence of the levels mixing.

In the case of d^7 ions as distinct from d^3 there occur both (1) the usual weak field $t_{2g}^5 e_g^2$ configuration corresponding to the $^4T_{1g}$ ground state, high-spin ($S=3/2$, $2S+1=4$), and (2) the strong field $t_{2g}^6 e_g^1$ configuration corresponding to the ^2E_g ground state, low-spin ($S=1/2$, $2S+1=2$). The right and the left parts of the diagram (see Fig. 104) divided by vertical line correspond to them. However, configuration of the strong field arises when $Dq/B=2.2$, that is, when $Dq>1500\,\text{cm}^{-1}$, that is, not carried for divalent ions having the lesser Dq values; hence only the left part of the diagram is important for Co^{2+}.

The position of the most intense band $^4T_{1g}(^4F) \rightarrow {}^4T_{1g}(^4P)$ determines the pink or crimson color of most Co compounds; only reduction of the Dq value leads to the shift of this band to the red part of the spectrum and thus to a blue (for example in Co_2SiO_4 or in $CoCl_2$) or even a green (in $CoBr_2$) color. (For determination of Dq, B and C see p. 84.)

Since Co^{2+} has the high value of the spin-orbit interaction constant (for a free ion $\xi=540\,\text{cm}^{-1}$), this interaction gives rise to the absorption band structure (which can be observed already at 77 K), besides, additional electronic-vibrational bands also occur (see Fig. 105). Lowering of the local symmetry leads to a further splitting of levels, which causes shift of the absorption bands in polarized light and dichroism.

Co^{2+} spectra in octahedral coordination can be subdivided as follows:

1. Spectra of Co^{2+} in the field of cubic (O_h) symmetry (for example in well-studied $MgO:Co^{2+}$) or in the fields close to cubic when the distortions of the octahedra are not great (in well-studied $CoCl_2$ with rhombohedral structure, and in hexagonal $CoBr_2$ and others). The characteristic structure of the absorption bands is related to spin-orbit splitting and phonon-assisted vibrational satellites of electronic transitions. See for example, the spectrum of $CoCl_2$ (Fig. 106) and the spectra of Co^{2+} in two positions in Co-analogs of monticellite (Fig. 107) and olivine [225, 290].

2. Spectra of minerals where Co^{2+} occurs usually in distorted octahedra; level splitting leads to dichroism.

The polarized absorption spectrum has been obtained for erythrite related to the structural type of vivianite (see above the description of Fe^{2+} spectra) with Co^{2+} in two very similar positions.

The energy level diagram for Co^{2+} in tetrahedral coordination is the same as for Cr^{3+}, $3d^3$ (since $d_{tetr} = d_{oct}^{10-n}$), even the C/B ratio (equal to 4.5) is similar, but here another order of Dq values occurs for Co^{2+} in the tetrahedra because Dq for M^{2+} is less than for M^{3+} and $Dq_{tetr} = 4/9 D_{oct}$.

The ground state of Co^{2+} in tetrahedra is $^4A_2$. Three regions of spin-allowed transitions $^4A_2 \rightarrow {}^4T_2$, $^4A_2 \rightarrow {}^4T_1$ and $^4A_2 \rightarrow {}^4T_1(^4P)$, consist each of four to six narrow bands arising from the spin-orbit interaction; more weak doublet

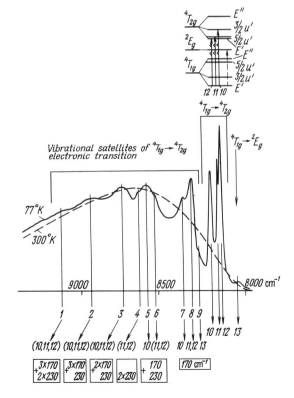

Fig. 105. Absorption band of Co^{2+} in MgO^4 $T_{1g} \rightarrow {}^4T_{2g}$ (in the infrared region) [279]. One discerns: pure electronic transition $^4T_1(E^1) \rightarrow {}^4T_{2g}$ (E, $5/2u$, $3/2u$) with three narrow lines due to allowed transitions into spin-orbital sublevels and phonon-assisted spectrum vibrational sidebands (analyzed in terms of two phonon modes of energy: 170 cm^{-1} and 230 cm^{-1}). Weak (spin-forbidden, quartet-doublet) transition $^4T_{1g} \rightarrow {}^2E_g$ is also observed

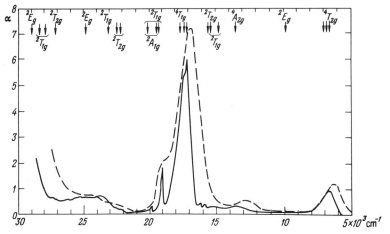

Fig. 106. Absorption spectrum of $CoCl_2$ [225]

transitions can be partially superimposed on them. The spectra have been studied extensively for $ZnAl_2O_4:Co^{2+}$, $ZnS:Co^{2+}$ (Fig. 108) [225, 231, 298], $ZnO:Co^{2+}$ [269], $CdS:Co^{2+}$ and other compounds of the $A^{II}B^{VI}$ type [298].

The pronounced Co^{2+} spectra have been observed in natural sphalerites (Fig. 108) with Co content of about 0.002% and in the presence of about 1% iron [276]; these can be used for analytical determination.

The interesting peculiarity of Co^{2+} spectrum in the tetrahedral coordination (without center of symmetry) is its very strong intensity. The oscillator strength for

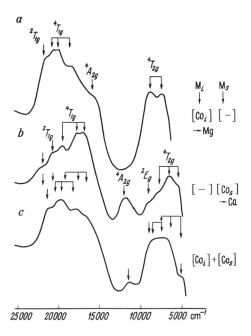

Fig. 107a–c. Absorption spectrum of Co^{2+} in synthetic monticellite (Mg, Co) $CaSiO_4$ (**a**), in cobalt analog of monticellite $Co(Ca, Co)SiO_4$ and the spectrum obtained by subtraction of the first one from the third (**b**) [290]

Fig. 108. Spectrum of Co^{2+} in sphalerite from Lovozero (0.03% Co) [276]

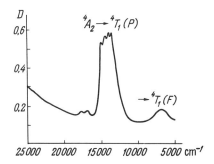

$^4A_2 \to {}^4T_1$ (^4P) in $ZnS:Co^{2+}$ is $f = 1.3 \cdot 10^{-2}$ [298] against $f = 1.6 \cdot 10^{-5}$ for $T_{1g}(E) \to {}^4T_{1g}(P)$ in $MgO:Co^{2+}$ [279], i.e., it is 1000 times greater. Even in comparison with other ions in tetrahedral coordination, the oscillator strength of Co^{2+} transitions is one or two orders of magnitude greater.

The components of $^4A_2 \to {}^4T_1$ (^4P) transitions have been observed in synthetic quartz with Co^{2+} content in the tetrahedral position to 0.00058 weight % [308]. Detection of about 0.0001 weight % of Co^{2+} is possible by absorption spectra.

In tetrahedral coordination, but with the reduction of tetrahedral symmetry to tetragonal S_4 (elongated along the S_4 axis tetrahedron), the spectra of Co^{2+} have been observed in synthetic Y–Al, Y–Ga and Y–Fe garnets. The superimposed spectra of the simultaneously occurring Co^{2+} in the tetrahedral, octahedral and dodecahedral positions and Co^{3+} in tetrahedral and octahedral positions [280] were resolved in these garnets.

Nickel. $3d^8:Ni^{2+}$. The ground term of free d^8 ion is ^3F, excited terms are ^3P, ^1D, ^1G, ^1S. The ^3F term splits in the octahedral crystal field into $^3A_{2g}$ (ground state, configuration $t_{2g}^6 e_g^2$), $^3T_{2g}$ and $^3T_{1g}$ (configuration $t_{2g}^5 e_g^3$), while the second triplet term ^3P transforms into $^3T_{1g}(t_{1g}^4 e_g^4)$. The behavior of these levels as well as the levels derived from singlet terms is shown in Figure 109. As for all ions with the ground F state (^3F in Cr^3 and Ni^{2+}, ^4F in V^{3+} and Co^{2+}), there is still only one term of the same multiplicity ^3P (or ^4P). This leads to the three spin-allowed transitions in the crystal field: two transitions between the levels of the split ^3F term ($^3A_{2g} \to {}^3T_{2g}$ and $^3A_{2g} \to {}^3T_{1g}$) and more transition to the level derived from ^3P (or ^4P) ($^3A_{2g} \to {}^3T_{1g}$). An estimation of Dq and B can be obtained from the equations:

$$^3A_{2g} \to {}^3T_{2g} = \Delta$$
$$\to T_{1g}(F) = 7.5B + 1.5\Delta - (b^-),$$
$$\to {}^3T_{1g}(P) = 7.5B + 1.5\Delta + (b^{-1}),$$
$$T_{1g}(F) - {}^3T_{1g}(P) = 2(b^-),$$

where

$$(b^-) = 1/2[(9B - \Delta)^2 + 144B^2]^{1/2}.$$

Weak and narrow spin-forbidden bands are observed as the result of transitions from $^3A_{2g}$ into ^1E_g and $^1A_{1g}(^1D)$; weak broad bands arise from the transition into

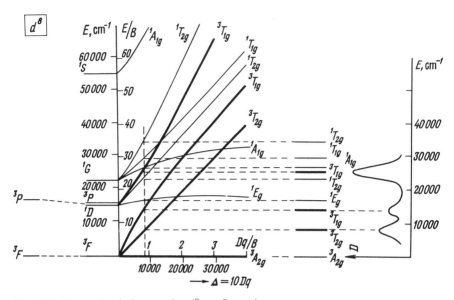

Fig. 109. Energy level diagram for d^8 configuration

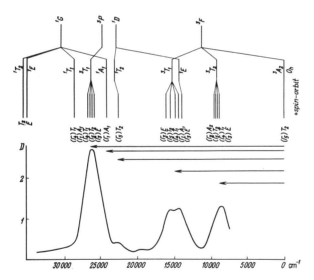

Fig. 110. Absorption spectrum of $NiSiF_6 \cdot 6H_2O$ [196]

the other singlet levels occurring in the UV. The large value of spin-orbit interaction constant ($\xi = 630\,cm^1$ for the free ion) leads to the further splitting of levels, as the case of Co^{2+} and Cu^{2+}.

Spectra of Ni^{2+} (see Fig. 110) have been studied in a great number of chemical compounds in various stereochemical forms. However, detailed studied spectra of Ni^{2+} in oxides (NiO, MgO, Al_2O_3), halides (NiF_2, MgF_2, $KMgNiF_3$, $NaNiF_3$, $NiCl_2$, $NiBr_2$) as well as spectra of $Ni(H_2O)^{2+}$ (in $NiSiF_6 \cdot 6H_2O$, $NiSO_4 \cdot 6H_2O$, $NiSO_4 \cdot 7H_2O$) have been interpreted in the octahedral field approximation and,

Fig. 111. Absorption spectra of Ni^{2+} in minerals [266]

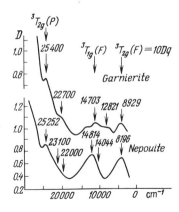

taking into account spin-orbit interaction, though in many of them Ni octahedra are distorted [268, 274, 278, 279, 298]. Among minerals the spectra of Ni^{2+} in octahedral coordination have been studied mainly [266] in Ni-containing silicates (Fig. 111).

The usual geochemical association of Ni^{2+} with Fe^{3+} and Fe^{2+} causes partial overlapping of spectra Ni^{2+} by the Fe^{3+} charge transfer band; besides Fe^{3+} and Fe^{2+}, crystal field bands can be observed together with the Ni^{2+} bands.

Spectra of impurity of Ni^{2+} have been found in lemon-yellow calcites [266]. The energy level diagram of Ni^{2+} in tetrahedral coordination is similar to the diagram of $V^{3+}(d^2)$ in octahedral coordination ($d^n_{tetr} = d^{10-n}_{oct}$). The ground state is $^3T_1$; three spin-allowed transition are observed here: $^3T_1 \to {}^3T_2$, $^3T_1 \to {}^3A_2$ and $^3T_1(F) \to {}^3T_1(P)$, but are shifted (in comparison with V^{3+}) to the longwave side (since $Dq_{tetr} = 4/9 Dq_{oct}$ and $DqM^{2+} < DqM^{3+}$). The absence of the symmetry center in tetrahedral positions removes parity forbiddeness that causes the rather strong increase of absorption intensity of Ni^{2+} in tetrahedra in comparison with octahedra.

Spectra of Ni^{2+} in tetrahedral coordination have been studied in detail in ZnO, ZnS, CdS, $MgAl_2O_4$, and Y–al garnets. Among natural minerals, the spectrum of Ni^{2+} in tetrahedra has been observed in sphalerites [276].

Copper. $3d^9:Cu^{2+}$. Electron configuration d^9 can be considered as configuration with one hole in filled d shell ($3d^{10-1} = 3d^9$), that is, with one unpaired electron.

Hence there is here only one (ground) term ^2D as in the case of d^1 configuration (see Ti^{3+}). The nearest excited free ion terms arise from the $3d^84s^1$ configuration and occur in the far UV region (Fig. 112).

Term ^2D splits into ground state ^2E_g and excited $^2T_{2g}$ in the octahedral crystal field, i.e., in the same states as in the case of the d^1 configuration, but with inverse ordering due to opposite charges: the positive hole in d^9 and the negative electron in d^1. The same levels, but with inverse ordering, are formed in the case of d^9 configuration in comparison with d^1 configuration in crystal fields of any symmetry; thus the splitting scheme of d^1 configuration can be used for d^9 ions (with the account of inverse level ordering). As in case of d^1 ions forbidden transitions (complicating spectra in the cases of all other configurations) are absent

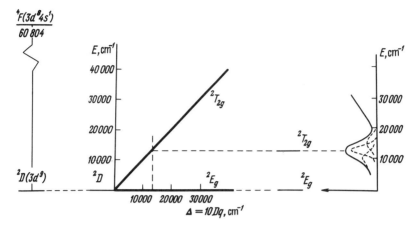

Fig. 112. Energy level diagram for d^9 configuration (Cu^{2+}) ($CuSO_4 \cdot 5H_2O$ spectrum; the absorption band $^2E_g \to {^2T_{2g}}$ is decomposed into three Gaussian components)

here, and all the broad spin-allowed absorption bands are related to transitions between the levels derived from $^2D \to {^2E_g} + {^2T_{2g}}$. Racah parameters B and C (as in d^1) are also absent here; d^9 ion states can be considered as one-electron states (as the states with one hole) and hence are designated either ^2E_g and $^2T_{1g}$, or e_g and t_{2g}, or $d_{x^2-y^2}$, d_{z^2} and d_{xy}, d_{xz}, d_{yz}.

However, the inverse ordering of the d^9 ion levels in crystal fields leads to their considerable difference from d^1 ions. It is related to the fact that among two states ^2E_g and $^2T_{2g}$, in the cubic field, ^2E_g state is subjected to the more strong action of the Jahn-Teller effect (i.e., to inner configuration instability) that causes splitting of ^2E_g.

In d^1 ions (Ti^{3+} and others), ^2E_g is excited state and the Jahn-Teller effect is shown in splitting of absorption bands due to the transition from unsplit (in the cubic field) ground $^2T_{2g}$ state into the split ^2E_g state.

In d^9 ions (Cu^{2+}), ^2E_g is the ground state and the Jahn-Teller effect which is shown in the ^2E_g state splitting leads to the stereochemistry of Cu^{2+} complexes never corresponding to the regular octahedron: instead of six-fold coordination, (4+2) coordination nearly always occurs in the form of an elongated octahedron. There are also square coordinations, tetragonal pyramidal, rhombically (or lower) distorted coordinations, rare trigonal distorted octahedra with additional rhombic or other distortions here.

However, since Cu^{2+} complexes are always in fields with symmetry lower than octahedral, thus the excited $^2T_{2g}$ state is also always split.

As distinct from a diversity of d^1 ion level splitting (Ti^{3+}, see Fig. 81), in the case of d^9 ions, tetragonal distorted octahedron is the most usual coordination (with possibly weaker rhombic distortion) which can pass to square coordination

Analysis and Experimental Survey of Transition Metal Ions Spectra 235

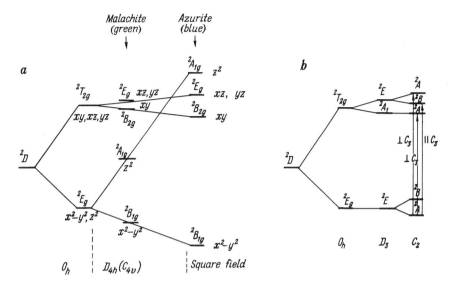

Fig. 113 a and b. Splitting of Cu^{2+} levels in octahedral coordination: **a** with increasing tetragonal distortion (elongation along C_4) and transition to square coordination; **b** in mostly trigonal field

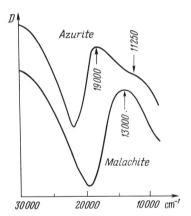

Fig. 114. Spectra of Cu^{2+} [276]

(Fig. 113). Hence two absorption bands must be observed in Cu^{2+} spectra (Fig. 114; tetragonal distortion), often nonresolved and maximum three–four bands (with rhombic or strong tetragonal fields or square fields plus rhombic distortion) [211, 265, 267, 276]. When there are two different positions of Cu^{2+} in a structure, the total spectrum will present the superimposition of two spectra, each of them can be represented by two–four broad absorption bands.

Tetragonal splitting is described by the D_s and D_t parameters; its meaning can be obtained from Fig. 98 showing the ^5D state levels of the Fe^{2+} ion.

6.4 The Nature of Colors of Minerals

Three periods may be distinguished in the history of studies of colors of minerals.

The first period was marked by qualitative observations and systematics. The culminating expression of this "romantic" period of the investigation of colors of minerals, which sprang from works of Goethe and Lomonosov, was the monograph of Fersman [227]. Here idiochromatic, allochromatic (connected with inclusions of other minerals) and pseudochromatic (interference) colors were distinguished and the concept of ion-chromophors was developed.

The second was the beginning of quantitative investigations of colors with the measurements of optical absorption spectra Absorption curves of some 60–70 minerals obtained in the 1930s–1950s [200] put the studying of this problem on an experimental basis, but without physical interpretations.

Quantitative measuring of color (with the help of color coordinates [236, 270, 272, 282, 283, 311]) is based on optical absorption and reflection spectra.

Contemporary understanding of the nature of colors of minerals in general and colors of concrete minerals in particular are based on the following position: understanding of color is the interpretation of the spectrum.

The intermediation of the spectra, determination of concentration dependences, orientation dependences (pleochroism), relations to valence states of ion coordinations, crystal structures, and chemical bonds can be performed in the framework of crystal field, molecular orbital, and energy band theories [276].

6.4.1 Types of Colors of Minerals

I. Color Related to Interband Transition in Different Points of the Brillouin Zone. The spectrum covers the UV visible region and is thus the color of opaque materials or in specular reflection. As distinct from absorption spectra, where absorption bands are separated from each other by the region with considerable transmission, in reflectance spectra there are broad bands which are only a little higher than the general high background. In the narrow visible region, reflection spectrum has the shape of slightly inclined or gently bent curves (Fig. 115).

For this reason only two colors can usually be observed in specular reflection: white (or grey with lower reflectivity value) with weak shades (blue or pink dependent on the slope of reflection curve in the visible region) and yellow.

Not only opaque minerals have this white or gray (or black) color with faint shades in specular reflection (observed usually by ore microscope in polished sections), but all minerals, including semi-transparent and colored in different colors in a sample, in diffuse reflection and absorption spectra, and in internal reflexes. Gray and white colors of specular reflection of colored (blue, green, red) samples of minerals are explained by the fact that the dispersion of reflectivity being a function of refraction and absorption (see Chap. 4.2) is determined in general by the dispersion of refraction for weak absorbent materials, while weak absorption bands are not revealed in it.

The dependence of the dispersion of reflectivity upon the dispersion of refraction explains the change of color of some minerals in reflected light with observations in air and in different immersion liquids (a typical example is covellite, which is dark blue in air, and red with different shades in immersion).

Fig. 115. Types of spectra in the visible region: $I-V$ are types of colors of minerals and inorganic compounds

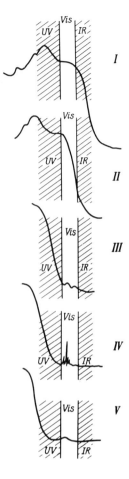

II. Color Connected with Absorption Edge Position in Visible Region, i.e., color of materials with the width of the forbidden bands corresponding to energies of optical transitions of visible region of spectrum ($3 \div 1.6$ eV). If the absorption edge cuts off parts of the spectrum neighboring with the UV region, its position at different wave lengths of the visible region causes bright and pure colors: yellow in greenockite (with energy gap 2.41 eV, corresponding to 514 nm) and orpiment (2.50 eV or 496 nm), red in cinnabar (near 600 nm), some sulfosalts of silver and others.

III. Colors Connected with Transitions Between d-Electron Levels. These are colors produced by transition group element ions entering compounds in the impurity form or as main component (ion-chromophors).

There are three subtypes here with their different characteristics:

1. Color connected with charge transfer spectra i.e., with transitions from bonding or nonbonding molecular orbitals into antibonding orbitals formed mainly from metal atomic orbitals (see Chap. 3.2).

Charge-transfer spectra are observed for all transition metal ions (including such ions as Ti^{4+}, V^{5+} and others which do not give rise to crystal field spectra) but for most of them, charge-tranfer bands occurs in the UV region and thus cannot be the cause of colors.

The importance of this mechanism of absorption spectra formation for minerals is connected with the fact that it is observed in the Fe^{3+} ion in the visible region. The maximum of the Fe^{3+} charge-transfer band occurs in UV region (230–250 nm for oxygen compounds), but its intensity is so great (near 7000 cm^{-1} in some silicate glasses in comparison with 0.1–10 cm^{-1} for crystal field spectra), that even the tail of the charge-transfer band occurring in the visible region and (with a considerable content of iron), even in the infrared region, is of central importance for the color of a mineral.

2. Color connected with crystal field spectra i.e., with the transitions between d electron levels split up by the crystal field as considered in detail above. It is the most usual cause of color in rockforming minerals.

The most widely occurring chromophors in minerals are Fe^{2+} and Fe^{3+} ions. Minerals of V, Cr, Mn, Co, Ni, Cu comprise more "localized" groups.

It is evident from the standpoint of the crystal field theory that each valence state of a transition metal in each coordination gives rise to principally different spectra, while with different ligands (F, Cl, H_2O, O, S, Se, Te) and with different distortion of structural position, the absorption band in these spectra can be shifted and split. This leads to different colors of compounds of one and the same transition element.

The crystal field theory enables determination of the dependence of color of this subtype (i.e., dependence of spectra of d electrons) on crystal structure, orientation, composition, and state of the chemical bond.

To determine the dependence of spectrum on the orientation of crystal, it is necessary to point out that the position of each transition does not change with the change of orientation (as distinct from EPR spectra). If a transition is allowed in one orientation and forbidden in an other, its intensity is different in these orientations, but the transition energy remains the same.

3. The change and increasing intensity of color connected with intervalent interactions is observed in the cases of Fe^{2+}–Fe^{3+}, Fe^{2+}–Ti^{4+} pairs and others in the structural positions which enable interaction between the cations.

IV. Color Connected with Transitions Between f Electron Levels, i.e., color caused by rare earth and actinide ions. Three subtypes are distinguished: (1) color caused by transitions within the f^k configuration, i.e., color of trivalent rare earth TR^{3+} and uranium ions U^{4+}, U^{3+}, U^{2+}; their spectra exhibit great numbers of narrow weak lines and produce very weak colors; (2) color caused by transitions $4f^k \rightarrow 4f^{k-1}5d$, i.e., transitions between the levels of the mixed fd configuration; the broad, intense absorption bands which determine colors are caused by divalent rare earths TR^{2+}; (3) color connected with transitions between molecular orbitals of uranyl UO_2^{2+} (bright yellow, green, orange colors of many uranium minerals).

V. Color Connected with Electron-Hole Centers, Molecular Ions, Free Radicals. (1) Color connected with impurity cations: (a) with d^1 configuration (Ti^{3+}, V^{4+}, Nb^{4+} and other, and also titanil, vanadil, niobil and others) and with d^n and f^k configuration, (b) with s^1, p^1, p^5 configuration (Pb^+ in green amazonite, Pb^{3+} in pink calcite and others); (2) colors connected with O^- and molecule ions O_2^-, S_2^- and others; (3) colors connected with free radicals SO_4^-, SO_3^- and others; (4) colors connected with F and V centers and F aggregate centers.

7. Structure and the Chemical Bond

7.1 Contemporary Methods of Description and Calculations of the Chemical Bond in Solids

A tendency to the regular arrangement of atoms in a compound and to the formation of crystal structure is one of the most important and characteristic manifestations of the chemical bond. The structure can be considered as a direct consequence of the mutual adjustment (self-consistence) of properties of atoms forming the compound. Molecular orbital and valence bond methods explain the stereochemistry of molecules and complexes. However, there are no isolated molecules in crystals. When passing from the electronic structure of the $FeCl_6^{3-}$ complex, for example, to $FeCl_3$ crystal (with similar octahedral groups $Fe\text{-}Cl_6$), the interaction of each atom with all lattice atoms is to be taken into account. Therefore, the description of the chemical bond in crystals cannot be restricted to molecular-orbital (or valence bond) methods.

The electronic structure of solids is described by the energy band theory. However, the usual versions of this theory employing such models as nearly free electron approximation, different Bloch functions at each point of the Brillouin zone etc. (see Chap. 4.1), place particular emphasis upon interpretation of physical phenomena, namely reflectance spectra, semiconducting, metallic and other properties of solids.

However, in this model description of the most important concepts of the chemical bond, such as effective charges, ionicity–covalency bond, participation of different atomic orbitals in the bond etc., i.e., of those concepts which most consistently result from the molecular orbital scheme, appears to be rather difficult.

Thus semiempirical methods were developed [121, 153], combining the advantages of molecular orbitals and energy band theories. On the one hand, they were a direct extension of the molecular orbital method, taking into account the characteristic features of the crystal structure, while on the other hand they represent one of the methods of the energy band theory (semi-empirical tight-binding method, see Chap. 4.1).

The molecular orbital scheme involves structure parameters and links them with parameters of the chemical bond: the molecular orbital scheme and group overlap integral are determined by coordination number and local symmetry; to obtain overlap integrals interatomic distances are used, which in turn, are obtained as equilibrium distances by the minimization principle; bond angles are related to the MO coefficient describing the amounts of s and p atomic orbitals (sc/pc).

This scheme lacks one structure parameter: the crystal lattice. Consider how its influence leads the MO scheme discussed earlier (see Chap. 3.2) to the semiempirical tight-binding methods [153].

In the molecular orbital calculation, one-electron wave functions of molecule or complex (ψ_{MO}) were constructed, Coulomb integrals H_{AA} were determined as valence states ionization energies for effective charges of atoms, and resonance integrals H_{AB} were also determined. Inserting ψ_{MO}, H_{AA} and H_{AB} into the secular equation, one obtains its roots corresponding to the energies of bonding and antibonding molecular orbitals and calculated coefficients of atomic orbitals in these molecular orbitals (Chap. 3.2).

The crystal orbital, that is, one-electron wave function in crystal, is obtained in two stages [153]:

1. Basic Bloch functions (see Chap. 4.1) are obtained from atomic orbitals (exemplified by crystals of the periclase MgO-type, with valence atomic orbitals of Mg 3s, and of O $2p_x$, $2p_y$, $2p_z$):

$$\psi_{2p_x} \to \psi_1(k) = (1/\sqrt{N}) \Sigma \exp(ikR) \cdot \varphi_{2p_x}(r-R),$$

$$\psi_{2p_y} \to \psi_2(k) = (1/\sqrt{N}) \Sigma \exp(ikR) \cdot \varphi_{2p_y}(r-R),$$

$$\psi_{2p_z} \to \psi_3(k) = (1/\sqrt{N}) \Sigma \exp(ikR) \cdot \varphi_{2p_z}(r-R),$$

$$\psi_{3s} \to \psi_4(k) = (1/\sqrt{N}) \Sigma \exp(ikR) \cdot \varphi_{3s}(r-R),$$

where $1/\sqrt{N}$ denotes the normalizing coefficient (N is the number of lattice atoms); Σ is summation over all lattice nodes; R is the lattice parameter.

2. Complete Bloch functions (crystal orbitals) are obtained as linear comation of these four basic Bloch functions

$$\Psi(k) = c_1 \psi_1 + c_2 \psi_2 + c_3 \psi_3 + c_4 \psi_4,$$

where c_1, c_2, c_3, c_4 are coefficients analogous to c_i in molecular orbitals. For this type of crystal one constructs also the linear combinations of three-centered A–B–A localized orbitals (see [153]).

In the case of tetrahedral crystals of ZnS or diamond type (or crystals with antifluorite structure), recourse was made to a semiempirical method of equivalent orbits developed for the $A^{II}B^{IV}$ [153] compounds: (1) at first hybrid orbits of atoms $\chi_A = sp^3$ and $\chi = sp^3$ are obtained from s and p orbitals of A^{II} and B^{IV}; (2) the wave functions of four A–B bonds (equivalent orbitals) are made up of these hybrid orbits of atoms

$$\varphi_i = 1/\sqrt{1+\lambda^2} \cdot (\chi_A + \lambda \chi_B),$$

where $\lambda = c_A/c_B$ is the covalency parameter and $1/\sqrt{1+\lambda^2}$ the normalizing coefficient; (3) multiplying by the structure factor $\exp(ikR)$ (see Chap. 4.1) one obtains Bloch functions (crystal orbitals)

$$\varphi(k/r) = 1/\sqrt{N} \cdot \Sigma \exp(ikR) \cdot \varphi_i(r-R).$$

It is then necessary to find the Coulomb integrals H_{AA} (denoted here as α_A) and $H_{BB}(\alpha_B)$:

$$\alpha_A = \alpha_A^0 + mZ/R, \quad \alpha_B = \alpha_B^0 - mZ/R,$$

where α_A^0 and α_B^0 are the Coulomb integrals for s and p electrons of A and B atoms in their valence state with effective charges determined just as when calculating molecular orbitals (see Chap. 3.2) or by other similar methods; Z denotes effective charges; R is the distance between A and B atoms; m is a semi-empirical constant taken to be equal to Madelung constant (see below) when calculating energy bands, but taken to be equal approximately to 1 for compounds of $A^{II}B^{VI}$ type with NaCl and ZnS structures when calculating effective charges [153]. The Madelung term mZ/R takes into account the interaction between the electron of A and B atoms and the lattice field.

If Bloch wave functions are obtained from hybrid orbits sp^3, then α^0 is determined as the energy equal approximately to 1/4 of the energy of the s orbit plus 3/4 of the energy of the p orbit. Resonance integrals H_{AB} (denoted here as β) are here just so as in calculations of molecular orbitals $\beta = \left(\dfrac{\alpha_A^0 + \alpha_B^0}{2}\right) \cdot S$, where S is the overlap integral; or the β values can be also determined empirically from thermochemical data [153].

Obtained α_A and α_B values (similar to Coulomb integrals in molecular orbital calculations), as well as the β values, are used for determination of the covalence parameter λ from the equation:

$$\lambda^2 + \lambda(\alpha_B - \alpha_A)/(\beta - 1) = 0$$

and of effective charge Z equal for compounds of $A^N B^{8-N}$ type to:

$$Z = \{N - (\gamma - N)\lambda^2\}/(1 + \lambda^2);$$

and for $A^{II}B^{VI}$(ZnS, PbS) to

$$Z = 2 - \frac{6\lambda^2}{1+\lambda^2} = \frac{2 - 4\lambda^2}{1+\lambda^2}.$$

In addition to the determination of these chemical bond parameters, the semi-empirical calculations can also be used for the band structure determination (see Chap. 4.1). To do this the α_A, α_B and β values are inserted into the secular equation (just as when calculating molecular orbitals):

$$\begin{vmatrix} \alpha_A - E & \beta \\ \beta & \alpha_B - E \end{vmatrix} = 0,$$

the roots of this equation yield energies at different points of the Brillouin zone, thus forming bands of the type of 2p, 3s etc. atomic orbitals. Analysis by this method of effective charges in compounds of NaCl, MgO, CaO, BaO, SrO, PbS type and of ZnS (sphalerite, wurtzite), diamond, Na_2O and Na_2S type and in compounds of some other types allowed to obtain values close to experimental ones [153].

7.1.1 Extension of the Bond Orbital Methods for Cristobalite and Quartz Structures

Applications of the energy band calculation methods (especially "first principles" methods) were restricted mainly by the simple metals and by structure types of diamond, zinc-blende, wurtzite and rock-salt. Even the pseudopotential theory was used most effectively for studies of the electronic properties and crystals structures of the same types of solid [146].

Thus the extension of the bond orbital approximation for studies of such important minerals and materials as silica polymorphs is of extreme importance [144, 460].

The bond orbital model is a method of electronic structure (energy band structure) calculation developed for tetrahedrally coordinated solids: diamond, zinc-blende, cristobalite, and quartz structure types, but which can be extended to a more general class of materials.

It is a tight-binding method based on a linear combination of atomic orbitals (LCAO) approach. It combines possibilities both for energy band calculation and chemical bond description, for interpretation of spectra and computation of the cohesive energy and wide range of properties. The method is limited in accuracy but is characterized by simplicity, clear physical and chemical meaning of all the steps of its procedure, and diversity of interpretations based on it.

Consider some main features of the bond orbital approximation as it applies to cristobalite and quartz structures. Taking into consideration a chain Si-O-Si, i.e., two silicon atoms and one oxygen:

1. Construct on the each Si atom four tetrahedrally directed sp^3 hybrid orbitals denoted $|h_1\rangle$ and $|h_2\rangle$.

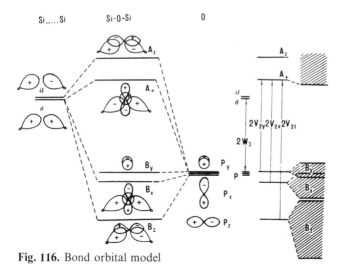

Fig. 116. Bond orbital model

The Dirac notation is useful here (see Chap. 1.7): $|h\rangle$ corresponds to ψ_{sp^3}; below we shall use also: $|p_z\rangle$ for ψ_{p_z}; $|b\rangle$ for ψ^b (bonding orbital); $|a\rangle$ for ψ^a (antibonding orbital); $\langle h_1 h_2 \rangle$ for $S = \int \psi_{h_1} \psi_{h_2}$ (overlap integral); $\langle h_1|H|h_1\rangle$ for H_{AA} (Coulomb integral); $\langle h_1|H|h_2\rangle$ for H_{AB} (exchange integral).

2. Valence orbitals used for oxygen are p_x, p_y, p_z (the oxygen 2s orbitals are essentially core states).

3. Form bonding and antibonding combinations of the two silicon Si...Si hybrid orbitals:

$$|b\rangle = \frac{1}{\sqrt{2(1+S)}}(|h_1\rangle + |h_2\rangle) \quad \text{and} \quad |a\rangle = \frac{1}{\sqrt{2(1-S)}}(|h_1\rangle - |h_2\rangle).$$

4. Now we have in all five orbitals: h_1, h_2, p_x, p_y, p_z, or equivalently, b, a, p_x, p_y, p_z.

Allowed interaction and mixing between these orbitals lead to new superbond orbitals (see Fig. 116):

$$|B_z\rangle = u_z(p_z) + u_a|a\rangle$$

$$|B_x\rangle = u_x(p_x) + u_b|b\rangle$$

$$|B_y\rangle = |p_y\rangle \quad \text{(nonbonding)}$$

and corresponding anti-bonding $|A_+\rangle$ and $|A_z\rangle$ orbitals (Fig. 116).

In the case of cristobalite with straight Si–O–Si chains there are the same orbitals, but B_x become also nonbonding.

In the case of zinc-blende-type compounds, one considers only two hybrid orbitals $|h_a\rangle$ and $|h_c\rangle$ where a stands for anion, the nonmetallic ion, and c for cation, the metallic ion; the bond orbitals are simply given by: $|b\rangle = u_a|h_a\rangle + u_c|h_c\rangle$ (i.e., without formation of the superbond $|B\rangle$ orbitals).

5. Define the fundamental parameters of the bond orbital model, which are basically only two, W_2 and W_3:

$$W_2 = -\langle h|H|p_z\rangle/(1-2S^2),$$

$$W_3 = (\langle a|H|a\rangle - \langle p_z|H|p_z\rangle)/(1-2S^2)^{1/2} = (\varepsilon_a - \varepsilon_z)/(1-2S^2)^{1/2},$$

where $S = \langle h|p_z\rangle$ (overlap integral), $\langle h|H|p_z\rangle$ is analogous to the exchange integral H_{AB} and $\langle a|H|a\rangle$ and $\langle p_z|H|p_z\rangle$ are analogous to Coulomb integrals H_{AA} in the molecular orbital model (see Chap. 3.3). The meaning of W_3 can also be seen from Fig. 116.

It is most remarkable that many properties may be computed quite simply in terms of only these two parameters, and measured values of the properties can then be used to determine the parameters [146, 460].

In quartz, in which the Si–O–Si chains are not straight, the parameters can be defined as follows:

$$W_{2z} = W_2 \left(\frac{1 - 2s^2}{1 - 2S^2 \cos \vartheta} \right) \cos \vartheta,$$

$$W_{2x} = W_2 \left(\frac{1 - 2s^2}{1 - 2S^2 \sin^2 \vartheta} \right) \sin \vartheta,$$

where $\vartheta = 1/2(180 - \varphi)$ and φ is the Si–O–Si angle. $W_{3z} = W_{3x} = W_{3y}$ can be taken to a good approximation. (In the considerations of diamond and zinc-blende structures one uses the designation

$$V_2 = -\langle h_a | H | h_c \rangle / (1 - S^2)$$

and

$$V_3 = (\langle h_c | H | h_c \rangle - \langle h_a | H | h_a \rangle)/(1 - S^2)^{1/2}$$

and one defines also the parameters

$$V_1 = -\langle h_a | H | h_a \rangle / (1 - S^2) = 1/4(\varepsilon_p^a - \varepsilon_s^a)$$

equal to one-fourth of the $s - p$ splitting of the atomic term value).

V_3 (or W_3) represents polar energy, V_2 (or W_2) covalent energy, and V_1 band-broadening energy.

6. One can now determine some characteristics of the bond and many properties using only the two parameters: W_2 and W_3 (without construction of the energy band scheme).

a) The definition of the polarity of the bond:

$$\beta_p = W_3/(2W_2^2 + W_3^2)^{1/2},$$

the covalency of the bond

$$\beta_c = W_2(2W_2^2 + W_3^2)^{1/2}.$$

One can determine the polarity of the each orbital $|B_z\rangle$, $|B_x\rangle$, $|B_y\rangle$ (see Fig. 116), using the values of W_{2z}, W_{2x}, W_{2y}, respectively. (For diamond and zinc-blende structures polarity and covalency are defined slightly different by: $\alpha_p = V_3(V_2^2 + V_3^2)^{1/2}$ and $\alpha_c = V_2(V_2^2 + V_3^2)^{1/2}$, metallicity is $\alpha_m = V_1(V_2^2 + V_3^2)^{1/2}$.)

b) The expression for $|B\rangle$ (see above) for SiO$_2$ can be written as:

$$|B\rangle = u_z |h_z\rangle + u_c |a\rangle = [1/2(1 + \beta_p)]^{1/2} |p_z\rangle + [1/2(1 - \beta_p)]^{1/2} |a\rangle.$$

For zinc-blende-type crystals:

$$|b\rangle = u_a|h_a\rangle + u_c|h_c\rangle = [1/2(1+\alpha_p)]^{1/2}|h_a\rangle + [1/2(1-\alpha_p)]^{1/2}|h_c\rangle.$$

c) In the latter case one identifies $|u_a|^2$ and $|u_c|^2$ as the fraction of each electron to be associated with the anion and cation and the effective charge is

$$Z_a^* = 8|u_a|^2 - Z_a \quad \text{and} \quad Z_c^* = 8|u_c|^2 - Z_c,$$

where Z_a and Z_c are the formal valences of the anion and cation.

For SiO_2 the effective charge is equal to the sum of the contributions of all three bond orbitals B_x, B_y and B_z:

$$Z_0^* = \beta_{px} + \beta_{py} + \beta_{pz} - 1 \quad \text{or, since} \quad \beta_{py} = 1,$$

$$Z_0^* = \beta_{px} + \beta_{pz}.$$

The value of the effective charge in quartz is 1.02.

d) The excitation energies denoted in Fig. 116 by $2V_{2x}$, $2V_{2y}$, $2V_{2z}$ and corresponding to the optical absorption peaks can be expressed through W_2 and W_3:

$$2V_{2x} = 2(2W_{2x}^2 + W_3^2)^{1/2},$$
$$2V_{2y} = W_3 + 2V_{2x},$$
$$2V_{2z} = (2W_{2z}^2 + W_3^2)^{1/2} + 2V_2 - 2W_{2z}S_z.$$

e) The two-parameter bond orbital method has also been used for calculations of the valence energy bands and density of states, total cohesive energy and a wide range of properties ([460] see also Chap. 8.2).

7. Thus far the interactions within the Si–O–Si chain (or within Zn–S or C–C groups of atoms) have been considered. In constructing the energy band structure it is necessary to take into account the interactions between the different bond sites [144, 189, 460].

For the diamond and zinc-blende types of structure these interactions are shown in Fig. 117. This leads to the six parameters K, A, C, T, S, G to be determined.

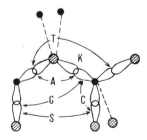

Fig. 117. Interactions between the bond orbitals and the corresponding parameters

In the high symmetry Γ, X, L of the Brillouin zone (i.e., truncated octahedron for diamond, zinc blende, and cristobalite, see Fig. 61a) these parameters are related to the energy values in these points by the simple expressions:

$$E(\Gamma_1) = 4A + 4C + 16G; \quad E(X_1) = -8T - 8S + 4A, \quad E(L_3) = -4T - 8S + 4G$$

etc. Then the bond orbital fundamental parameters (V_1, V_2, V_3 or W_2, W_3) can be determined from the parameters K, A, C, T, S, G, which are computed from the measured values of the interband transition energies.

7.2 Lattice Energy of Ionic Crystals

This problem is considered in many works on solid states physics [149], crystal chemistry [369] and geochemistry [5, 8, 11]. Until recently, an extremely important part of interpretations of geochemical processes has been assigned to lattice energy. Over the last years better understanding of the limitations and possibilities of different approaches to the determination of the lattice energies and their uses in geochemistry has been gained. Methods of lattice energy calculations have been developed by use of special programs and of present-day computing facilities.

In order to discuss all the aspects of this problem, bearing in mind the place of each of them in the contemporary picture of the chemical bond, and realizing their logical hierarchy, one has first to compare the different approaches to the determination of lattice energy.

1. Determination of the lattice energy from interactions between atoms and electrons in crystals (Fig. 118).

A. Quantum-mechanical determinations.

a) From electron–electron, electron–nucleus, nucleus–nucleus interactions. In the case of molecules these are carried out in terms of the molecular orbital method, in the case of crystals within the framework of various methods of the energy band theory. The binding energy (the cohesion energy of molecules and crystals) is determined in these cases as the difference between the electron energy in crystal (or molecule) and the mean energy of valence electrons of free atoms.

These are the most rigorous methods, but their principles make it difficult to obtain accurate values of the cohesion energy. This is because the latter are the small quantities obtained here from the difference of great quantities of electron energies in crystals and in free atoms. Moreover, all methods of molecular orbitals and of energy band theory are based on the one-electron approximation, while the cohesion energy is essentially a collective quantity and therefore it is necessary to introduce the correlation energy. It is, nevertheless this direction which holds the greatest promise for obtaining the most reliable values, using (as in the case of the energy band scheme construction) empirical parameters inserted at different stages of the calculation. Recently, the calculations of cohesion energy have been developed by means of pseudopotential [146] and by using the bond orbital approximation [460].

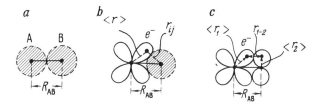

Fig. 118a–c. Interaction models in calculations of the lattice energy. **a** Model of ionic-ionic interaction (Born model): pure ionic M^+X^- state; point charges; **b** interaction model in the crystal field theory: point ligands (one ligand ion is shown) acting on atomic d orbitals of the central ion; **c** quantummechanical description: electron–electron, electron–nuclear, nuclear–nuclear interactions

b) From interactions between the electron of the atom in crystal and other atoms taken as point charges, i.e., the model similar to that of the crystal field (see Chap. 2.1). The possibility of obtaining numerical values which can be related to spectroscopical (optical, EPR, Mössbauer) parameters, as well as calculations of crystal field stabilization energy for transition metal ions are the principal advantages of these methods. However, covalency in the model of point charges is either taken into account empirically (using spectroscopical parameters) or is replaced by the concept of polarizibility borrowed from the electrostatic theory.

B. Nonquantum determinations, i.e., from interactions between ions referred to as point charges equal to formal charges. The cohesion energy between such ions is determined by electrostatic (Coulomb) interactions. This Madelung-Born-Meyer model is naturally suitable only for calculations of the lattice energy of ionic crystals: in pure covalent crystals, the charges are equal to zero, and in partly covalent crystals this model can be used to the extent to which these crystals are ionic. This model is suitable only for the compounds of monovalent ions of M^+X^- type (alkali halides) but not for compounds of other valence states because their effective charges differ strongly from formal charges. If it were possible to take into account the covalency and variety of interactions, then this model would have a very important advantage; namely, cohesion energy thus obtained would not be a small, relative value as compared to the high values of valence electron energies as in the case of quantum–mechanical calculations.

2. The determination of the lattice energy from the processes accompanied by disruption of linkage (experimental thermochemical determinations). In molecules, the cohesion energy is determined by their dissociation energies [322]. In crystals the lattice energy U is determined as the heat of decomposition of crystal into gaseous ions (for example $TiO_2 \rightarrow Ti^{4+} + 2O^{2-}$). As a rule it is not obtained directly, but by summation of the energies of successive processes (Haber-Born cycle; Figs. 118 and 119): heat of crystal decomposition to standard elements $TiO_2 \rightarrow Ti + 2O$ (enthalpy: $-\delta H_{st}$) or oxides (enthalpy: $-\delta H_{ox}$); sublimation energy of metal $\delta H_{subl} Ti_{sol} \rightarrow Ti_{gas}$ and dissociation energy of nonmetal $\delta H_{diss} O_2 \rightarrow O + O$; ionization energies of both elements: $Ti \rightarrow Ti^{4+}$ (ionization potential) and of $2O \rightarrow 2O^{2-}$ (electron affinity F)

$$U = \delta H_{st} - \delta H_{subl} - \delta H_{diss} - IP + F.$$

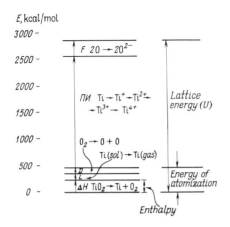

Fig. 119. Components of the lattice energy of TiO_2 ($U = 2874$ kcal mol^{-1}): enthalpy: $-\Delta H = 218.0$; sublimation heat of Ti-metal $L = 101.0$; dissociation energy of O_2 $D = 117.2$; ionization potential $Ti \rightarrow Ti^{+4}$ IP $= 2100$; electron affinity of $2O^{2-}: -F = 169 \cdot 2 = 338$ kcal mol^{-1}

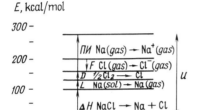

Fig. 120. Components of the lattice energy of NaCl ($U = 184$ kcal mol^{-1}): $-\Delta H = 98$, $L = 26$, $D = 29$, IP $= 118$, $-F = 87$ kcal mol^{-1} (note that the scale of energies here is increased five-fold as compared with Fig. 119)

The crystal decomposition to gaseous atoms is called the atomization energy (E_{at}); then $U = E_{at} - IP + F$, i.e., the lattice energy differs from the atomization energy by the sum of the metal ionization energy and the electrons affinity of nonmetal.

All these thermochemical values are, of course, dependent on the bond strength of atoms in crystal, but it is only the lattice energy which is comparable to the value obtained from calculations of atom interactions. However, these experimental, but in essence, calculated thermochemical values are obtained relative to dissociation to completely ionic states. Predetermination of such calculation is related to the fact that the energies of ionization to the states with formal charges corresponding to pure ionic states ($Ti \rightarrow Ti^{4+}$ and $O \rightarrow O^{2-}$) are included into the lattice energy. For multicharge ions (see, for example, TiO_2, Fig. 119) the sum of ionization potentials is 10 to 20 times larger than the energy of other components included in the lattice energy calculation, and accounts for 70 to 80% of the lattice energy. For TiO_2 ΣIP of

$$Ti \rightarrow Ti^+ \rightarrow Ti^{2+} \rightarrow Ti^{3+} \rightarrow Ti^{4+} = 2100 \text{ kcal},$$

electron affinity of $O \rightarrow O^{2-} = 169 \cdot 2 = -338$ kcal, while $\delta H_{st} = 218$ kcal, $\delta H_{subl} = 101$ kcal and $\delta H_{diss} = 117.2$ kcal. It is a drastic increase of the ionization potentials of multicharged ions $M^+ \rightarrow M^{2+} \rightarrow M^{3+} \rightarrow M^{4+}$, which gives rise to sharp changes of the lattice energy, when passing from M^+X^- to $M^{2+}X^{2-}$, $M_2^{3+}X_3^{2-}$, $M^{4+}O_2^{2-}$.

It is the enormous ionization energy values of multicharge ions which make the thermochemically obtained lattice energies an order of magnitude greater than formation enthalpies, dissociation energies, and atomization energies (Fig. 120). The atomization energy, not including ionization energies, has often been used in the last years as characteristic of the bond strength [335]. It is, however, only an indirect characteristic of binding energy and its meaning is, in fact, no more than the sum of decomposition heat (enthalpy) and the sublimation and dissociation energies, but not of cohesion energy of ions in lattice.

Calculations of the Lattice Energy of Ionic Crystals [149, 162, 353, 354, 358, 361]. The system of ions taken as rigid spheres (i.e., without consideration of electronic structure of ions) not overlapping (i.e., without consideration of the covalency), and with positive and negative charges (being equal to the valence) at the points of the crystal lattice, is exposed to two types of force.

1. Attractive forces between ions of opposite charges and repulsive forces between ions with charges of the same sign. The Coulomb potential energy of the interaction between two ions A and B with charges $Z_A e$ and $Z_B e$ and interatomic distances R_{AB} is equal to

$$-\frac{Z_A Z_B e^2}{R_{AB}}$$

i.e., is inversely proportional to $1/R_{AB}$ (i.e., the first power of R_{AB}).

2. Internuclear repulsion-limiting approach of ions at short distances. This is expressed according to Born by the simple empirical function reflecting the rapid increase of interaction with decreasing interatomic distance, i.e., by inverse power function $1/R_{AB}^n$, where n is an empirically determined number (usually between 4 and 12) taken from compressibility experiments (i.e., repulsion is in inverse proportion to the n-th power of R_{AB}).

Then the interaction energy of the ion pair is

$$-\frac{Z_A Z_B e^2}{R_{AB}} + \frac{1}{R_{AB}^n}.$$

Let us pass now from the ion pair to crystal.

1. Determine the interaction of a given ion, not only with one neighboring ion, but with all ions of the crystal. To do this one multiplies each term of the above expression by structure coefficients A, Madelung constant, and B expressed in terms of A (their meaning will be explained below).

2. Summarize this for all crystal ions. The interaction energy of one ion with all crystal ions is to be multiplied by the number of ion pairs (the number of molecules) in the gram-molecule of the compound (of the crystal), i.e., by the Avogadro number N:

$$U = -N\frac{A Z_A Z_B e^2}{R_{AB}} + N\frac{B}{R_{AB}^n}.$$

Attraction and repulsion forces in crystal being equal

$$\frac{AZ_AZ_Be^2}{R_{AB}} = \frac{nB}{R_{AB}^{n+1}}.$$

Then B can be expressed as

$$B = \frac{AZ_AZ_Be^2 R_{AB}^{n+1}}{R_{AB}^2 n} = \frac{AZ_AZ_Be^2 R_{AB}^{n-1}}{n}.$$

When substituting this value into the expression of the lattice energy, one obtains the Born formula

$$U = \frac{NAZ_AZ_Be^2}{R_{AB}}\left(1 - \frac{1}{n}\right).$$

Thus, the lattice energy per mol of crystal is expressed in terms of ion charges $Z_A e$ and $Z_B e$, the distance between them (R_{AB}), Madelung constant, and empirically determined coefficient n.

The Madelung Constant and the Lattice Energy Calculation Exemplified by NaCl. Figure 121 shows a lattice plane of NaCl. Let us calculate the interaction of one ion (Na$^+$) with other lattice ions: e/r_{ij}, where r_{ij} is the distance between the given ion and each of the other ions of Na$^+$ and Cl$^-$ [162]. The main point of the calculation is that the distance to each group of ions is expressed in terms of the shortest interatomic distance R_{Na-Cl}, which in turn can be expressed in terms of the lattice parameter $R_{Na-Cl} = r = a/2$. Thus, equal distances to 12 ions with coordinations $\langle 110\rangle$, $\langle 1\bar{1}0\rangle$, $\langle 101\rangle$ etc. are $r\sqrt{2}$, and the interaction energy of the Na$^+$ with these Na$^+$ ions is $-12e^2/r\sqrt{2}$, i.e., is equal to $-12/\sqrt{2}$ in terms of e^2/r. For eight Cl$^-$ ions (they lie off the plane of Fig. 121) with $\langle 111\rangle$ coordinates, the Na–Cl distance is equal to $r\sqrt{3}$, and the energy is $+8e^2/r\sqrt{3}$, or in terms of e^2/r it is $+8/\sqrt{3}$. Thus, we sum positive and negative energies (Table 39) until additional contributions do not change (within the accuracy of the calculation) the resulting small positive value. It is this value which is the Madelung constant, i.e., the sum of inverse distances expressed in terms of interatomic distance $1/r_{ij}$ and multiplied by the number of ions with such distance.

The energy (proportional to $1/r_{ij}$) decreases with increasing distance, but when passing to the far coordination spheres, the number of ions with the given distance increases and, therefore, the convergence is attained slowly, i.e., the Coulomb interaction is a long-range interaction.

The Madelung constant in cubic crystals depends only on the structure type (individual features of different crystals are taken into account, when multiplying A by $Z_A Z_B$, i.e., by ion charges and interatomic distance). For noncubic crystal the dependence of the Madelung constant on the ratio of the lattice parameters is easily established, for example, c/a for hexagonal, trigonal and tetragonal symmetries [364].

Fig. 121. The calculation of NaCl lattice energy (see Table 39) [162]

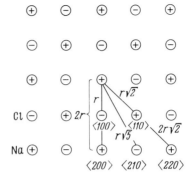

Knowing the Madelung constant for the given structure type (for NaCl $A = 1.747558 \approx 1.75$), the lattice parameter (for NaCl $a = 5.62$ Å, $R_{Na-Cl} = a/2 = 2.81$ Å $= 2.81 \cdot 10^{-8}$ cm) and the coefficient n (for NaCl $n = 9$), one can calculate the lattice energy:

$$U = NA \frac{Z_A Z_B e^2}{R_{AB}} \left(1 - \frac{1}{n}\right) = 6.02 \cdot 10^{23} \cdot 1.75 \cdot \frac{(4.8 \cdot 10^{-10})^2}{2.81 \cdot 10^{-8}} \cdot 0.9 = 77.67 \cdot 10^{11} \text{ erg}$$

or

$$(1 \text{ kcal} = 4.18 \cdot 10^{10} \text{ erg}): U = \frac{77.67 \cdot 10^{11}}{4.18 \cdot 10^{10}} = 184 \text{ kcal} \cdot \text{mol}^{-1}.$$

Mathematical methods of calculation of Madelung energy (Bertaut, Evjen and Ewald methods) have been considered in a number of papers [312, 314, 316, 337, 364]. The development of electronic computors led to direct methods of lattice sum calculations by using their special programs [314, 338, 362].

Table 39. NaCl lattice energy calculations (see Fig. 119; [162])

1	2	3	4	5
Cl	$\langle 100 \rangle$	-6	$\sqrt{1}$	$(e^2/a) \cdot (-6) \cdot (2/\sqrt{1}) = -12.000\,000$
Na	$\langle 110 \rangle$	12	$\sqrt{2}$	$(e^2/a) \cdot (12) \cdot (2/\sqrt{2}) = 16.970\,563$
Cl	$\langle 111 \rangle$	-8	$\sqrt{3}$	$(e^2/a) \cdot (-8) \cdot (2^a\sqrt{3}) = -9.237\,604$
Na	$\langle 200 \rangle$	6	$\sqrt{4}$	$(e^2/a) \cdot (6) \cdot (2/\sqrt{4}) = 6.000\,000$
Cl	$\langle 210 \rangle$	-24	$\sqrt{5}$	$(e^2/a) \cdot (-24) \cdot (2/\sqrt{5}) = -21.466\,252 \times 1/2$
Na	$\langle 211 \rangle$	24	$\sqrt{6}$	$(e^2/a) \cdot (24) \cdot (2/\sqrt{6}) = 19.595\,918$
Na	$\langle 220 \rangle$	12	$\sqrt{8}$	$(e^2/a) \cdot (12) \cdot (2/\sqrt{8}) = 8.485\,282$
Cl	$\langle 221 \rangle$	-24	$\sqrt{9}$	$(e^2/a) \cdot (-24) \cdot (2/\sqrt{9}) = -16.000\,000 \times 1/4$
Na	$\langle 222 \rangle$	8	$\sqrt{12}$	$(e^2/a) \cdot (8) \cdot (2/\sqrt{12}) = 4.618\,802] \times 1/8$
				$-3.503\,538 : 2 = 1.75$

The first column denotes the atom type, the second the coordinates of the point in NaCl lattice, the third the number of atoms of the given position, the fourth the distance from the reference point and the fifth the contribution into potential energy.

Quantum-Mechanical Calculations of Lattice Energy of Ionic Crystals. These were performed for some simple compounds using the same pure ionic structure approximation as in the Born theory. The total lattice energy is considered here to consist of not only the electrostatic term obtained here from interaction between nuclei, electrons, and between electrons and nuclei, but also of the exchange term and the overlap energy. Empirical parameters are not employed in the calculations.

The data obtained for NaCl [333, 334, 342] and other halides [360] are close to the results of the simple electrostatic calculations. Thus the more rigorous methods justify the use of the simple model in treatment of the compounds close to pure ionic ones, composed of the ions with filled shells and with effective charges approaching the formal ones as alkali halides. However, for MgO the quantum-mechanical calculations differ essentially from the simple electrostatic ones [101, 320].

Simplified formulas. By substitution of numerical values of $N = 6.021 \cdot 10^{23}$ and of $e^2 = (4.8 \cdot 10^{-10})^2$ and by conversion of R_{AB} from cm units to Å ($1 \text{ Å} = 10^{-8}$ cm), i.e., $6.02 \cdot 10^{23}(4.8 \cdot 10^{-10})^2/10^{-8} = 138 \cdot 10^{11}$ erg or $138 \cdot 10^{11} : 4.8 \cdot 10^{10} = 329.7$ kcal, the lattice energy becomes

$$U = Ne^2 A \frac{Z_A Z_B}{R_{AB}}\left(1 - \frac{1}{n}\right) = 329.7\, A \frac{Z_A Z_B}{R_{AB}}\left(1 - \frac{1}{n}\right).$$

The magnitude of the Madelung constant varies little within compounds with a similar type of chemical formula: for NaCl (coordination number 6) $A = 1.75$, for CsCl (coordination number 8) $A = 1.76$ for ZnS (coordination number 4) $A = 1.638$ (sphalerite) and $A = 1.641$ (wurtzite). Therefore, in compounds of one type (for example M^+X^-) the lattice energy is determined mainly by interatomic distances (increasing with decreasing R_{AB}). The choise of n has comparatively little influence on the lattice energy magnitude. For ions with the configuration of Ar shell n is taken to be equal to 9, Ne shell $n = 7$, He shell $n = 5$, Kr shell $n = 10$, Xe shell $n = 12$.

When passing from M^+X^- to $M^{2+}X^{2-}\ M^{4+}X_2^{2-}$ etc., the lattice energy increases four times, eight times etc., only due to the difference in charges. Moreover, when the stoichiometry is changed and the number of atoms in formula is increased, the Madelung constant increases by jumps. Thus in simplified estimations of the lattice energy it was believed in earlier papers [3] that it is possible to exclude the Madelung constant and substitute for it the sum of the atoms (Σm) in compound's formula. One takes as basis compounds of the M^+X^- type represented by the NaCl structure with $A = 1.75$. Then the energy formula becomes

$$U = 329.7\, A \frac{Z_A Z_B}{R_{AB}}\left(1 - \frac{1}{n}\right) = 329.7 \cdot 1.75/2 \frac{\Sigma m Z_A Z_B}{R_{AB}}\left(1 - \frac{1}{n}\right)$$

$$= 287.2 \frac{\Sigma m Z_A Z_B}{R_{AB}}\left(1 - \frac{1}{n}\right).$$

Taking $n=9$ and $(1-1/n)=0.9$ and substituting interatomic distance by the sum of cation radius (r_c) and anion radius (r_a) one obtains:

$$U = 256.1 \frac{\Sigma m Z_A Z_B}{R_{AB}} = 256.1 \frac{\Sigma m Z_A Z_B}{r_c + r_a}$$

which is known as the Kapustinsky formula [3, 5]. The seeming advantage of the structureless simplified formula is that it needs only the chemical composition data for the calculation (ion radii are taken from the tables). However, in such a case, the same values are obtained for crystal, for rock of the same chemical composition, and for the mixture of the corresponding oxides (i.e., the formation heat of the crystal from oxides–enthalpy, is lost here). They express properly the order of magnitude when passing from one type of compound to another, as well as the succession of increase of the lattice energies in a series of compounds of the same type. However, this information is readily apparent from the analysis of the formula, while use of these and similar simplified expressions is not advisable for obtaining numerical values.

Geochemical Applications of the Lattice Energy Concepts. It is necessary to distinguish between the use of the concept of "lattice energy" itself and the use of the lattice energy values obtained from the calculations [317, 326, 359]. In geochemistry one distinguishes between two aspects of the lattice energy concept: (1) as the value characterizing the energy decrease in the system due to the transition to the thermodynamically equilibrium crystalline state, (2) as the characteristic of the bond strength, i.e., of the energy of ion interaction with all lattice ions. In considering the energetical aspects, one needs to take into account two features of the lattice energy concept. First, thermodynamically this value is a summary one: the process of crystal decomposition to gaseous ions ($NaCl \rightarrow Na^+_{gas} + Cl^-_{gas}$) does not usually occur while the actual processes involved as successive steps are melting (or solution), decomposition to oxides, sublimation and dissociation. The energies of these individual processes should not obligatorily vary symbatically with the lattice energy variation from compound to compound. Thus the stability of compound is to be determined with respect to the concrete processes of its decomposition and to be characterized by the heats of decomposition, sublimation, dissociation, and reaction, respectively.

Second, the lattice energy does not include parameters of conditions related to the composition and the formation, namely temperature, pressure, volume, and masses or chemical potentials of components. Thus, the lattice energies do not represent energy characteristics of stability, crystallization sequences and of change of paragenetic associations; these are provided by thermodynamical potentials such as internal energy $f(S, V, n_1 \cdots, n_i)$, enthalpy $f(P, S, n_1, ..., n_i)$, Helmholtz' free energy $f(T, V, n_1, ..., n_i)$ and Gibbs' free energy $f(T, P, n_1, ..., n_i)$. Unfitness of the lattice energy for determination of crystallization sequences can be illustrated, for example by consideration of diopside–anorthite eutectic: either diopside or anorthite can be crystallized earlier, depending on the melt composition.

At the same time the thermochemical characteristics of minerals, particularly heat capacity and enthropy, are correlated with the state of the chemical bond.

It seems that the two directions of this aspect of the use of the lattice energy are apt to be important in geochemistry, namely, in intra- and intercrystalline distribution of cations [319, 321, 323, 329–331, 336, 340, 355–357] and probably, isotope fractionation, i.e., enrichment of the states with greater binding energy by a heavier isotope [315].

As a measure of the bond strength, the lattice energy can be considered only in the case of compounds with the bond of one type. In the case of complex compounds the measure of the bond strength is the sum of binding energies of two or more atomic groups (for example, $Mg-O_6$ and $Si-O_4$ in Mg_2SiO_4), or the sum of different bond energies (for example, $Mo-S_6$ bonds and the bonds between $Mo-S_6$ layers in molibdenite MoS_2). The binding energies may be of great importance when considering the mechanism of crystallization processes (as a part of crystal chemical analysis of magmatic crystallization), when describing the mechanism of processes of melting, solution, inorganic reactions etc. The structure of melts and solutions, and the forms of element transport in solutions are also determined by the binding energies.

The most direct relations exist between the binding energy and the mechanical properties, strength of crystals, elastic and plastic properties, and surface energy.

Numerical values of the lattice energies only for alkali halides are likely to be of real significance. The step-wise increase of the lattice energies when passing to compounds of divalent ions (oxides and chalkogenides), tri- and tetravalent ions does not correlate with small changes in formation heat and other thermochemical values.

Similarly, the great difference in the lattice energies of SiO_2 and Mg_2SiO_4 obtained from the simple electrostatic theory does not mean that Mg-O and Si-O bond strength in Mg_2SiO_4 differs greatly from Si-O bond strength in SiO_2, or that their melting temperatures and the thermochemical parameters are quite different, or that there is a great difference in the molecular orbital schemes of $Si-O_4$ clusters in quartz and forsterite.

Thus, relative correlations of series of compounds of the same type based on the molecular orbital schemes or on the semiempirical tight-binding approximations, as well as on improved electrostatic model calculations (see Chap. 7.3) seem to be more rigorous nowadays.

7.3 Lattice Sums, Crystal Field Parameters, Spectroscopical Parameters, and Intracrystalline Distribution

In the crystal field theory the coefficients called lattice sums (the Madelung constant is a particular case of them) are used, which can be calculated exactly from atom coordinates and provide those general characteristics of the charge distribution in the lattice. They also enter into all spectroscopical parameters as the substance characteristics irrespective of different mechanisms of obtaining electronic, nuclear and spin spectra. The same coefficients represent the method of analytical description of the actual coordination polyhedra distortions in struc-

Crystal Field Potential Expression in Terms of Lattice Sums. The crystal field is a potential created at the point of each ion site (described by valence atomic orbitals) by all other lattice ions presented as point charges; i.e., in the crystal field model the ligand ions only are taken to be the point ions while the central ion (of transition metal) is presented by atomic s, p, and d orbitals (see Fig. 29) in contrast to the simple electrostatic model taken when calculating the lattice energy (see Chap. 7.2) with the both ions taken to be the point charges.

Consider first the interaction between two point charges: between the ligand ion L with the charge Ze (or q_L) and the electron located at some point P (Fig. 122). The potential created by the ligand ion is $V = \dfrac{Ze}{r_{ij}}\left(\text{or }\dfrac{q_L}{r_{ij}}\right)$; the energy $E = eV = \dfrac{Ze^2}{r_{ij}}$. Here r_{ij} is the distance between the ligand ion and the electron of the central ion (note that neither electron radius r, nor interatomic distance R_{ML} participate here).

Thus, the problem is reduced to determining not $1/R_{AB}$ (as in Chap. 7.2) but the value $1/r_{ij}$, which is neither crystal constant (as interatomic distance R_{AB} or R_{ML}), nor electron characteristic (as its radius r). Thus it is necessary first of all that r_{ij} be expressed in terms of R_{ML} and r.

1. From Fig. 122, from the right-angled triangle $P-L-(R-z)$, one can obtain: $r_{ij}^2 = (x^2 + y^2) + (R-r)^2$, but $x^2 + y^2 + z^2 = r$ and $(x^2+y^2) = r^2 - z^2$, then $r_{ij}^2 = r^2 - z^2 + R^2 - Rz + z^2$; $r_{ij} = \sqrt{R^2 - 2Rz + r^2} = R\sqrt{1 - 2z/R + (r/R)^2}$. The same is obtainable if the position of the point P is expressed in spherical coordinates:

$$r_{ij}^2 = [(r^2 - (r\cos\vartheta)^2] + (R - r\cos\vartheta)^2 = R^2 - 2rR\cos\vartheta + r^2.$$

Then $1/r_{ij}$ is finally expressed in terms of r and R_{ML}.

2. Replacing r/R by q and $\cos\vartheta$ by t gives $\dfrac{1}{\sqrt{1-2qt+q^2}}$. However, this expression has a solution available in all mathematical reference books because it presents the Legendre polynom:

$$\frac{1}{\sqrt{1-1qt+q^2}} = \sum_{l=0}^{\infty} P_l^m(t)q^l = \sum P_l^m(\cos\vartheta)(r/R)^l$$

and

$$\frac{1}{r_{ij}} = \frac{1}{R}\sum P_l^m(\cos\vartheta)(r/R)^l.$$

3. As far as the powers of $q^l = (r/R)$ are concerned, everything is clear; being multiplied by $1/R$ they take the form r^l/R^{l+1}.

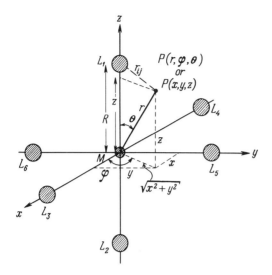

Fig. 122. Distance r_{ij} between an electron at a point P (with spherical coordinates, r, ϑ, φ, or rectangular coordinates x, y, z) of transition metal ion M and the point charge of ligand ion L_1 in terms of electron radius r and interatomic distance R

4. $P_l^m(\cos\vartheta)$ can be expanded according to l powers

$P_0(\cos\vartheta) = 1$,

$P_1(\cos\vartheta) = \cos\vartheta$,

$P_2(\cos\vartheta) = 3\cos^2\vartheta - 1$,

$P_3(\cos\vartheta) = 1/2(5\cos^3\vartheta - 3\cos\vartheta)$,

$P_4(\cos\vartheta) = 1/8(35\cos^4\vartheta - 30\cos^2\vartheta + 3)$ etc.

Moreover, before each term there is here a normalizing multiplier, $\left[\dfrac{2l+1(1-m)!}{2(1+m)!}\right]^{-1/2}$, cumbersome in the general form, but since l and m are small integers, it acquires simple values in the particular cases, for instance, $1/\sqrt{3/2}$ is obtained if $l=1$ and $m=0$; $1/\sqrt{5/2}$ if $l=2$ and $m=1$ (see Table 2).

Therefore the solution of $1/r_{ij}$ is

$$\frac{1}{r_{ij}} = \frac{1}{R\sqrt{1-2qt+q^2}} = \frac{1}{R}[P_0(t) + P_1(t)q + P_2(t)q^2 + P_3(t)q^3 + \cdots].$$

5. The position of the point $P(r, \vartheta, \varphi)$ is determined not only by the angle ϑ, but also by the angle φ (dependence on this angle is determined by $\exp(im\varphi)$ function). The product of two angular functions, in this case of $\Theta_l^m(\vartheta)$ and $\Phi_m(\varphi)$, is a spherical function Y_l^m (spherical harmonic) of the power l and order m:

$$Y_l^m(\vartheta, \varphi) = N \cdot \Theta_l^m(\vartheta) \cdot \Phi_m(\varphi) = P_l^m(\cos\vartheta) \cdot \frac{1}{\sqrt{2\pi}} \cdot \exp(\pm im\varphi).$$

6. Finally, the position of the electron is not a point position, its distribution (angular dependence) is determined by the boundary corresponding to s, p, d_{xy}, d_{z^2} etc. atomic orbitals, which themselves represent spherical harmonics, i.e., products of $\Theta_l^m(\vartheta_j)$ and $\Phi_m(\varphi_j)$ functions (see Tables 1 and 2).

Thus, we obtain finally

$$\frac{1}{r_{ij}} = \sum_l \sum_m \frac{4\pi}{2l+1} \left(\frac{r^l}{R^{l+1}}\right) Y_l^m(\vartheta_i, \varphi_i) \cdot Y_l^m(\vartheta_j, \varphi_j),$$

where the first term is the normalizing coefficient, the second corresponds to $1/R\left(\frac{r}{R}\right)^l$ with r as the average electron radius of the given atomic orbital and R as interatomic distance, the third and the fourth terms are two similar spherical harmonics, each of which represents a product of two angular functions: one product describes the angular position of ligand ions (ϑ_i and φ_i angles), another describes the electron angular distribution (ϑ_j and φ_j angles).

Here the expression for spherical harmonics $Y_l^m(\vartheta, \varphi)$ for the most important instances with $l = 0, 2, 4$ and $m = 0$:

$Y_l^m = \Theta_l^m \cdot \Phi_m \longrightarrow \Phi_m = \frac{1}{2\sqrt{\pi}} \exp(\pm i m\varphi)$

$\Theta_l^m = \left[\frac{(2l+1)(1-m)!}{2(1+m)!}\right]^{-1/2} \cdot P_l^m(x); \quad x = \cos\vartheta$

$P_l^m(x) = \frac{(-1)^l}{2l \cdot l!} (1-x^2)^{m/2} \frac{d^{l+m}(1-x^2)^l}{dx}$

Y_0^0	$1/\sqrt{2}$	1	1	$1/2\sqrt{\pi} \cdot 1^*$
Y_2^0	$\sqrt{5/2}$	1/8	$4(3\cos^2\vartheta - 1)$	$1/2\sqrt{\pi} \cdot 1$
Y_4^0	$\sqrt{9/2}$	$1/16 \cdot 24$	$48(35\cos^4\vartheta - 30\cos^2\vartheta + 3)$	$1/2\sqrt{\pi} \cdot 1$

From this, general expressions for spherical harmonics in spherical and rectangular coordinates are obtainable:

Y_0^0	$1/2\sqrt{\pi}$	$1/2\sqrt{\pi}$
Y_2^0	$\sqrt{5/4}\sqrt{\pi}(3\cos^2\vartheta - 1)$	$\sqrt{5/4}\sqrt{\pi}(3z^2 - r^2)$
Y_4^0	$3/16\sqrt{\pi}(35\cos^4\vartheta - 30\cos^2\vartheta + 3)$	$3/16\sqrt{\pi}(35z^4 - 30z^2 r^2 + 3r^4)$.

To convert from the expression in spherical coordinates to the expression in rectangular coordinates, one has simply to substitute $\cos\vartheta$ for z/r. Expressions for other harmonics are given in many works [49] etc.

---

* $\exp(\pm im\varphi) = \cos(m\varphi) \pm i\sin(m\varphi)$; if $m = 0$ $\sin(m\varphi) = 0$, $\cos(m\varphi) = 1$, $\exp(\pm im\varphi) = 1$

In the expression of potentials, two kinds of term are distinguished, namely, terms describing the position of ligand ions and those describing the distribution of electron density:

$$V = Ze \sum \frac{1}{r_{ij}} = Ze \sum \frac{4\pi}{2l+1} \cdot Y_l^m(\vartheta_i, \varphi_i) \cdot \left(\frac{r^l}{R^{l+1}}\right) \cdot Y_l^m(\vartheta_j, \varphi_j)$$

$$= Ze \sum \underbrace{\frac{4\pi}{2l+1} \cdot \frac{1}{R^{l+1}} \cdot Y_l^m(\vartheta_i, \varphi_i)}_{\text{ligands}} \cdot \underbrace{r^l \cdot Y_l^m(\vartheta_j, \varphi_j)}_{\text{electrons}} = Ze \sum A_l^m \cdot r^l \cdot Y_l^m(\vartheta_j, \varphi_j).$$

The coefficients A_l^m here embrace characteristics of the position of ligands, namely, interatomic distance, R, and $Y_l^m(\vartheta_i, \varphi_i)$ i.e., the product of angular functions describing the arrangement of ligands in terms of ϑ_i and φ_i angles; they are called lattice sums.

The second part of the expression, i.e., $r^l Y_l^m(\vartheta_j, \varphi_j)$, embraces characteristics of the distribution of electron density, namely, electron radius r^l and $Y_l^m(\vartheta_j, \varphi_j)$, i.e., the angular part of the electron wave function in the form of spherical harmonics.

Since A_l^m and $r^l Y_l^m$ enter into each term V_l^m with the same l and m, the expressions for Y_l^m describing the angular arrangement of ligand ions and the angular distribution of electron density are completely identical, with the same normalizing coefficient, but they have a different meaning. In A_l^m the angles describe the position of ions, while in r^l, the angular distribution of electron density for the wave function $\psi_{lm} = r^l \cdot Y_l^m$ is represented, i.e.,

$$Y_2^0 = d^0 = d^{z^2}(l=d=2; m=0); \qquad Y^{2\pm 1} = d_{\pm 1};$$

$$Y_2^{\pm 2} = d_{\pm 2}[d_{xy} = 1/2(d_1 - d_{-1})] \qquad d_{yz} = -1/i\sqrt{2}(d_1 + d_{-1}),$$

$$d_{xy} = 1/i\sqrt{2}(d_2 - d_{-2}), \qquad d_{x^2-y^2} = 1/\sqrt{2}(d_2 + d_{-2}).$$

A_l^m values in particular cases are simply numbers. Here is for instance, the A_4^0 value in the octahedron (see Fig. 122).

Ions	ϑ_i	φ_i	$\cos \vartheta_i$	$(35\cos^4\vartheta - 30\cos\vartheta + 3)$
L_1	0	—	1	35−30+3=8
L_2	180	—	−1	35−30+3=8
L_3	90	0	0	0− 0+3=3
L_4	90	180	0	3
L_5	90	90	0	3
L_6	90	270	0	3
				28

Lattice Sums, Crystal Field Parameters

Ions	ϑ_i	$\cos \vartheta_i$	$3\cos^2\vartheta - 1$
L_1	0	1	4
L_2	180	−1	2
L_3	90	0	−1
L_4	90	0	−1
L_5	90	0	−1
L_6	90	0	−1
			$4-4=0$

$$A_4^0 = \frac{4\pi}{2l+1} \cdot \frac{1}{R^{l+1}} \cdot Y_l^m = \frac{4\pi}{9} \cdot \frac{1}{R^5} \cdot Y_4^0(\vartheta_i, \varphi_i)$$

$$= \frac{4\pi}{9R^5} \cdot \frac{3}{16\sqrt{\pi}} (35\cos^4\vartheta - 30\cos^2\vartheta + 3) = \frac{4\pi}{9R^5} \cdot \frac{3}{16\sqrt{\pi}} \cdot 28 = \frac{7\sqrt{\pi}}{3R^5}.$$

$$A_2^0 \frac{4\pi}{2l+1} \cdot \frac{1}{R^{l+1}} \cdot \frac{5}{4\sqrt{\pi}} (3\cos^2\vartheta - 1) \text{ in the octahedron (and in other coordi-}$$

nations with cubic symmetry) becomes zero:

As the symmetry is reduced to the axial one (and lower) angles deviate from right angles or R_1 and R_2 differ from R_{4-6}; as a result field gradient arises and the term A_2^0 appears. Thus, the term A_l^m in the expression of the potential $V_l^m = Ze\Sigma A_l^m \cdot r^l \cdot Y_l^m$ presents the lattice sum determined by the position of ligand ions and expressed in particular instances simply by numbers. Thus, for the octahedron:

$$V_4^0 = Ze\Sigma A_4^0 r^4 Y_4^0 = 6 \cdot Ze \frac{7\sqrt{\pi}}{3R^5} r^4 Y_4^0.$$

In binary compounds, A_l^m is a purely geometric value, i.e., it is dependent only on coordinates of atoms, in the case of triple and more complex compounds, however, when the potential is formed from contributions of ions with different charges, one can determine A_l^m for each sublattice of similar ions, but in order to obtain the total potential, it is necessary to know the ion charges.

Thus, we have obtained the expression for the terms of the crystal field potential having unfolded the value A_l^m but not yet unfolding the value $Y_l^m(\vartheta_j, \varphi_j)$ for electron density distribution.

Consider now the total potentials for some concrete types of crystal field which are obtained as the sum of the terms:

cubic field

$$V_{cub} = V_4^0 + V_4^4 + V_4^{-4};$$

tetragonal field

$$V_{\text{tetr}} = V_2^0 + V_4^0 + V_4^4 + V_4^{-4} \quad (\text{i.e.,} \ V_4^0 + V\text{cub});$$

trigonal field

$$V_{\text{trig}} = V_2^0 + V_4^0 + V_4^3 + V_4^{-3} \quad (\text{i.e.,} \ V_2^0 + V_{\text{cub}},$$

but with fixed three-fold symmetry axis);

rhombic field

$$V_{\text{rhomb}} = V_2^0 + V_2^0 + V_4^0 + V_4^2 + V_4^{-2} + V_4^4 + V_4^{-4} \quad (\text{i.e.,} \ V_2^2, V_4^2, V_4^{-2} + V_{\text{tetr}}).$$

l and m in V_l^m (as well as in A_l^m and Y_l^m) take on the following values: (1) for d electrons $l = 2, 4$; cubic symmetry $l = 4$, lower symmetry $l = 4, 2$; for f electrons $l = 2, 4, 6$; cubic symmetry $l = 4, 6$, lower symmetry $l = 2, 4, 6$; (2) m in Y_l^m is dependent on φ angle as $\exp(im\varphi)$, therefore for d electrons $m = 0$ in the case of axial symmetry, $m = \pm 2$ at rhombic symmetry, $m = \pm 3$ at trigonal symmetry $m = \pm 4$ at tetragonal, $m = 0$ and ± 4 in the case of cubic symmetry; for f electrons terms with $l = 6$ appear.

Now, write the potential of the octahedral field (dropping for the time being Ze):

$$V_{\text{oct}} = V_4^0 + V_4^4 + V_4^{-4} = A_4^0 r^4 Y_4^0 + A_4^4 r^4 Y_4^4 + A_4^{-4} r^4 Y_4^{-4},$$

and since $A_4^{\pm 4} = \sqrt{\frac{5}{14}} A_4^0$ (for the octahedron), then

$$V_{\text{oct}} = A_4^0 r^4 Y_4^0 + \sqrt{\tfrac{5}{14}} A_4^0 r^4 Y_4^4 + \sqrt{\tfrac{5}{14}} A_4^0 r^4 Y_4^{-4} = A_4^0 r^4 [Y_4^0 + \sqrt{\tfrac{5}{14}} (Y_4^4 + Y_4^{-4})].$$

When expressing Y_4^0, Y_4^4 and Y_4^{-4} in rectangular coordinates we have:

$$V_{\text{oct}} = A_4^0 r^4 \left[\frac{15}{4\sqrt{\pi}} (x^4 + y^4 + z^4 - 3/5 r^4) \right]$$

or, when substituting $A_4^0 = \frac{7\sqrt{\pi}}{3R^5}$,

$$V_{\text{oct}} = \frac{7\sqrt{\pi}}{3R^5} r^4 \left[\frac{15}{4\sqrt{\pi}} (x^4 + y^4 + z^4 - 3/5 r^4) \right]$$

$$= \frac{35 r^4}{4 R^5} (x^4 + y^4 + z^4 - 3/5 r^4) = D r^4 (x^4 + y^4 + z^4 - 3/5 r^4),$$

where $D = \dfrac{35}{4R^5} = \dfrac{15}{4\sqrt{\pi}} A_4^0$.

Now, consider the value of the term characterizing the electron wave function: $r^4(x^4 + y^4 + z^4 - 3/5 r^4)$, i.e., the expression obtained by summing three cubic harmonics Y_4^0, Y_4^4 and Y_4^{-4} in rectangular coordinates. Let us find its solution for the mean radius $\bar{r}$ and mean coordinates of the electron.

A special method known as the method of equivalent operators has been elaborated for this purpose. We shall not consider it, because ready solutions for all required states are available in the tables ([81]; Tables 5–9) whence we obtain:

$$\sum(x^4+y^4+z^4-3/5r^4)$$
$$=\beta\bar{r}^4[I_x^4+I_y^4+I_z^4-1/5I(I+1)(3I^2+3I+1)]$$
$$=\frac{\beta\bar{r}^4}{20}[35I_z^4-30I_z(I_z+1)I_z^2+25I_z^2-6I(I+1)+3I^2(I+1)^2]$$
$$+\frac{\beta\bar{r}}{8}[I_+^4+I_-^4].$$

$I=2$ for d electrons; we find coefficient $\beta=2/63$ (for D states in Table 5 in [181], the total coefficient $F=12$ for $I=2$ in Table 7a in [181]:

$$r^4(x^4+y^4+z^4-3/5r^4)$$
$$=\frac{\bar{r}^4\beta}{20}(35I_z^4-30I(I+1)I_z^2+\cdots)$$
$$=\frac{\bar{r}^4\cdot 2}{20\cdot 63}\cdot 12=\frac{2}{105}\bar{r}^4=q;$$

then

$$Dr^4(x^4+y^4+z^4-3/5r^4)=Dq=\frac{35}{4R^5}\frac{2r^4}{105}=\frac{1}{6}\cdot\frac{\bar{r}^4}{R^5}.$$

For d electrons split into two states t_{2g} and e_g $I_z=1$ for t_{2g} and $I_z=0$ for e_g. In the same Table 7a in [181] we find for these values multipliers 6 for e_g and -4 for t_{2g}. Hence for e_g

$$r^4(x^4+y^4+z^4-3/5r^4)=\frac{\bar{r}^4\beta}{20}(35I_z^4-\cdots)=\frac{2}{105}\bar{r}^4\cdot 6=6q$$

and

$$V_{oct}=Dr^4(x^4+y^4+z^4-3/5r^4)=6Dq;$$

for t_{2g}: $V_{oct}=-4Dq$.

Then the crystal field strength (the separation between t_{2g} and e_g) is

$$\Delta=6Dq-(-4Dq)=10Dq=\frac{10\bar{r}^4}{6R^5}=\frac{5}{3}\cdot\frac{r^4}{R^5}.$$

Lattice sums A_l^m enter into spectroscopical and energy parameters in the following forms:

A_0^0 → interaction energy of ions with filled shells (Madelung energy);
ΣA_l^m → crystal field stabilization energy (for ions with d and f electrons) ($l \neq 0$);
$A_0^0 + \Sigma A_l^m$ → total energy of interaction of ions with d and f electrons;
A_4^0, A_4^4 → Dq: crystal field strength in the case of cubic symmetry for d ions;
A_2^0 → axial splitting by crystal fields of trigonal and tetragonal symmetry (in optical absorption spectroscopy); crystal field gradient in NMR and NQR spectra; quadrupole splitting in Mössbauer spectra; "axiality" in EPR spectra;
A_2^2 → splitting by rhombic and lower fields in optical spectra; asymmetry parameter in nuclear spectra (NMR, NQR and NGR spectra); "rhombicity" in EPR spectra;
$+A_6^{0,2,4,6}$ → for ions with f electrons.

The coefficient A_0^0 is of special significance:

$$V_0^0 = Ze\Sigma A_l^m r^l Y_l^m = Ze\Sigma \frac{4\pi}{2l+1} \cdot \frac{1}{R^{l+1}} \cdot Y_l^m \cdot [r^l \cdot Y_l^m]$$

$$= Ze \frac{4\pi}{1} \Sigma \frac{1}{R} \cdot \frac{1}{2\sqrt{\pi}} \cdot 1 \cdot \frac{1}{2\sqrt{\pi}} = Ze\Sigma \frac{1}{R},$$

i.e., A_0^0 is the Madelung lattice sum. Energy $E = eV_0^0$ is much larger than the crystal field stabilization energy determined by other lattice sums A_l^m with $l \neq 0$; A_0^0 is inversely proportional to the first power of interatomic distance $R\left(A_0^0 \approx \frac{1}{R}\right)$, whereas other A_l^m are inversely proportional to higher powers of $R\left(A_l^m \approx \frac{1}{R^{l+1}}\right)$ (Fig. 123). A_0^0 does not enter into spectroscopical parameters, because the potential related to it causes only shift of all levels.

A_4^0 acquires different values depending on coordination (octahedral, tetrahedral, cubic), thus determining different values of the crystal field strength:

$$A_{4(oct)}^0 = \frac{35}{4R^5}; \quad A_{4(tetr)}^0 = -\frac{35}{9R^5}; \quad A_{4(cub)}^0 = -\frac{70}{9R^5};$$

i.e., $Dq_{oct} = \frac{9}{4}Dq_{tetr} = \frac{9}{8}Dq_{cub}$.

A_2^0 reflects axial distortions, especially pronounced in EPR spectra $B_2^0 = A_2^0$ $\cdot r^2 \sqrt{5/4\sqrt{\pi}}$, i.e., including normalizing coefficient of $Y_2^0(\vartheta_j, \varphi_j)$ (describing the spin function and electron radius r), as well as in nuclear spectra (NMR, NQR, NGR) where A_2^0 enters into expressions for the crystal field gradient with which the nucleus quadrupole moment interacts.

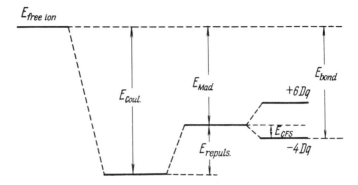

Fig. 123. Contribution of Madelung energy (E_{Mad}) and crystal field stabilization energy (E_{CFS}) into binding energy (E_{bond}); E_{Coul}, energy of Coulomb interaction between the central ion and ligands; E_{repuls}, repulsion energy

The lattice sum calculations were performed for garnet [318], zircon [347], andalusite, kyanite, sillimanite [344], spodumene [345], hydrated iron oxides [325] and in great detail for corundum considering the charges not as point charges, but as dipole, quadrupole and octupole charges [327, 328]. It has been possible to attain a good agreement with experiment when calculating the crystal field gradient in Al_2O_3, Fe_2O_3 etc., as well as in ludwigite-vonsenites with regard to different exchange components [351, 352].

7.4 Atomic and Ionic, Orbital, and Mean Radii

The concept of atomic sizes is a natural component of the physical picture of the world, a part of the atomic model of matter. Structure models of crystal chemistry, close packing of atoms, stability rules for coordination polyhedra with different cation–anion size ratios, limits of isomorphic substitutions, and estimation of interatomic distances are related to the atomic sizes. In geochemistry, the history of atoms of the earth, the atomic sizes also form an essential and the most pictorial part of the concepts of atoms, all related crystal chemical propositions are at the same time geochemical propositions and are moreover when estimating volumes of elements in the earth's crust, discussing the phenomena of migration diffusion etc.

Numerous attempts to compile the systems of atomic or ionic radii have been made for more than 50 years: beginning from Bragg in 1920 [367] then Wasastjerna [397], Pauling [113], Goldschmidt [6], Zachariasen [367], Ahrens [366], Belov and Bokij [369], Batsanov [367] and in recent years Shannon and Prewitt [390], Whittaker and Muntus [399], Van Vechten and Phillips [369] etc. [366, 367, 386].

However, in spite of the different approaches to the system of radii, all these works use the same conception, the decisive features of which are an assumption of constancy and additivity of radii and approximation of the pure ionic (or pure covalent) bond.

It should be noted that the system of orbital radii based on quantum-mechanical calculations of the wave functions (see below) introduced by Slater [393] and extended by Shchukarev and Lebedev [380] belongs to the same crystal chemical conception of constant and additive radii, but uses an approximation of the covalent bond for all crystal types (even for the most ionic). This long period of development, mostly crystal chemical, is nearing its completion.

New and more real possibilities in understanding and estimating atomic sizes are offered by applying quantum-mechanical analysis of the electron density distribution directly in molecules and crystals.

7.4.1 Ionic Radii and Molecular Orbitals

In order to understand the main point of atomic size dependencies on different factors, one should consider the modern model of the chemical bond i.e., the molecular orbitals method (see Chap. 3.2) as well as results of calculations of free atom wave functions available for all atoms (see Chap. 1.5), to wave functions of atoms in crystals calculated at present only for some compounds [371, 383, 387], and to molecule topography [77, 121].

In previous sections (Chap. 7.2 and 7.3) the angular parts of atomic orbitals were considered when constructing and analyzing the molecular orbital schemes. Consider now the role of the radial part of the wave functions (atomic orbitals) in the formation of molecular orbitals and their relation to the atomic sizes.

As applied to free atoms, the radial parts of atomic orbitals have already been discussed in detail (see Chap. 1.5). In addition, if should be noted that all quantum-mechanical calculations (for instance, lattice sum calculations discussed in Chap. 7.3) include not orbital radii but mean radii r_{mean} (Fig. 124), which are 0.1–0.2 Å greater than r_{orb} for the given atomic orbital, because the shape of curves of radial distribution is asymmetric and their external (relative to the maximum) part is extended much farther than the internal part.

At first, as a zero approximation, superimpose curves of radial distribution of electron density of two atoms (Fig. 125), for example, those of Na and Cl taken at the distance equal to the interatomic distance in crystal ($R_{NaCl} = 2.81$ Å). Ionic radius can be determined thereby from the overlap minimum or from the electron density minimum. In both cases the ionic sizes are determined from the total electron densities and not from individual valence atomic orbitals.

Now, consider molecular orbital schemes (see Fig. 48). From these it follows first of all that the bond is formed not from one pair of orbitals but from a number of molecular orbitals obtained as linear combination of ligand s and p orbitals and atomic s, p, and d orbitals. And even empty (excited) atomic orbitals of the central atom can participate in formation of molecular orbitals.

Each molecular orbital covers the whole molecule or complex, and boundaries of each atomic orbital can be determined in it only conditionally, i.e., from the electron density minimum in the topography of this molecule; in the case of high covalency this minimum may be not observable.

Ionic Radii and Molecular Orbitals

Fig. 124. Mean, orbital and atomic radii in case of free atom

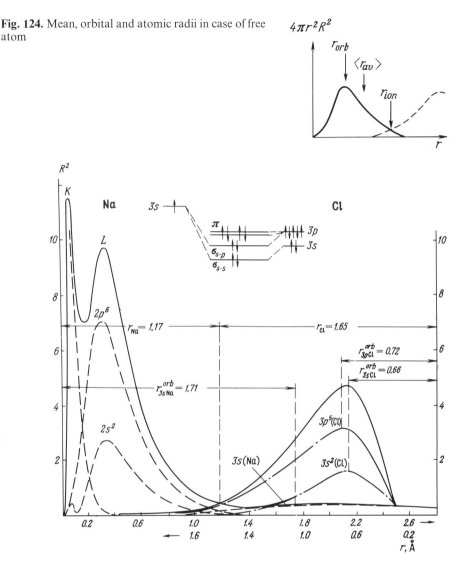

Fig. 125. Superposition of atomic orbitals and total electron densities of Na^+ and Cl^- ions in NaCl

However, the separation of atomic orbital boundaries in molecular orbitals is possible, as it is possible to determine the fraction of atomic orbitals in molecular orbitals by c_A, c_B coefficients.

It should be taken into account that one atomic orbital can participate ("be exchanged") in a number of molecular orbitals. In addition to σ-bonds, the overlap yielding π-molecular orbitals, as well as the presence of antibonding molecular orbitals, has to be taken into consideration. Simple superposition of atomic orbitals (as in Fig. 125) ignores the differences in their relative energies (their

VSIE = H_{AA}) influencing (together with H_{AB}) on effective charge and effective electron configuration.

All this concerns valence electrons. The wave functions of inner electrons (core electrons) are characterized by small values of radii of electron density maxima (r_{orb}), but since the values of electron density for inner shells (especially those of cations) are much higher than those for valence shells, so even tails of these wave functions have a high density. Therefore the electron density of core electrons also influences the atomic sizes.

Thus, the atomic sizes in compounds are determined by their total electron density (and not by individual valence atomic orbitals) including the core electrons, but anylysis of its dependencies requires consideration of the overlapping of individual pairs of atomic orbitals.

The sizes of free atoms are much greater than those of free cations (at the expense of the loss of valence electrons where only inner core electrons are left, see Chap. 1.5), but only slightly smaller than the anionic sizes.

To estimate the relative anionic and cationic sizes in molecules and crystals, notice that: (1) bonding molecular orbitals in ionic-covalent compounds are mainly anionic (ligand) ones; c_A coefficients (i.e., those of cations) in molecular orbitals are always less than c_B (those of ligands); (2) the number of cation valence electrons is always less than that of anion (s^1, s^2, s^2p, s^2p^2 in M^I, M^{II}, M^{III}, M^{IV} as against s^4p^4, s^2p^5 in B^{VI} and B^{VII}) and cation coordination is greater than or equal to anion coordination; (3) the wave function of cation valence electrons differs greatly from the wave functions of electrons of inner shells; it extends very far, but the slope is extremely gently; in contrast, the wave functions of anionic electrons are condensed and much more intensive. The valence electrons of the anion are essentially its core electrons.

Therefore, the valence electrons of cations in compound do not directly determine their sizes, because these electrons enter the molecular orbitals, which are mainly of ligand character. In most ionic compounds, the ionic sizes are determined mainly by the overlap of the wave functions of cation core with valence orbitals of anion. The contribution of atomic orbitals of cation in molecular orbitals describing the total electron density distribution of the compound increases with increasing covalency.

Because of this, the cationic sizes in compounds decrease sharply in comparison with the atom (similar to the free cation sizes), while anion sizes are almost unchanged with respect to the atom.

Consider now whether the bond length of the same ions, for example, Si–O, varies in different compounds.

First of all, the bond length varies with the coordination number (CN), not only of the cation, but also of the anion. The variation of the CN of the cation is responsible for the change of the type of molecular orbitals' scheme, for the different expression for group overlap integral, different number and shape of each of the molecular orbitals, and for the atomic orbital overlap at different angles. Variation of the CN of the anion causes the change of its effective charge and effective electron configuration. A greater CN of cation and anion leads to the increase of their sizes primarily because of the greater number of neighbours with which the electron density is shared, but the bond length between them is dependent on both the CN of the cation and that of the anion, determining the relative number of electrons participating in the bond on both sides.

The coordination being the same, the length of one and the same bond varies according to second neighbors: Si–O distance in Si–O–Si and in Si–O–M (bridging and nonbridging oxygen in silicates) is different, owing to differences in the effective charge of oxygen and hence, in the distribution of its electron density.

Consider the two of variation of mean distances as well as individual bond variation within a coordination polyhedron (in Å):

Si–O in SiO molecule $R_{av} = 1.44$,

Si–O_4 in quartz $R_{av} = 1.61$,

Si–O_6 in stishovite $R_{av} = 1.77$.

In 3T muscovite

Si_1–O_a = 1.622 $\quad$ Si_2–O_b = 1.6192

O_c = 1.690 $\quad$ O_c = 1.5452

O_d = 1.675 $\quad$ O_d = 1.5622

O_e = 1.698 $\quad$ O_e 1.6862.

That is, in 3T muscovite the individual Si–O lengths vary from 1.698 to 1.545 Å (i.e., by 0.15 Å) and bonds in the Si–O_4 polyhedron vary only between 0.05 and 0.07 Å depending on mean CN of oxygen (1.604 at CN of oxygen two and 1.651 Å at CN four).

The lengths of individual Fe^{2+}–S bonds in tetrahedra vary between 2.28 and 2.36 Å and in octahedra between 2.26 and 2.65 Å.

Consider now the sizes of the same cation in combination with different anions (Fe^{2+} in Fe–O, Fe–S, Fe–Cl ...) or the sizes of the same anion in combination with different cations (O^{2-} in Na–O, Fe–O, Si–O, As–O ...). When passing from the Na–O bond to Fe–O or Si–O or other bonds, different atomic orbitals with a different number of electrons, overlap integrals and ionization energies of valence states become involved in the bond (even for the same CN); it causes not only sharp changes in equilibrium interatomic distances, belonging principally to self-consistent values in the calculation of molecular orbitals, but also the principal change in molecule or complex topography. In zero approximation, it can be estimated from superposition of the total electron densities of atom pairs at the distance equal to interatomic distance in the crystal or molecule.

Thus, ionic sizes, like all properties of atoms in compounds, are variable quantities; the ionic sizes are the sizes for each individual pair of ions (for instance silicon and oxygen sizes in SiO_2 differ from the silicon size in SiF_4 and oxygen sizes in MgO) and for each coordination number. They vary from compound and from one nonequivalent bond to another, according to the second neighbors and structure geometry. There are no ionic sizes in general, but there are the sizes with respect to one or another of the ions; this value is always to be determined for a pair of ions.

7.4.2 Systems of Additive Ionic and Atomic Radii

Two steps in compiling any of the systems of radii can be distinguished: (1) determination of reference values of radii of some ions; (2) construction of the system of radii.

1. Radius values were determined based on the following considerations:

a) From consideration of interionic distances, measured by means of X-ray diffraction methods, presenting ions as adjoining spheres.

Bragg (1920) determined the sulfur radius as 1.03 Å from S–S distance in pyrite (where the atoms are bound covalently) equal to 2.05 Å. Then $r_{Zn} = d_{ZnS} - r_S = 2.35 - 1.03 = 1.32$ Å, $r_O = d_{ZnO} - r_{Zn} = 1.97 - 1.32 = 0.65$ Å. Thus, the atomic radii were obtained where the cations were found to be large and the anions small. Later Bragg determined the radius of the oxygen ion from the mean O–O distance in crystals being 2.7 Å, and obtained $r_{O^{2-}} = 1.35$ Å [367]. Lande (1920) assumed from the interatomic distances of pairs of compounds MgSe = 2.73 Å and of MnSe = 2.73 Å, or MgS = 2.60 Å and MnS = 2.59 Å etc., assumed that the anions are in contact with each other then $r_{Se^{2-}} = 2.73$ Å $\cdot \sqrt{2}/2 = 1.93$ Å and $r_{S^{2-}} = 2.60$ Å $\cdot \sqrt{2}/2 = 1.83$ Å and $r_{Mg^{2+}} \leq 0.76$ etc (the same for LiI and Li$_2$Te; see [379]). If the ions are not assumed to be rigid adjoining spheres, then (1) S–S, O–O, etc. distances (not only in these, but in any structures) give sulfur or oxygen boundaries along S–S or O–O directions as overlap minimum of their electron densities, but only for each particular structure, (2) the cationic sizes are not determined thereby, because the overlap of its electron density with the anion is quite different from S–S or O–O overlap, and the overlap minimum of Mg–S or Mn–S etc. is not related to S–S overlap minimum.

b) From refractometry measurements, Wasastjerna (1923) found molar refraction to be

$$R_{mol} = \frac{n^2-1}{n^2+2} \cdot \frac{M}{d} \frac{\varepsilon-1}{\varepsilon+2} = \frac{4}{3}\pi N\alpha = 0.6023 w(r_{cat}^3 + r_{an}^3)^c,$$

where n is the refraction index, M the molecular weight; d the density; ε the dielectric constant; α the polarizability; coefficient 0.6023 is obtained from the Loschmidt number N (number of molecules per 1 cm^3); w is the valence, and c the empirical constant (for $\lambda = 589$ nm, $c = 1.365$). The ratio of interatomic distances is then taken as $\sqrt[3]{\alpha_{cat}} : \sqrt[3]{\alpha_{an}}$ under the assumption that $r^3 \approx \alpha$ and that all polarizing electrons belong to the anion, i.e., under assumption of the pure ionic bond [397].

Kordes used the relation $\lg R_{mol}^I : \lg R_{mol}^{II} = (r_{cat}^3 + r_{an}^3)^I : (r_{cat}^3 + r_{an}^3)^{II}$ (where R_{mol} are molar refractions) for compounds with one common ion. Having measured molar refractions and interatomic distances, for example, for NaCl and KCl, the three radii of Na$^+$, K$^-$ and Cl$^-$-ions can be determined, but not four radii (r_{Cl} in NaCl and r_{Cl} in KCl), i.e., the equality of one of the radii in two compounds has to be assumed [374, 375].

Wasastjerna determined the reference radii to be of F$^- = 1.33$ Å, of O$^{2-} = 1.32$ Å etc., according to Kordes the radius of F$^- = 1.32$, of O$^{2-} = 1.35$ Å.

c) From the ratio of nuclear charges of isoelectronic ions.

Pauling (1927) noted that Na^+ in NaF had the electron configuration $1s^2 2s^2 2p^6$, i.e., the same as F^-, but the charge of Na nucleus was equal to 9 and that of the F nucleus to 11, therefore (taking into account the shielding of nuclei by inner electrons, see Chap. 1.5) $r_{Na^+} : r_{F^-} = (9-4.52)(11-4.52)$. Then, knowing the interatomic distance of NaF (2.31 Å), one obtains: $r_{Na^+} = 0.95$ Å, $r_{F^-} = 1.36$ Å, i.e., $r = c/Z^*$, where c is the constant, and $Z^* = Z - \sigma$ the effective charge of nucleus, i.e., r is proportional to the shield. Slater charge for outer electrons (see Chap. 1.5). In such a way one obtains "monovalent" radii from which ionic radii are deduced by means of multiplication by correction for contraction of ions in crystal [113]; this is an ionic model, but even in NaF the effective charge of ions is about 0.9 and electron configuration is $Na^{+0.9} 2p^6 3s^{0.1}$ and of $F^{-0.9} 2p^{5.9}$.

2. When compiling all the types of crystal chemical systems of ionic radii, the Goldschmidt method (1929) is used. If the reference values of r_{F^-} and $r_{O^{2-}}$ are assumed to be 1.33 Å and 1.32 Å respectively (according to Wasastjerna measurements), the radii of other ions are obtained from the interatomic distance using the additivity principle. Agreement between interatomic distances of different compounds calculated from the radii and those obtained from experimental X-ray data is considered as a criterion of the correctness of the tables of ionic radii. In the systems of ionic radii by Zachariasen (1931, 1954), Ahrens (1952), Belov and Bokij (1952), and Batsanov (1962) the same reference values of reference radii of F^- (1.33 Å) and Cl^- (1.81 Å) are taken, but the systems differ only by the reference value of the oxygen radius ranging between 1.32 and 1.40 Å.

In consequence of this, the tables differ only slightly and in principle are methodically quite analogous to the Goldschmidt system. The Pauling system of radii (with O^{2-} 1.40 Å and F^- 1.33 Å) is numerically also close to the crystal chemical systems.

Analysis of crystal chemical data [394] showed that it is impossible to present interatomic distances in all types of crystals and molecules by means of one system of radii.

Therefore one constructed the different systems for ionic radii (considered above), for covalent radii [146, 396] and for molecular radii [376, 378]. In each of the systems one distinguishes also the radii for different coordination numbers.

The most detailed system of ionic radii is that proposed by Shannon and Prewitt [390], designed especially for oxygen and fluorine compounds, with due regard for the coordination number of both cation and anion, for high and low spin states (where necessary), yielding good agreement between calculated and measured interatomic distances for about 1000 compounds. There are two parallel variants: (1) "effective" ionic radii based on O^{2-} radius in coordination equal to 1.40 Å and of $F^- = 1.33$ Å, (2) "crystal" radii on the basis of O^{2-} in coordination equal to 1.26 Å and $F = 1.19$ Å.

The variant of this system is given in the table of ionic radii proposed by Whittaker and Muntus [399]. They took the interatomic distances of O^{2-} and F^- (in six-fold coordination) to be equal to 1.32 and 1.25 Å, respectively, and calculated the other radii in the light of the fact that they are to conform with the stability limits of different coordinations depending on r_{cat}/r_{an} ratio after Magnus-Goldschmidt.

Van Vechten and Phillips [396] calculated covalent radii for compounds with chemical formula $A^N B^{8-N}$ (or sphalerite, wurtzite and diamond structure).

For the tetrahedral coordination (in Å):

Be	B	C	N	O	F	
1.157	0.963	0.835	0.747	0.681	0.672	
Mg	Al	Si	P	S	Cl	
1.406	1.292	1.201	1.130	1.127		
Ca	Cu	Zn	Ge	As	Se	Br
Sr	Ag	Cd	In	Sn	Sb	Te	I
1.859			1.105			

For the octahedral coordination, the covalent radii of O and F are 0.737, Si 1.239, P, S, Cl 1.23, Cu ... Br 1.343, Ag ... I 1.541.

7.4.3 Appraisal of the Systems of Additive Radii

A general estimation of the physical meaning of the systems of additive radii can be obtained by analysis of the atomic size concept itself from the standpoint of the molecular orbital method (see earlier). Since the atomic radii are not additive and vary from compound to compound, there cannot be a unified system for atomic sizes. The conception of the system of radii is related to the idea of invariable properties of atoms, such as electronegativity.

Due to variation of effective charges, schemes of molecular orbitals, energies of valence states and overlap integrals from one compound to another, the sizes of both atoms participating in the chemical bonding change nonadditively. Consideration of interatomic distances in crystals reveals a number of types of compounds where these distances cannot be deduced either from ionic or from covalent radius systems. Even if the consideration is restricted to one type of compound, for example, an oxygen compound, the additivity will not occur, because the size not only of cations but also of anions varies essentially (for oxygen from 1.40 Å in the most ionic compounds to 0.60 Å in the O_2 molecule).

Thus, the systems of ionic radii allow an estimation of interatomic distances, but do not give the sizes of ions. At the same time the order of magnitude of these sizes is likely to be realistically estimated at least for the most ionic and most covalent compounds.

Discrepancies between the systems considered are not fundamental, and all the systems give the same order of magnitude of atomic sizes.

7.4.4 Orbital Radii (see also Chap. 1.5)

Slater [392, 393] noticed that the sums of orbital radii of the outermost atomic orbitals of free atoms allow an approximate estimation of the interatomic distances between every pair of atoms. These are naturally only approximate values, because interatomic distances themselves vary for a given pair of atoms depending on the coordination, on the second neighbors, and on the molecular or crystal state.

It is evident from the discussion of the concept of ionic radii from the point of view of the contemporary model of the chemical bond, namely of the molecular orbital model, that in the general case orbital radii do not determine the sizes of ions [392, 380]; the values for ions are determined by overlap of the total electron densities but not of the individual atomic orbital, there are actually no orbital maxima in the molecular orbital formed from atomic orbitals; in ionic-covalent crystals valence electrons of cations are in molecular orbitals mainly of a ligand nature but not vice versa, etc.

The determination of approximate interatomic distances for ionic-covalent crystals from orbital radii is no more than a formal procedure (taking different initial sizes of a reference ion, one can compile a system of radii by means of which it would be possible to estimate interatomic distances approximately); the idea of large cations and small anions in compounds conforms neither with calculations of wave functions nor with experimental data.

7.4.5 Experimental X-Ray and Electron Diffraction Determinations of Atomic Sizes [370, 373, 381, 382, 385, 388, 398, 400]

X-ray and electron diffraction studies are based on the scattering of X-rays or electrons by atomic electrons in crystal. Electron density maps are the first experimental results of structure analysis. The electron density distribution on them corresponds to the total electron densities obtained from calculations of wave functions, and the density minima correspond to atomic radii. However, atoms are revealed in X-ray structure analyses, on the one hand, and in the chemical bond and electron spectroscopy, on the other, in different ways. Atomic coordinates are determined from electron density maxima corresponding to core electrons, while the chemical bond is realized by outer parts of wave functions of valence electrons including from the cation part only one to five electrons. Therefore direct estimations of electron density between ions yield extremely low values: for KF 0.08, NaF 0.06, NaBr 0.1 el/Å [382] and even for covalent diamond the bond "bridges" are estimated to be only 0.2–0.5 el/Å [370, 398].

Thus in structure analysis, the electron densities in a crystal can be related to the superposition of spherical atoms, and observation of deviations from the sphericity and electron density distribution in the space between atoms is within the possibilities of this method, requiring high precision measurements, analysis of structure factor variations and consideration of the temperature factor, describing heat motions of the atom which also lead to electron density distortion.

The few special measurements of electron density distribution (Table 40) reveal first great variations (nonadditivity) of ionic sizes (Na from 1.09 to 1.31 Å; Cl from 1.16 to 1.71 Å); secondly considerable deviations from the values in the tables of ionic radii. At the same time, they conform with the order of magnitude of the ionic sizes assumed in them.

At present the calculations of atomic sizes in crystals are carried out in two directions.

Table 40. Atomic radii according to X-ray and electron diffraction data (NaF, KF, NaBr)

Compound	Atom	r, Å	Ref.	Compound	Atom	r, Å	Ref.
NaCl	Na	1.17	[381]	NaBr	Na	1.31	[382]
	Cl	1.64			Br	1.69	
NaF	Na	1.09	[382]	RbCl	Rb	1.58	[373]
	F	1.22			Cl	1.71	
KF	K	1.40	[382]	AgCl	Ag	1.46	[381]
	F	1.28			Cl	1.16	

Table 41. Crystal radii in NaCl-structure alkali halide crystals obtained from the Born model [372]

	Li	Na	K	Rb	Cs
F	0.83	1.15	1.48	1.61	1.77
	1.19	1.16	1.19	1.20	1.23
Cl	0.90	1.20	1.51	1.65	(1.80)
	1.67	1.62	1.63	1.64	(1.67)
Br	0.93	1.22	1.52	1.66	(1.81)
	1.83	1.77	1.78	1.79	(1.81)
I	0.45	1.25	1.54	1.68	(1.82)
	2.05	1.99	1.99	2.00	(2.01)

Table 42. Atomic sizes in europium sulfides estimated from augmented plane wave method

Compound	Atom	r, Å
EuS	Eu	1.562
	S	1.416
$Eu^{2+}S^{2-}$	Eu	1.568
	S	1.410

For the special case of alkali halides (for which the ionic model is taken to be a good approximation) Fumi and Tosi extended the ionic theory of Born using the monovalent and "basis" radii of Pauling, this theory taking account of the repulsion of ions with nearest and next nearest neighbors [372, 395]. Their "crystal" radii (Table 41) have been used by Shannon and Prewitt [390, 391] and by Whittaker and Muntus [399] to compile the systems of additive radii.

The most general and direct method of estimation of the atomic sizes in crystals is provided by the calculation of the wave functions of atoms in crystals.

Petrashen et al. [387] and Maslen [383] calculated the wave functions of Ca^+ and F^- for fluorite, from which the sizes of these ions can be obtained.

An estimation of atomic sizes (Table 42) has been made also from calculations within the framework of the energy band theory by the augmented plane wave

Fig. 126. Determination of atom sizes by energy band theory methods (by the augmented plane wave method) for EuS [371]

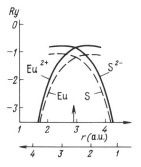

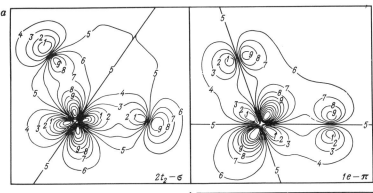

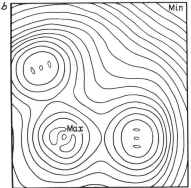

Fig. 127 a and b. Contours of constant electron density of the molecular orbitals of MnO_4^- calculated by the cluster method [98]. **a** Normalized $2t_2-\sigma$ and $1e-\pi$ bonding orbitals in O–Mn–O plane; **b** total electron density in O–Mn–O plane

method in EuS [371] where the atomic boundaries are determined from intersection of potential energy curves (Fig. 126).

Atomic boundaries can also be determined from electron density calculations (Fig. 127) by the "cluster" method [98, 121].

8. Chemical Bond in Some Classes and Groups of Minerals

8.1 Diversity of the Aspects of a Complex Phenomenon of Chemical Bond in Solids

With the background of all the concepts, models and parameters discussed in the previous sections, we may proceed to apply them to the consideration of actual classes and groups of minerals and of individual compounds and minerals. This means approaching minerals and inorganic compounds according to the following two principles:

1. Let us proceed not from the valence bond method with all its consequences; let us renounce the use of the concept of electronegativity. The understanding and interpretation of the electronic structure of compounds, their crystal structure and properties will be based on the molecular orbital and energy band theories. By way of a compromise in this transitional period (from VB to MO and energy band) some mingling of concepts from the different methods will be tolerated.

2. As a result of this approach, instead of rather general descriptions of the ionicity-covalency of compounds and correlation of all the properties with the same electronegativity, one takes into consideration the availability for the compounds of the MO calculation data, of the energy band schemes obtained by the different methods, of the particular theories of the certain parameters developed for the actual classes of compounds, the interpretations of spectroscopical and crystal structure data and parameters describing electrical, magnetic and optical properties from the standpoint of electronic structure.

The diversity of the crystal structure types and the extreme variability of properties of compounds, including many technically important properties, call for a much more concrete and complete description of the electronic structure (i.e., the chemical bonding) of solids.

For each class of compound, the special system of schemes of calculation and experimental parameters describing their electronic structure must be elaborated. At present we have only the beginnings of these particular systems and only some few calculations and measurements have been performed till now.

However the general directions of these descriptions appear to be outlined by the following components.

1. Theoretical calculations: the MO scheme and ligand fields; effective charges and configurations; intracrystalline field calculations and related spectroscopical parameters; the energy band structures; bond orbital approximation and the calculations based on it of the physical properties and estimations of the chemical bonding characteristics; valence bond and hybridization.

2. Parameters of solid state spectroscopy as the most direct information on atoms in compounds (described in the framework of the molecular orbital and energy band theories). For different classes of compounds, the different methods of spectroscopy are applicable and the parameters thus obtained have different significance (see Chap. 5).

3. Crystal structures. Following the determinations of the crystal structure of all the most important minerals and the special stage of precision refinement of these structures, the new stage, related to the comprehensions of the data, obtained emerged: at first in the form of certain empirical rules and of valence bond method correlations, and then from the standpoint of recent theories of electronic structure. Coordination polyhedra in crystal structures are of central importance both for spectroscopy of solids and for electronic structure of solids since all their theories involve symmetry transformations of wave functions. Structure motif is a direct result of the tendency of complexes in crystals to form stable electron configurations; on the other hand, it determines electric field gradients and exchange interactions in crystals.

The refinements of atomic coordinates appear to be specially intended for the intracrystalline field and the related spectroscopical parameters calculations. The individual interatomic distances and valent angle variations are explained by the pecularities of the binding of the atoms in crystal structures.

Partitioning of the atoms between nonequivalent sites within mineral structures, order and disorder, are determined by the energies of these sites.

4. Electrical, magnetic, and optical properties. Various combinations of conduction types and magnetic ordering types, determined from the electrical and magnetic parameters and their temperature dependence, are of vital importance in a description of the exchange interactions between atoms in crystals, and represent their consequences. In reflectance and absorption spectra one distinguishes various types of transition representing the MO, crystal field and energy band schemes.

5. Thermochemical data (energies of dissociation and atomization, enthalpies etc.). Of interest are the calculations of the bond energies based on the energy band methods and their relation to thermochemical and spectroscopical data.

These more detailed and concrete presentations of the chemical bond enable nonempirical explanation of phase relations, existence of limited or continuous solid solutions, isomorphous substitutions, phase transformations at high pressures etc.

Let us consider the bases for systematics of inorganic compounds (especially of minerals) suitable to discussion of their chemical bonding (since classification into ionic, covalent, metallic and molecular crystals is too general).

The classifications of minerals are now crystallochemical. However, in many cases they prove to be crystal structure classifications rather than crystallochemical ones and do not correspond to the natural division of minerals according to their properties and genetical positions. For example, corundum Al_2O_3 and hematite Fe_2O_3 belong to the same crystal structure type, but have quite different optical, magnetic, electrical, and mechanical properties.

Classifications of minerals and inorganic compounds can be based now on two bases of equal importance.

1. Crystal structure base. Classification according to electron structure types is very close to that according to crystal structure types.

a) It is convenient to take the coordination polyhedron for the unity of the classification. The classes of compounds must then be distinguished: (1) with cation polyhedra MO_m^{n+} oxydes, MS_m^{n+} sulfides, $MHal_m^{n+}$ halides, etc. (2) with both cation and anion polyhedra–silicates, carbonates, sulfates etc. (3) without coordination polyhedra-elements (diamond, sulfur etc.), metals, alloys.

The type of coordination polyhedron determines (for a given electron configuration of the atoms) the type of molecular orbital scheme and, in simple compounds, the type of energy band scheme. Local point symmetry of the coordination polyhedron determines symmetry properties of central importance in concepts of molecular orbitals, energy bands, crystal fields, and spectroscopical parameters.

b) The crystal structure motif represents the next step in the description of the electronic structure of compounds. The mode of joining of cation polyhedra (by vertices, edges or faces) determines the exchange interactions and consequently the electrical, magnetic, and optical properties. The linkage of anion polyhedra leads to the polymerization and formation of complex radicals and larger clusters (in particular in silicates and sulfosalts). One distinguishes according to the state of bonding between the bridge and nonbridge ions of oxygen and sulfur, i.e., connecting two radicals or belonging to radical and cation polyhedron.

c) Precision refinements of atomic coordinates are the base for the calculation of energies of sites, intracrystalline fields and related spectroscopical parameters.

2. Type of electron configuration of the atoms and relative energies of the atomic orbitals in a given compound.

Oxides, sulfides, and halides can be divided according to the type of electron configuration of the cations: oxides (sulfides, halides) of alkali metal with ns^1 configuration, metals with ns^2 configuration, nontransition elements with p^n configuration, transition metals, rare earths, actinides. Relative energies of the atomic orbitals change strongly from one period of the system of elements to another, and vary essentially by going from one element to another one within a period.

Similarly one can classify cation polyhedra in complex compounds. However, for example, the $Mg-O_6$ polyhedra in periclase MgO and in diopside $CaMgSi_2O_6$ differ first of all in the state of bonding of the oxygens, in exchange interactions and in the interaction of Mg with all atoms of the lattice. As a special case, one distinguishes the oxycations titanile, vanadile etc., and among them, as a special type, uranile.

At a present there are not enough experimental and theoretical data for systematic description of the chemical bond in all these types of compound (but they are increasing rapidly). Consider some of the most important classes and groups of minerals for which these concrete data are already available. It is important that already now one can outline the tentative schemes of description of the chemical bonding in these classes of compounds in the framework of the above concepts and experimental data.

8.2 The Chemical Bond in Silicates

Let us first consider the tetrahedral oxyanion SiO_4^{4-}, characterizing the general features of the electronic structure of silicates, then the cation polyhedra MO_m^{n+} (mainly octahedra $Mg-O_6$, $Fe-O_6$ etc. and more complex polyhedra Na, $K-O_{7,8,9}$ etc.) and finally the electronic structure of silicates as a whole.

8.2.1 Description of the Chemical Bond in SiO_4^{4-} in Terms of the Calculated Molecular Orbital Diagram

Molecular orbital calculations have been made for the SiO_4^{4-} by the self-consistent Wofsberg-Helmholz (or extended Hückel) method for regular tetrahedral symmetry (see Chap. 3.2).

The distortions of the tetrahedra are small enough to neglect the splittings of the t and e_2 states in the lower symmetries.

The consideration of the SiO_4^{4-} in silicates such as isolated tetrahedra proves a good approximation of the chemical bond investigations, supported at this stage by the agreement of the MO calculations with the X-ray spectra of the silicates.

Table 43. Molecular orbital diagram calculations for SiO_4^{4-} (after [428])

MO		E, a.u.	Si			O		
			C_{3s}	C_{3p}	C_{3d}	C_{2s}	$C_{2p\sigma}$	$C_{2p\pi}$
I	$1a_1$	−0.327	0.38	—	—	0.92	0.13	—
	$1t_2$	−0.308	—	0.29	0.23	0.93	0.03	0.04
II	$2a_1$	0.270	0.74	—	—	−0.39	0.55	—
	$2t_2$	0.364	—	0.66	0.33	−0.32	0.45	0.39
III	$1e$	0.432	—	—	0.35	—	—	0.94
	$3t_2$	0.468	—	0.10	0.49	−0.12	0.13	−0.85
IV t_1		0.579	—	—	—	—	—	1.00

I. Lower bonding molecular orbitals formed mainly from ligand (oxygen AO) orbitals and composed mainly by oxygen 2s atomic orbitals ($1a_1$ and $1t_2$).
II. Strongly bonding molecular orbitals ($2a_1$ and $2t_2$) formed from silicon 3s and 3p orbitals (with admixture of 3d Si orbitals) and oxygen 2p orbitals; it is these molecular orbitals which are mainly responsible for the chemical bonding.
III. Weakly bonding molecular orbitals (essentially d–p bonding) formed mainly from oxygen 2p orbitals with small amount of silicon 3d and 3p orbitals ($1e$ is pure π molecular orbitals, $3t_2$ is σ, π orbital).
IV. Nonbonding molecular orbital t_1 formed from oxygen 2p orbital; it is the last occupied orbital and its energy determines the ionization potential of SiO_4^{4-}.
V. Not shown in the table. Antibonding molecular orbitals formed essentially from silicon atomic orbitals ($4t_2, 2e_2, 3a_2, 5t_2$).

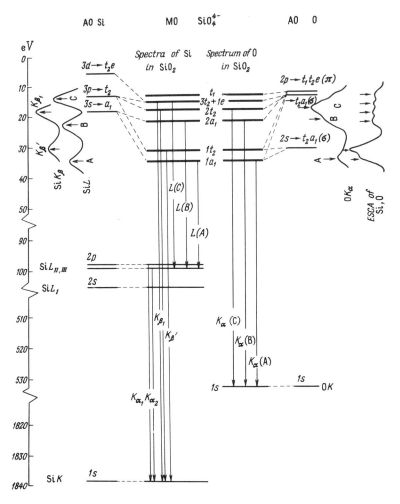

Fig. 128. Molecular orbital diagram for SiO_4^{4-}, X-ray spectra SiK_α, SiK_β, $SiL_{II,III}$, OK_α and ESCA spectra

The molecular orbital description of the chemical bond in SiO_4^{4-} is presented in Fig. 128 (cf. Fig. 48) and in Table 43. The calculations are believed to provide a correct qualitative molecular orbital diagram, although more thorough calculations change the eigenvalues and the orbital compositions for SiO_4^{4-}. The MO diagram shows in particular the involvement of the Si $3d$ atomic orbitals in the chemical bond: in the $1e_1(s-p_\pi)$, $2t_2$ and $3t_2(d-p_{\sigma,\pi})$ molecular orbitals. Note that the relative Si $3d$ percent character of the chemical bond in silicates is very small (being determined not by c_i but by c_i^2 coefficients), though it should not be underestimated.

Mulliken bond overlap populations (see Chap. 3.2) computed for T–O bonds (T = B, Al, Si) in anorthite, albite, microcline and reedmergnerite correlate with the observed bond lengths and O–T–O and T–O–T angles [431, 432, 451].

8.2.2 Molecular Orbital Diagram for the SiO_4^{4-} According to X-Ray and ESCA Spectra

The quantitative MO results (Fig. 128) are used to assign the transitions in X-ray and X-ray electron spectra [423, 426, 477]. The $SiK_\alpha(2p\to 1s)$, $SiK_\beta(3p\to 1s)$, $SiL_{II,III}(3d\to 2p)$, $OK_\alpha(2p\to 1s)$ emission spectra are produced by the transitions from the corresponding p-like and d-like molecular orbitals of the SiO_4^{4-} to the inner shell levels 1s and 2p of Si and 1s of O. The X-ray electron spectra have no selection rules and therefore arise from the transitions at all the atomic and molecular orbital levels.

The combination of all these experimentally determined X-ray emission and electron spectra made it possible to compile a full MO diagram for the SiO_4^{4-} (Fig. 128). The relative intensities of the bands in these spectra reflect the contributions of the corresponding atomic orbitals to the molecular orbital and the contribution of the molecular orbital to the chemical bond.

Another description of the bonding in the SiO_4^{4-} from the X-ray and electron spectra is provided by the "chemical shift" for SiK_β and AlK_β, $SiL_{II,III}$, SiK_α and AlK_α, OK_α emission bands, and for ESCA transitions from Si 1s and 2p levels [457, 473].

The results obtained can be treated in two different ways:

a) The transitions from the molecular orbitals (i.e., SiK_β, $SiL_{II,III}$, OK_α spectra) can be interpreted only in terms of the MO diagrams: the shifts of the corresponding bands reflect the changes of the MO energy levels, and the variations in the band intensities reflect the changes in the contributions of the atomic orbitals into the molecular orbital. The changing coordinations result in changes in the MO diagram, hence the transitions arising from the different MO and the corresponding bands in the X-ray spectra are not compared directly. Only empirical correlations of the SiK_β band shifts with the interatomic distances are possible.

b) The transitions between the inner shell electron levels (SiK_α and AlK_α arising from 2p to 1s transitions), and ESCA transitions from Si and Al 1s and 2p levels, are not connected directly with the MO diagrams and their interpretations are based on the concepts of the effective charge.

8.2.3 Effective Charges of Si and Al in the Silicates and Aluminosilicates

It has already been determined by X-ray crystal structure investigations that the total number of electrons at the Si atom is not 14 ($1s^2 2s^2 2p^6 3s^2 3p^2$) and not 10 (Si^{4+} without $3s^2 3p^2$) but 12, and at the oxygen atom not 8 ($1s^2 2s^2 2p^4$) or 10 ($O^{2-} 1s^2 2s^2 2p^6$), but 9. From these observations the effective charges were determined as $+2$ for Si and -1 for O. One obtains the same order of magnitude from the detailed analysis of electron density distribution in the X-ray diffraction pattern for quartz, calculations of electrostatic energy for garnets, olivine, zircon by fitting the various charge values on Si and O (see Chap. 7.3), estimations of the effective charge from thermochemical data [318, 326].

However, the most direct methods to describe the states of atoms in crystals are spectroscopical and, particularly for the characteristics of effective charges, X-ray spectroscopy methods are the most suitable.

The effective charges of Si in a number of minerals determined from the K_α line shifts in X-ray emission spectra are equal to $+1.4 \pm 0.2$ (in forsterite $+1.42$, in quartz, beryl, topaze $+1.40$, in opal $+1.38$, in orthoclase $+1.37$, in diopside $+1.35$, in anorthite, muscovite $+1.34$, in zircon $+1.25$ [176, 188]).

The effective charge of Al in aluminosilicates is around $+1.20$, that of O is from -0.98 to -1.14 [176].

It should be noted, however, that the interpretation of the shifts of the K_β band in the Si and Al spectra, as well as the shift of the oxygen K_α band (arising as a result of the transition from the $2p$-like molecular orbital to the $1s$ level of oxygen), needs consideration of the molecular orbital diagram.

Correlations exist between K_α and K_β line shifts in X-ray emission spectra and coordination number, interatomic distances, Al/Si ratio in the aluminosilicate minerals, but for the understanding of these correlations it is necessary to use the MO diagram.

The order of magnitude of the effective charge of Si, Al, O in silicates is approximately determined, but a more accurate treatment calls for further development of the very concept of effective charge. Its uncertainty is connected mostly with the determination of the volume within which one takes into account the fraction of electron density of a given atom. For SiF_4, taking the conventional sphere of the radius 2 a.u. (i.e., 1.06 Å) and using in the calculation the pd atomic orbital basis of Si, one obtains the charges $Si^{+2.75}$ and $F^{-0.69}$ [117].

A discrepancy in the effective charge values obtained by the different methods is also explained by the different physical content of the mechanisms of phenomena determined as effective charge.

Electron spectroscopy determinations of the shifts of the energies of Si, Al, Mg, Na $2p$ levels which are also connected with effective charges enables a comparison of variations in a series of related compounds [457].

Characteristics of effective charges ($Si^{+1.40}$, $O^{-0.98}$ etc.) must not eliminate the concept of valency (see above, Chap. 3.3). There are few determinations of effective charges of cations in silicates [176], but they indicate the order of magnitude for this class of mineral: $Na^{+0.75}$ in Na_2SiO_3, $Mg^{+1.01}$ in $MgSiO_3$.

8.2.4 Silica Polymorphs: Energy Band Schemes; Bond Orbital Model and Calculations of the Electronic Structure and Properties

A further approach in the description of the chemical bond in silicates must be related to consideration of the energy band.

Thus far it has been done for the relatively simpler case of the SiO_2 polymorphs, with only one type of coordination polyhedron.

The silicon atom (such as carbon and germanium) belongs to the "tetrahedral" atoms in which four sp^3 hybrid orbitals directed towards neighboring atoms easily arise from the s^2p^2 electron configuration.

For this type of solid the bond orbital method exists (see Chap. 7.1), introduced first for diamond and zinc-blende type of crystals, then generalized to an arbitrary tetrahedrally coordinated solid, and more recently applied to crystalline and amorphous SiO_2 and GeO_2 [460].

The simplest forms of SiO_2 are β-cristobalite with the Si atoms forming a diamond lattice, and β-tridymite with the Si atoms in a wurtzite lattice.

Hence the band structure of β-cristobalite can be used as a starting point in tackling the problem of the electronic structure of all the crystalline forms of silica and $Si-O_4$ clusters in silicate crystals, as well as of the electronic structure of glasses.

The first Brillouin zone for β-cristobalite is the same truncated octahedron as for diamond and zinc-blende structures (Fig. 61a).

The valence energy bands and density of states have been calculated for β-cristobalite by the two-parameter bond orbital method (see Chap. 7.1) and by the three-dimensional extended Hückel method, which is an extension of the band theory of one-dimensional polymers [423, 449].

In terms of the bond orbital model, the theory of the electronic structure of quartz and other forms of SiO_2 as well as vitreous silica has thus been presented: (1) the parameters W_2 and W_3 (a covalent and a polar energy) were calculated; ionicity and effective charges were determined; (2) these parameters were fitted to the optical absorption spectra as well as to the X-ray emission and photoemission spectra; (3) variation of the bond angle, bond length, and lattice distortion were interpreted in terms of the polarities and effective charge on the oxygens as well as in terms of the parameters related to the energy band scheme and corresponding to the optical-absorption peaks; (4) the refractive indices and dielectric constants were calculated for α- and β-quartz, α- and β-cristobalite, β-tridymite, coesite, keatite, and vitreous silica [460].

8.2.5 Cation Polyhedra in Crystal Structures of Silicates

In discussing the chemical bond in silicates it is convenient to distinguish the three types of cation polyhedra: octahedral $Mg-O_6$, $Fe-O_6$, $Al-O_6$ etc. and close to them $Ca-O_6$, $Ca-O_7$; polyhedra $K-O_{7,8,9}$, $Na-O_{7,8,9}$; tetrahedral $Li-O_4$, $Be-O_4$, $Al-O_4$ etc.

Among the octahedral cations one distinguishes the cation of nontransition elements (first of all Mg) and that of transition metal (first of all Fe). They are very close to each other from the crystal structure aspect, but differ in electron configuration and state of bonding.

The MO diagrams of the octahedral cation polihedra are principally the same as the typical MO schemes shown earlier in Fig. 48. The actual MO diagrams have been calculated by the self-consistent-field X_α scattered wave (SCFX_αSW) method for TiO_2, FeO and Fe_2O_3 [479] and for SiO_6^{-8} (in stishovite), AlO_6^{9-} and MgO_6^{-10} [478].

Lowering of the symmetry leads to splitting of the molecular orbitals, but the distortions of the octahedra are usually small and the splitting can be neglected in some chemical bond discussions. However, these splittings are clearly revealed in electron paramagnetic resonance spectra, Mössbauer spectra and in some optical absorption spectra.

8.2.6 Degree of Ionicity-Covalency in Cation Polyhedra According to Superfine Structure of Electron Paramagnetic Resonance (EPR) Spectra

The parameter of superfine structure reveals the strongest dependence on the state of the chemical bond among the EPR parameters (Fig. 129).

This superfine structure (sfs) can be observed in the spectra of the impurity transition metal or lantanide ion. It is very often represented in silicate minerals by the ion Mn^{2+} which is one of the best EPR ions. The state of the chemical bonding of the impurity Mn^{2+} ion and the Mn^{2+} EPR sfs parameter depend on the type of ligand ions (decreasing rapidly in a series with the ligands F–Cl–O–S–

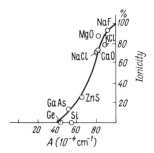

Fig. 129. Ionicity in cation polyhedra of silicates plotted against superfine structure of EPR spectra

Table 44. Superfine structures of EPR spectra of Mn^{2+} parameter in some minerals (in 10^{-4} cm^{-1} units)

Coordination polyhedron	Mineral	A, B, C or A_{av}
Mg–O$_6$	Periclase	81.4
	Forsterite	78.8
	Diopside	
	Tremolite, Talc	
	Magnesite	86.3; 87.3
	Dolomite	86.8; 85.7
Ca–O$_6$	CaO	84.5
	Calcite	87.8; 87.7
	Dolomite	86.8; 87.6
	Apatite	91.4; 89.6; 84.0
Ca–O$_7$	Apatite	72.0; 81.6; 80.2
	Sphene	89.9
	Apophyllite	84.9
Ca–O$_8$	Diopside	85.3; 85.6
	Tremolite	87.0; 85.4
	Datolite	88.8
	Sheelite	88.9
Al–O$_6$	Corundum	79.6; 78.8
	Spodumene	83.6; 81.7
Al–O$_4$	Petalite	70.0
Li–O$_4$	Petalite	69.0

Se–Te) and on the peculiarity of the bonding of the cation being replaced by Mn^{2+}.

Greater values of the sfs correspond to greater ionicity: for example, in $NaF:Mn^{2+}$ sfs parameter is $91.5 \cdot 10^{-4} cm^{-1}$, in $NaCl:Mn^{2+}$ $82.1 \cdot 10^{-4} cm^{-1}$ while in $ZnS:Mn^{2+}$ it is only $64.0 \cdot 10^{-4} cm^{-1}$.

The sfs parameters of Mn^{2+} in the $Mg-O_6$, $Ca-O_6$ and other polyhedra in silicate and some other minerals are listed in Table 44.

Comparison of these values shows: (a) in the Mg and Ca polyhedra in the silicate minerals the chemical bonds are more ionic than in NaCl, but less ionic than in NaF; the degree of ionicity is here about 80–85% (Fig. 129); (b) the states of the ionicity of the bond in $Ca-O_6$, $Ca-O_7$ and $Ca-O_8$ are close each to other and more ionic than in $Mg-O_6$; (c) the bonds $Mg-O_6$ and $Ca-O_6$ in silicates are more ionic than in MgO and CaO; (d) the degree of ionicity of $Mg-O_6$ and $Ca-O_6$ in silicates is close to that in carbonates, sulfates, and phosphates; substitution of oxygen by fluorine (and in some cases OH) increases the sfs value; (e) in each of the polyhedra $Mg-O_6$ and $Ca-O_6$ in different subclasses and groups of silicates, the states of the bonding are very similar; more subtle differences can be revealed by more accurate measurements and by analysis, taking into account both crystal structure and electronic structure factors in these crystals; (f) in tetrahedral cation polyhedra $Li-O_4$ and $Al-O_4$, the chemical bonds are essentially more covalent, in fact the most covalent among oxygen complexes.

8.2.7 Energies of the Structural Sites, Energies of Stabilization, and Intracrystalline Fields in Silicates

A number of factors has in the last years favored the development of calculations and measurements of the energies and potentials in silicate minerals.

1. Mössbauer, EPR, infrared and optical absorption spectra, as well as detailed X-ray diffraction data, have been used for determinations of the distribution of Mg–Fe, Mg–Mn, and Ca–Fe between nonequivalent sites in crystal structures of many minerals; this distribution can be written in the chemical formula of the minerals, it can be ordered and disordered.

2. Spectroscopical parameters used in determining the cation distribution between different structural positions contain information about the crystalline fields in these positions; the assignment of spectra to certain positions needs the calculation of relation between the distortions of the different non-equivalent polyhedra, crystalline fields in these polyhedra and related spectroscopical parameters; different lattice sum coefficients A_l^m represent those common features which enter all the spectroscopical parameters from the geometrical characteristic of the cation position.

3. With the availability of computers and special programs the simplified expressions for the calculations of lattice energy and other properties have become unnecessary, and calculations for the most complex silicate structures as well as the inclusion in the calculation (in programs) of more and more types of interaction has become feasible.

4. These calculations need precision data on the coordinates of atoms; the refinements of the crystal structures of the most important rock-forming silicates carried out in 1960s, seen to have been directed to this aim.

There were two directions of these calculations.

1. Calculations of the energies (potentials) of the structural positions.

a) Crystal field stabilization energies (CFSE) can be obtained from the optical absorption spectra (see Chaps. 6.3 and 7.3). Site preference of the transition metal ions is determined by the CFSE: the ions are enriched in the positions where they have higher CFSE, and those from the two ions enter a given position which have in this position higher CFSE.

b) The cohesive energies for the ions in certain structure positions were computed for many silicate minerals. The major contribution to the energy of a position in crystal structure is provided by the electrostatic (Coulomb) interactions of the charges giving the Madelung energy. The calculations of the Madelung energies were made by direct summation for amphiboles [362] and by the method of Ewald (which is more efficient for complex structures) for a number of most important silicate minerals [337, 339, 346]. Use of the Ewald method with the availability of special programs and computers completely resolves the problem of computing Madelung energy. However, these calculations are performed in the ionic model of solids and the total cohesive energy is the sum of the Coulomb interactions, which can be computed quite precisely within the framework of the ionic model and a phenomenological core–core repulsive interaction determined by use of empirical coefficients.

The magnitude of CFSE is equal to about 0.5–1.0 eV (~ 4000–$8000\,cm^{-1}$) while that of the cohesive energy of an ion in the lattice is about 7–9 eV. This means that the site preference is determined mostly by the Madelung energy alone, but when the transition metal ions enter the same positions, the relative values of CFSE become important.

2. Electric-field gradient calculations have been performed taking into account many types of interactions, including both the short-range effects (overlapping of the wave functions) and long-range interactions with the lattice. It is very important that these calculations can be compared with the experimental parameters: quadrupole splitting of Mössbauer spectra of iron measured in many silicate minerals, electric (crystalline) field gradient eQq in nuclear magnetic resonance spectra of the Li, Be, B, Na, Al nuclei and parameter B_2^0 in electron paramagnetic resonance spectra.

There are two approaches in these calculations. (1) The Li, Al, etc. atoms are taken as point charges determined by formal valency, while oxygen ions are considered as dipole, quadrupole and octupole moments with the corresponding polarizabilities; there is good agreement between the calculated electric field gradients and that measured by nuclear magnetic resonance for spodumene, andalusite, sillimanite, kyanite, zoisite, corundum [344, 345]. (2) Other approaches applied to iron [351, 352] take into account overlapping of the wave functions; comparison with the quadrupole splitting in the Mössbauer spectra of hematite gave a good agreement.

8.2.8 Mössbauer Characteristics of the Bonding of Iron and of the Site Population in Silicate Minerals

Mössbauer spectra, besides such mineralogical applications as determination of oxidation state of iron, analysis of Fe^{3+}/Fe^{2+} ratio and of quantitative site population, studies of isomorphous substitutions and of magnetic behavior, identification of microcrystalline and ultra fine particles and so on, supply information on certain aspects of the chemical bonding of iron (and tin) in minerals (see Chap. 5.2).

The two principal Mössbauer parameters, isomer shift (IS) and quadrupole splitting (QS), reflect different aspects of the state of iron in silicates: IS is related mainly to the s electron density at the iron nuclei and increases with the degree of ionicity, while QS is a measure of the electric-field gradient at the Fe sites.

In most Mg–Fe rock-forming silicates, the IS for Fe^{2+} is usually 1.20 ± 0.10 mm s^{-1} (relative to stainless steel) and for Fe^{3+} it is usually 0.50 ± 0.10 mm s^{-1}. These values are typical for essentially ionic compounds and are close to the IS values in other oxygen compounds. Quadrupole splittings for iron in silicates have always rather large values; for Fe^{2+} of the order of $1.50 \div 2.50$ mm s^{-1} (increasing up to maximum observed for Fe^{2+} value of 3.50 mm s^{-1} in garnets) and for Fe^{3+} around 0.50–1.0 mm s^{-1} (up to 2.15 mm s^{-1} in epidote). This corresponds to the large anisotropy of the electric fields in the low-symmetry silicates.

Quadrupole splitting is mainly used to determine the structure position of iron and estimate intra- and intercrystalline distribution and cation ordering in silicate minerals.

8.2.9 Crystal-Chemical Meaning of the Nuclear Magnetic Resonance (NMR) Parameters in Silicates

Electric-field gradients measured in nuclear magnetic resonance studies is the same gradient which causes quadrupole splitting in Mössbauer spectra of iron, but in NMR it is determined for other nuclei, for example Li^7 in spodumene, Be^9 in beryl, B^{11} in danburite, Na^{23} in albite, Al^{27} in feldspars, andalusite, kyanite, stillimanite, zoisite, spodumene, Si^{29} in beryl etc.

Just as in Mössbauer spectroscopy, these measurements do not give the values of the crystal (electric) field, but only its gradient. The relation between the crystal field and its gradient can be compared to the relation between the refractive indices N_g, N_m, N_p and the optic axial angle: small changes of N_g, N_m or N_p can lead to great changes of the optic axial angle, similarly small distortions of the coordination polyhedron can cause great changes of the crystal field gradient.

Interpretations of the electric-field gradient parameter measured from NMR spectra are more complete when compared with the calculated values. The agreement between the experimental and theoretical values indicates to what extent the different interactions composing the complex phenomenon of the chemical bond were taken into account.

8.2.10 Bond Length and Angle Variations; Bridging and Nonbridging Oxygens

Relations between structural features and chemical bonding in silicates have a double meaning.

On the one hand, type and distortion of the coordination polyhedron (its local symmetry) determine the splitting of d electron levels by crystal fields, types of the molecular orbital and energy band schemes, symmetry of the parameters of EPR, NMR, optical and Mössbauer spectra; the atomic coordinates, interatomic distances and angles are used in calculations of lattice sums, energies of structural positions, molecular orbital and energy band quantitative diagrams.

On the other hand, structural features such as individual and average bond length and angle variations in one structure and in different crystal structures must be understood from the standpoint of recent concepts in chemical bonding.

The statement of this problem becomes possible after accumulation of the crystal structure refinements with interatomic distance measurements with average standard deviations of 0.004–0.002 Å.

The sizes of the atoms themselves are more variable in combination with other atoms (for example, the sizes of Si in the pairs Si–F, Si–Cl, Si–O, Si–S, Si–Si or the sizes of O in the pairs Na–O, Ca–O, Si–O, S–O, Cl–O etc.) than the interatomic distances (for example Si–O) in different structures. However, the refinements of the many silicate structures demonstrated quite definitely rather significant variations of Si–O distances and Si–O–Si bond angles.

Individual Si–O distances in Si–O_4 tetrahedra vary from 1.55 to 1.720 Å [405, 406]. The mean Si–O_4 distances range from 1.608 Å (for oxygen coordination number 2) to 1.622 Å (c.n.3) and 1.638 Å (c.n.4) [419]. The mean Si–O distances in the different linkage of silicon–oxygen tetrahedra are 1.605 Å in framework silicates, in sheet silicates 1.62 Å and in orthosilicates 1.63 Å (for Al–O_4 respectively 1.757; 1.77 and 1.80 Å). These empirical data naturally require interpretation.

First, the empirical correlations between Si–O distances and other structural features were established, which one could also attempt to understand from the standpoint of one or another model of the chemical bonding.

In silicate structures (as well as in other polymeric structures) one distinguishes bridging oxygen atoms (i.e., each shared by two silicon atoms) and nonbridging oxygens (i.e., linked to one silicon atom and to nontetrahedral cation) [409].

In orthosilicates (olivines, garnets etc.) with individual silicon–oxygen tetrahedra there are only nonbridging (nb) oxygens: $[SiO_4^{nb}]$. In pyroxene chains there are in each Si–O_4 tetrahedron two bridging oxygens shared each by two tetrahedra and two nonbridging: $[SiO_2^b O_2^{nb}]$.

In amphibole double chains there is one nonbridging oxygen in each Si–O_4 tetrahedron, while the other three oxygens (shared each by two tetrahedra) are bridging: $[Si_4 O_5^b O_6^{nb}]$ (for a cluster of six tetrahedra).

In sheet silicates three oxygens (by means of which silicon–oxygen sheet is formed) in each tetrahedron are bridging, and only one oxygen, through which this sheet links to the octahedral cation, is nonbridging: $[Si_2 O_3^b O_2^{nb}]$.

In framework silicates and quartz, all oxygens are bridging: $[SiO_2^b]$.

Bonding of bridging oxygen (Si–O^b–Si) is essentially covalent but that of nonbridging (Si–O^{nb}–M) has a double character: its binding with the octahedral cation is essentially ionic (according to EPR and other data), while the binding with Si is very close to the binding of Si with bridging oxygens.

This difference in the state of bonding is revealed in the variation of the Si–O distances: in the presence of both bridging and nonbridging oxygen, the bond lengths Si–O^b are longer than those of Si–O^{nb}.

For example, in diopside the Si–O^{nb} distances are 1.602 Å and 1.585 Å, mean distance 1.594 Å, while the Si–O^b are 1.664 Å and 1.687, mean 1.676 Å.

However the bond length variation is revealed also in structures with a single type of oxygen, for example, in framework anorthite structure the individual distances Si–O^b vary from 1.559 to 1.646 Å and the mean bond length depends on the number of the nearest neighbor Ca atoms: it is 1.632 Å with the two Ca, 1.622 with one Ca and 1.588 without the neighbor Ca atom [419].

Thus besides the bridging or nonbridging characteristics of oxygen in silicates its average coordination number is also significant.

More detailed correlations showed that the Si–O distances depend also on the bond angle Si–O–Si and the kind of adjoined nontetrahedral cation [405, 406, 419, 476]. These Si–O bond length variations have been discussed on the basis of the different concepts of the chemical bond.

1. The Si–O distance in silicates has been compared with the sum of covalent radii of Si and O corrected by the degree of ionicity expressed as the Si and O electronegativity (EN) difference [419]:

$$\text{Si–O} = r_{\text{Si}}^{\text{cov}} + r_{\text{O}}^{\text{cov}} - 0.09(\text{EN}_{\text{O}} - \text{EN}_{\text{Si}}).$$

Since instead of the observed mean Si–O distance in SiO_4^{4-} 1.62 Å, one obtained by this calculation 1.76 Å, another correction was suggested for the $3d - 2p$ π-bond contribution [425].

The involvement of the Si 3d orbital in σ, π – molecular orbitals $1t_2, 2t_2, 3t_2$ and in the π-molecular orbital $1e$ is evident from the MO scheme of SiO_4^{4-} (see Fig. 126, and cf. Fig. 48) and is supported by the experimental X-ray spectroscopy observations [108]. However it is doubtful whether the $3d\text{Si}-2p\text{O}$ π-bond can explain the correction in the above expression, since the $d-p_\pi$ bond description is based on the MO theory, but the electronegativities and additivity of radii originate from the valence bond method. Besides this, the variation of individual and mean Si–O distances must be dependent not only on the amount of the $d-p_\pi$ bond, but on the populations of all the molecular orbitals of SiO_4^{4-}.

2. The correlation between Si–O distances and Si–O–Si bond angles has been related to $s-p$ hybridization of the oxygen atomic orbitals: the wider Si–O–Si angles correspond to greater amounts of s orbital in the sp hybrid state of the oxygen; the Si–O lengths are shorter when the amount of s orbital is greater.

3. The influence of nontetrahedral cations on the Si–O distances has been estimated by using the electronegativities of the cation [418].

4. However, the mean coordination number, correlation with bond angle and role of nontetrahedral cations does not unequivocally explain Si–O distances. To take into account more crystal-chemical factors influencing variation of the

individual Si–O bond lengths, one considers the valence forces for the each oxygen atom [405, 406]:

$$\text{Si–O} = (\text{Si–O}_{av}) + b \cdot \Delta p,$$

where b is the empirical constant and Δp is the difference between the valence force for a given oxygen atom according to electrostatic valence rule and mean valence force for all the oxygen atoms of a given SiO_4 tetrahedron; examples of calculations see [405, 406].

5. The extended Hückel molecular orbital calculations (see Chap. 3.2) of Mulliken bond overlap populations for the SiO_4^{4-} tetrahedra in feldspars, olivines, and sodamelilite correlate with the experimental Si–O distances: shorter bonds are associated with the larger populations and with oxygen atoms with smaller charges [431].

6. The most complete recent theory of the electronic structure of quartz, cristobalite, and coesite is now the bond orbital approximation (see above and also Chap. 7.1). Most quantities of this method, such as total energy, polarities, charge on oxygen, and parameters related to optical absorption peaks, are expressed as simple analytical functions of Si–O–Si angle, Si–O distance and two fundamental parameters [460].

8.2.11 Interlayer Bonding and Surface Energy Calculations in Sheet Silicates

The energy of the bonding between the composite layers in sheet silicate structures consists mainly of ionic and van der Waals (intermolecular) components (see Chap. 8.4). Variations in the amounts of the contributions of the two components depend on the type of the structure [433, 434, 435].

1. In the micas the composite layers have a net negative charge which essentially determines the ionic bonding between the layers.

2. In sheet silicates with electrically neutral layers the interlayer bonding involves (a) long hydrogen bonds between the hydroxyls and oxygens of the adjacent layers (in the 1:1 phylosilicates such as kaoline and serpentine) or (b) van der Waals and ionic bonds (in the 2:1 phyllosilicates such as talc and pyrophyllite). The ionic contributions in the bonding between the electrically neutral layers originates from the long-range electrostatic (Madelung) energy.

The estimation of the energy of the ionic interlayer bonding was made by calculating the electrostatic energy for the normal structure and for hypothetical structures which contain the same layers but have increasing distances between them. The difference between the energy of the structures with separated layers and that of the normal mineral is the surface energy and corresponds to the ionic contribution to the cohesion energy.

8.2.12 Mantle Properties, High Pressure Spectroscopy, and Electronic Structure of Silicates

Speculations on the composition and structure of the mantle minerals and interpretations of geophysical data on the earth's interior are now based on results of high pressure experiments and theories of the electronic structure of minerals.

It is now accepted that the minerals of the earth's mantle are mainly ferromagnesian silicates; in the upper mantle: olivine, pyroxene, and pyrope; in the transition zone (350–1000 km): phases of olivine composition with spinel-type structure and then strontium plumbate and calcium ferrite-type structure as well as complex garnet solid solutions, jadeite and different phases with ilmenite, perovskite, and calcium ferrite-type structures; in the lower mantle: magnesiowüstite and stishovite [203].

However, besides these phase transformations continuous changes in the electronic structure and properties of the mantle minerals also exist. The calculations of lattice and site energies, as well as of overlap integrals for the minerals under conditions approximating those of the earth's crust and mantle need the data of high temperature and high pressure crystal structure determinations. Structure refinements of olivines, pyroxenes, amphiboles, feldspars, garnets, quartz etc. at high temperatures are now available [468]. Recent technical advances also permit high-pressure crystallographic study [195, 429] and one may hope that the simultaneous application of high pressure and high temperature in crystal structure refinements will become possible.

High pressure spectroscopy includes mainly optical absorption and Mössbauer spectra. It was the spectroscopic studies which suggested that continuous changes in mantle properties are related to the changes in the electronic structure of the constituent minerals. Pressure dependence of the isomer shift and quadrupole splitting [429], d–d transition and charge transfer bands, Racah and $10Dq$ parameters permit these changes to be traced. It was shown that the edge of the charge transfer bands in ferromagnesium silicates moves toward the infrared with pressure and thus blocks radiative heat transfer at the condition of the earth's mantle [255, 285]. The high-spin state of the ferrous cations in silicates appears to be retained to the depths of their stability [203].

Variation of the $10Dq$ parameter (related to the interatomic distances and hence to the volume of the coordination polyhedra) with pressure provides information on site compressibilities [255].

Descriptions of the electronic structures of the mantle minerals with increased covalency of the bonding must be given in terms of energy band theories, probably those of the bond orbital type.

8.3 The Chemical Bond in Sulfides and Related Compounds

The understanding of sulfide and related minerals with structural and compositional features reflecting peculiarities of ore deposition processes has advanced over the last decades mainly owing to the three factors.

1. Many of them became the subject of intensive studies in consequence of the discovery of properties of these types of compound which have found most diverse use in technical applications and physics. These are semiconductors with the structure types pyrite, marcasite, and arsenopyrite; ferroelectrics semiconductors of the antimonite, Sb_2S_3, type; thiospinels and selenospinels combining ferromagnetic with semiconducting properties; superconductors (some dichalcogenides); materials for quantum electronics and acousto-optics (of proustite,

Ag_3SbS_3, type); luminophors (ZnS, CdS and others); IR radiation detectors (PbS and others) etc. The crystal where synthesized which expanded enormously in comparison with the known representatives of natural minerals this class of compounds: sulfides of transition metals and rare earths, sulfosalts, thiospinels, selenospinels. This enables more complete understanding of sulfide minerals as a part of the vast class of compounds with a great diversity of properties.

2. Development of the new physical methods of investigation especially affected sulfide minerals, since crystal optics and microscopy, having for so many years provided such detailed characteristics of rock-forming minerals, offer less possibilities for opaque sulfide minerals. Of great importance were the developments of reflectance spectroscopy with measurements of spectra from vacuum UV to far IR and with interpretations of peaks based on the energy band theory, X-ray and electron spectroscopy, Mössbauer spectroscopy (for sulfides of iron, tin, and later of Te, Au and others), nuclear quadrupole resonance (sulfides and sulfosalts of As, Sb, Bi), some data from EPR and NMR of sulfides.

Diversity of electrical properties (insulating, semiconducting, with metallic-type conductivity, superconducting; Hall effect, thermoelectromotive force etc.), magnetic properties (diamagnetic, paramagnetic, ferro- and antiferromagnetic, with Pauli paramagnetism), and diversity of the combination of these properties in the same crystal are described by the energy band schemes and represent the characteristics of important aspects of their electronic structures.

3. Many mineral species of the sulfide class found to be included in the list of model systems which are the subject of most detailed theoretical considerations, calculations and experimental measurements: ZnS and CdS (sphalerite–wurtzite and greenockite), PbS–PbSe–PbTe (galena–PbSe–altaite), $CuFeS_2$ (chalcopyrite), $FeAs_2$ (löllingite), MoS_2 (molybdenite), SnS_2 (stanisulite), NiAs (nickeline), FeS (troilite pyrrhotite), thiospinels of $FeCr_2S_4$ type (dobreelite) and many others, AsS and As_2S_3 (realgar and orpiment), Sb_2S_3 (antimonite), Bi_2S_3 (bismuthine), sulfosalts of As, Sb, Bi (proustite, pyrargyrite and many others).

8.3.1 Diversity of the Aspects of the Chemical Bond in Sulfide and Related Compounds and the Theoretical Schemes

This group of compounds includes sulfides, selenides, tellurides, arsenides, and sulfosalts. Unlike oxides and halides, which are more ionic than covalent, this group belongs essentially to covalent compounds. Another characteristic feature is that many compounds of this group exhibit some amounts of metallic bond (metal–metal interactions) or, especially in disulfides, S–S bond (or As–As, As–S bonds in sulfoarsenides etc.). However, besides these general features extreme diversity of the chemical bonding exists within this group, and peculiar crystal structures and most different states and forms of the electronic structure are found.

The degree of covalency of the bonding varies to a great extent, increasing in the series of S–Se–Te compounds and in disulfides with complex $[S_2]^{2-}$ radicals, but decreasing in cation groups of the sulfosalts. The contribution of metallicity to the chemical bonding has a different meaning for transition metal sulfides (delocalized, or collective, d electrons causing metallic conductivity) and for nontransition metal compounds like PbS (where it corresponds to the amount of nearly free electrons in crystal orbitals of the energy band scheme). The chemical bonds in sulfosalts with the complex radicals $[AsS_3]^{3-}$, $[SbS_3]^{3-}$, $[BiS_3]^{3-}$ etc. differ from the bonding in sulfides no less than the bonds in silicates, sulfates, etc. from those in oxides.

With respect to the energy gap width, and corresponding to its absorption edge, there occur compounds ranging from insulators and semiconductors to crystals with metallic conductivity and from transparent and colorless to colored and opaque for visible region crystals.

Directed covalent bonds suggest more specific exhibition of each type of the electron configuration of metals. Exchange interactions between metals depend on the relative position of d electron levels with respect to energy bands.

The close position of d and p orbitals of the atom S leads to diversity of its coordinations.

Low-spin states of transition metals have been observed in minerals only in sulfides and related compounds (parallel with more common high-spin states).

In many types of sulfide the donor-acceptor bonds occur. In some compounds defect structures are stable.

Understanding of these versatile aspects of the chemical bond sulfides, of the causes of realization of one or another structure motif so diverse in sulfides and related minerals, interpretations and calculations of spectra and properties with all their extensive technical applications, inferences concerning isomorphous substitutions, stability and phase transformations are based on the corresponding aspects of consistent models of the molecular orbital – ligand field – energy band theories.

Because of the predominant covalent component in the chemical bond in sulfides, the lattice energy calculations, without taking into account overlapping of atomic orbitals, which are already insufficient for oxygen compounds, are inadequate. The crystal field theory provides a satisfactory qualitative description of optical absorption and EPR spectra of transition metal sulfides with essentially large energy gap width, but any estimations of the crystal field strength, superfine structure etc. need more adequate theories.

In qualitative discussions of certain crystal-chemical features of sulfides, one can already draw some essential conclusions from the consideration of electron configuration of atoms and their tendency towards formation of the stable (filled or half-filled) electron shells. In these discussions the concept of donor-acceptor band is also useful (see below).

From the valence bond model one uses the pictorial representations of the bonds by means of hybrid atomic orbitals. However it also entails using the valence structure concept (see below).

However, the most complete and rigorous description of the electronic structure of sulfides can be obtained on the basis of the energy band theory especially using the pseudopotential and bond orbital methods (see Chap. 7.1). In particular, the estimations of covalency (or ionicity) can be made in the framework of the tight-binding methods (see Chap. 7.1) and by using the reflectance spectra and their interpretations based on the calculated energy band diagrams (see Chap. 4.2).

Different theoretical schemes employed in interpretations of the aspects of electronic structure, spectra and properties of sulfides are compared in Fig. 128.

8.3.2 Energy Gaps in Sulfides, Types of Crystal and Types of Optical Transition; Ionicity and Band Scheme

In the computed energy band schemes showing curves of the crystal orbital energy dependence on the point position in a Brillouin zone (Fig. 130d), it is important to see the chemical origin of the electronic states denoted by the symmetry types (Γ, Δ, L, etc.), i.e., to see from which atomic orbitals in the main these states are made (from $3s$, $3p$ etc.). For example, in sphalerite and wurtzite quite different sets of points exist in the Brillouin zones (because of different symmetry: cubic and hexagonal) and different energy band schemes (see Figs. 71, 72). However, in both of them the valence band is composed mainly of S $3p$ orbitals, while the conduction band is of Zn $3s$ orbitals (it should be noted, however, that this is only an indication of the predominant contributions in these bands; a more reliable description takes into account mixing in both bands of the metal and sulfur orbitals, where the more covalent bond corresponds to the greater mixing).

Simplified schematic energy band diagrams (Fig. 130c) showing only total widths of the bands with their lineage (i.e., with the indication of predominant contributions of the different metal or sulfur atomic orbitals) are often used in descriptions of the general type of band structure and in discussions of some properties.

Types of band scheme (Fig. 131) are distinguished by the predominant state of the valence and conduction bands, existence of a narrow band or levels of d electrons (in transition metal sulfides), the widths of these bands, and primarily by the energy gap width. According to the width of the energy gap, sulfide minerals can be considered as semiconductors with large gaps (for example, sphalerite) and semiconductors with small gaps (for example, galena). Transition metal sulfides can be related to semiconductors and to compound with metallic conductivity.

The width of the energy gap also determines the position of the absorption edge, i.e., the position of the optical transition with minimum energy (with maximum wave length). If the absorption edge occurs in the IR region, the crystal is opaque (for example, galena), if it is in the visible, the crystal is colored and its color is determined by the part of visible region cut by the absorption edge (for example, cinnabar) and if it is in the UV, the crystal is transparent and colorless (for example, sphalerite without any impurities, which is extremely uncommon for natural sphalerites).

Reflectance spectra of sulfides in visible and UV are very complicated in nature, because of the complexity of the band structure. However, individual peaks in the reflectance spectra overlap and form bands consisting of several peaks and corresponding to the transitions between different crystalline orbitals belonging to the same bands (for example, in galena between $3p$S-like levels of the valence band and $6p$Pb-like levels of the conduction band, see Chap. 4.2). In transition metal sulfides one also observes the optical transitions connected with d electron levels occurring in the forbidden band; the number of these transitions depends on the filling of these split levels.

In interpreting the total energy gap E_g for such essentially covalent but partially ionic crystals as sulfides, one distinguishes two components of the E_g: (1)

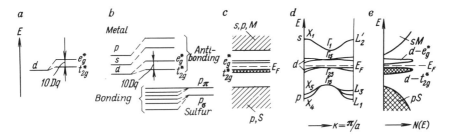

Fig. 130a–e. Energy level and energy band schemes used in descriptions of electron structure of transition metal sulfides. **a** Crystal field model (see Fig. 30); **b** molecular orbital model (see Fig. 48). **c–e** Three aspects of the energy band model; **c** valence band and conduction band total widths only are shown; in the forbidden bands occurs narrow d electron bands; **d** crystalline orbital level energy variation in different point of Brillouin zone (i.e., variation with wave vector k); in the forbidden band shown the d electron $\Gamma_{25}(t_{2g})$ and $\Gamma_{15}(e_g)$ states; **e** density of states $N(E)$ plotted against energy; in **d** and **e** only there is horizontal coordinate

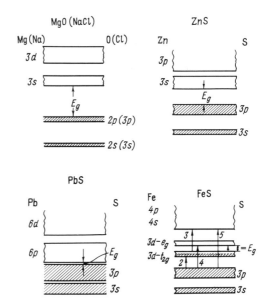

Fig. 131. Types of energy band scheme

covalent, or homopolar, E_h, and (2) additional ionic, or heteropolar, component C_{AB} [465, 466].

The method of estimation of the homopolar energy gap E_h was suggested for the crystals $A^N B^{8-N}$, that is for the semiconductors A^{IV}, $A^{III}B^V$, $A^{II}B^{IV}$ and dielectrics $A^I B^{VII}$ with tetrahedral diamond and zinc blende-type structures [465]. Interatomic distances and heats of formation for these compounds are approximately equal when A and B belong to the same row of the periodic system. Since E_h depends only on the bond length, then this homopolar component of the total energy gap of heteroatomic crystals can be determined from the energy gap values of the homoatomic crystals of the same row of the periodic system. Thus as for the

E_h for BN (second row of the periodic system) the E_g value of carbon is taken, for AlP it is the E_g of silicon, for GaAs it is the E_g of germanium. In the case of ZnS, where Zn belongs to fourth row, while S to the third row, one takes E_h as the mean with respect to E_g of Ge (fourth row) and E_g of Si (third row).

The experimentally determined total energy gap E_g is related to covalent, E_h, and ionic, C_{AB}, energy gaps by the expression:

$$E_g^2 = E_h^2 + C_{AB}^2.$$

Thus, after experimentally measuring E_g and evaluating E_h, one can determine C_{AB}. These values are used to estimate the ionicity f_i (the fraction of ionic or heteropolar character in the bond) and the covalency f_c (the fraction of covalent or homopolar character):

$$f_i = C_{AB}^2/E_g^2; \quad f_c = E_h^2/E_g^2; \quad (f_i + f_c = 1).$$

For example, for ZnS (sphalerite): $E_g = 3.6$ eV; $E_h = (E_g\text{Ge} + E_g\text{Si}):2 = (0.74 + 1.2):2 = 0.49$ eV and $f_c = 0.27$.

8.3.3 Interactions M–M and M–S–M in Transition Metal Sulfides and Their Relation to Properties and Structures

Only sulfides of transition elements (i.e., elements typically chalcophile) represent the most widely distributed species among sulfide minerals.

They resemble other sulfides (those of nontransitional metals) first of all by the small energy gap, but the presence of d electrons imparts to them features common to other transition metal compounds. Even semiconducting properties are exhibited in them in another way. Moreover, among sulfides and oxides, only those of transition metals can in some cases reveal metallic properties, but again following a scheme other than in metals.

Consider the most general peculiarities of electronic structure of transition metal sulfides as based mainly on energy band schemes, but compared with the molecular orbital and crystal field approaches.

In the energy band schemes of transition metal sulfides one distinguishes (Fig. 130): (1) broad valence bands (upper band composed mainly of $3p$ orbitals of S) and broad conduction bands (lower band is made mainly of ns and np orbitals of transition metal) and (2) narrow bands of d electrons lying either in the forbidden band or in the conduction or valence band.

The d electron states in the band scheme as, in the crystal field model, are split: in the octahedral field there are a lower narrow t_{2g}^*-like band (i.e., Γ_{25}, $X_3 + X_5$, $\Sigma_1 + \Sigma_2$ ... band) and a higher e_g^*-like band (i.e., Γ_{12}, $X_1 + X_2$, $\Sigma_1 + \Sigma_4$ ... band). These narrow bands can be separated from each other, or overlap, depending on the structural type and interactions with sulfur and with the lattice [154].

As in the MO model, both $t_{2g}^*(\pi)$ and $e_g^*(\sigma)$ states are antibonding.

Consideration of the relative position and broadening of the d electron bands permits a description of some important properties and structural features of transition metal sulfides (Fig. 132).

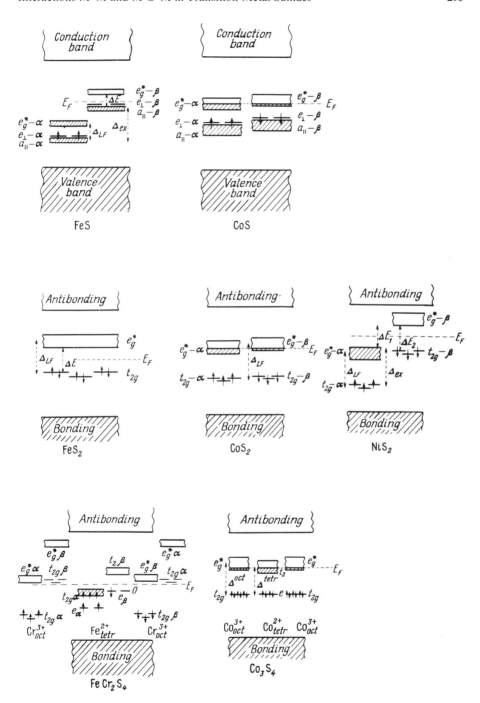

Fig. 132. Simplified energy band schemes for transition metal monosulfides, disulfides and thiospinels [417, 421]

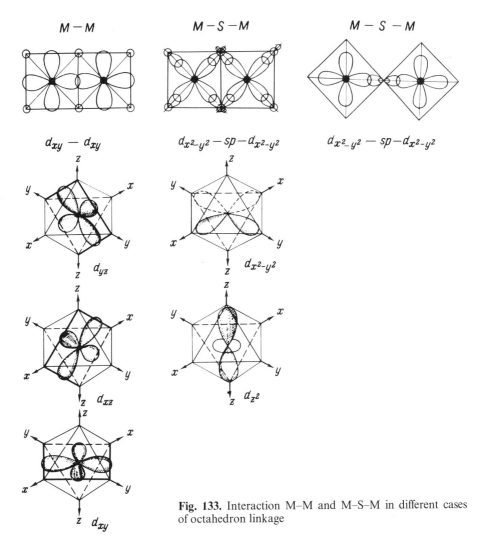

Fig. 133. Interaction M–M and M–S–M in different cases of octahedron linkage

Two types of d electron in crystals: localized and collective; two mechanisms of the delocalization: "covalent" (M–S–M) and "metallic" (M–M). Two extreme types of outer d electron behavior exist in solids:

1. Localized electrons: this means first of all that interaction between neighboring transition metal cations is weak enough; the corresponding states are described by discrete levels in energy band schemes and can be treated within the framework of the crystal field theory.

2. Collective electrons: the d electrons are distributed over all the lattice atoms; this demands the use of translation symmetry and the energy band theory for their description.

In octahedral coordination d electron states are split into t_{2g} orbitals (d_{xy}, d_{xz}, d_{yz}) directed between the corners of octahedron and e_g orbitals ($d_{x^2-y^2}$, d_{z^2}) pointed to the vertices, i.e., towards ligand atoms.

When the octahedra joints in a structure by edges or faces (Fig. 133) the direct interaction metal–metal can be realized through overlapping of the t_{2g} orbitals of transition metal cations of the neighboring octahedra. This M–M interaction of t_{2g} orbitals of the neighboring cations leads to the contribution of the metallic component in the chemical bond. The M–S–M interaction through the e_g orbitals of the cations is realized in the cases of essentially covalent bonding.

The cation–cation interaction corresponds to the "metallic" mechanism of delocalization of the d electrons while the cation–anion–cation interaction is described as the covalent mechanism of the delocalization.

The critical values of the overlap integrals Δ_c (for cation–cation d or t_{2g} orbitals) were suggested as the criterion for the localized or collective behavior of the d electrons [463].

If the overlap integral of d orbitals of the cation in a given crystal Δ_{cc} is greater than the critical value Δ_c, the electrons cease to be localized and form the bands; for their description one has to pass from the crystal field theory to the energy band theory.

Since the overlap integral Δ_{cc} is connected with the distance R_{cc} between these cations (the shorter the distance the larger the overlapping) one also uses the critical cation–cation distance R_c: if the observed R_{cc} is shorter than the critical value R_c, then $\Delta_{cc} > \Delta_c$, and one passes to the energy band description.

For the transition metal sulfides and oxides [463], these critical cation–cation distances were determined semi-empirically (in Å):

	oxides	sulfides
Fe	2.95	3.15
Co	2.87	3.07
Ni	2.77	2.97.

That is in sulfides R_c is approximately by 0.2 Å longer than in oxides.

In the case of cation–anion–cation interaction, one uses the critical overlap integral of the σ orbitals (overlap integrals of σ, π and $\bar{\pi}$ orbitals are lesser than that of σ orbitals).

In addition to the crystal field splitting it is necessary take into account electron–electron exchange interaction within the d electron subshell. This leads to the splitting of the each d electron level into spin α and spin β states (Fig. 134).

Consider some examples of the energy band analysis for certain transition metal sulfides.

1. In the series of the compounds FeS–CoS–NiS– which all form with NiAs structure, the difference in electronic structure of the metals reveals only an addition of one more d electron to each next compound ($Fe^{2+} - d^6$, $Co^{2+} - d^7$, $Ni^{2+} - d^8$), but by this their properties are very different.

In the structure type NiAs the octahedral polyhedra $M-S_6$ (distorted along the L_3 axes) are superimposed on each other by their octahedral faces along the c axis. The distances between the cations along the c axis (R_{cc}) and in basal plane (R_{bb}) are (in Å):

	FeS	CoS	NiS
R_{cc}	2.90	2.58	2.67
R_{bb}	3.45	3.36	3.43.

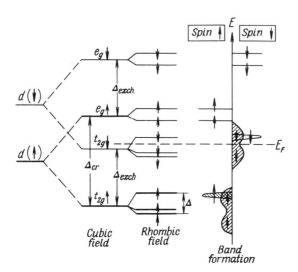

Fig. 134. Splitting of d electrons levels with the spins α and β [437]

In all these compounds R_{cc} is less than the critical distance $R_c = 2.15$ (for sulfides of Fe), $R_c = 3.07$ (Co), $R_c = 2.97$ (Ni) while R_{bb} is greater than R_c.

From the two types of metal d orbitals t_{2g} orbitals are responsible for the M–M interaction and e_g orbitals for the M–S–M interaction.

The t_{2g}^* orbitals are split in this structure type (with the trigonal local symmetry of M) into two states: $a_\parallel$ related to the M–M interaction along the c axis and $e_\perp$ related to the M–M interaction in the basal plane. Since $R_{cc} < R_c$ the $a_\parallel$ states form bands, while since $R_{bb} > R_c$, the $e_\perp$ states remain as discrete levels (see Fig. 132).

The e_g^* orbitals, because of the covalency of sulfides, are always represented by bands and not by discrete levels.

Further peculiarities of the energy band schemes of these compounds [402] are revealed by taking into account their electrical and magnetic properties (see Table 45).

From the μ_{eff} and n_{eff} values one determines the high-spin state of Fe^{2+} in FeS ($t_{2g}^4 e_g^2$, i.e., two electrons in the $a_\parallel$ band, two electrons in the e_g^* band and two electrons in $e_\perp = t_{2g}$ discrete levels), the low-spin state of Co^{2+} in CoS($t_{2g}^6 e_g^1 \to a_\parallel^4 e_\perp^2 e_g^{*1}$) and the low-spin state of Ni^{2+} in NiS ($t_{2g}^6 e_g^2 \to a_\parallel^4 e_\perp^2 e_g^{*2}$).

In high-spin FeS the strong exchange interaction Δ_{ex} (see Fig. 132) provides the splitting into states with α spin and β spin which is greater than the ligand (crystal)

Table 45. Energy gap, electrical and magnetic properties of some NiAs-type transition metal sulfides [402]

Sulfides	Conductivity type	E_g	μ_{eff}	n_{eff}	Magnetic properties
FeS	Semiconducting	0.27	5.18	4.27	Antiferromagnetic
CoS	Metallic	—	1.70	0.95	Antiferromagnetic
NiS	Metallic	—	2.66	1.84	Antiferromagnetic

E_g is the energy gap (eV); n_{eff} is effective number of the unpaired electrons.

Table 46. Energy gaps, electrical and magnetic properties of some pyrite type transition metal sulfides [412, 463]

Sulfides	Conductivity type	E_g, eV	μ_{eff}, (μ_B)	Magnetic properties	R_{cc}, Å	State
FeS_2	Semiconducting	0.37	0.0	Diamagnetic	3.83	$t_{2g}^6 e_g^0$
CoS_2	Metallic	—	1.85	Ferromagnetic	3.91	$t_{2g}^6 e_g^1$
NiS_2	Semiconducting	0.12	3.19	Paramagnetic	4.06	$t_{2g}^6 e_g^2$
		0.32	2.48			

field splitting Δ_{LF}. The energy gap of FeS $E_g = 0.27$ eV is too small to correspond to the transition of d electrons directly into the conduction band. Thus that E_g value corresponds to the distance between the top of the higher filled narrow band $a_\parallel - \beta$ and the bottom of the empty narrow band $e_g^* - \beta$ of the d electrons.

Localized electrons of the discrete $e_\perp - \alpha$ levels polarize unpaired electrons in the narrow bands: parallel in $a_\perp - \alpha$ and $e_g^* - \alpha$ and antiparallel in $a_\parallel - \beta$ (in all four unpaired electrons). This determines the antiferromagnetic properties of FeS (and in the same manner of CoS and NiS).

In CoS and NiS, metallic conductivity is explained in the band scheme by the occurrence of an unpaired electron in the e_g^* band and the overlapping of the $e_g^* - \alpha$ and $e_g^* - \beta$ bands.

2. In FeS_2–CoS_2–NiS_2 (Table 46, Fig. 132) with the structure type pyrite, the direct M–M interaction does not occur ($R_{cc} < R_c$), t_{2g}^* states are represented by discrete levels while e_g^* states are broadened into bands owing to covalent mixing of the M–S–M orbitals (with overlap integrals $\Delta_{c-a-c} > \Delta_c$) [412, 416, 420, 463, 484].

All these compounds exist in the low-spin state. If the bands are partially filled one observes metallic conductivity (as in CoS_2), if they are empty (as in FeS_2) the compounds are semiconductors; in NiS_2 semiconducting properties are determined by the forbidden band between $e_g^* - \alpha$, $t_{2g}^* - \beta$ and $e_g^* - \beta$.

3. In thiospinels AB_2S_4 (where A are tetrahedral and B octahedral cations) the distances between the cations R_{AA}, R_{BB} and R_{AB} are longer than the critical value R_c. Thus the t_{2g}^* states (for octahedral cations) and e^* states (for tetrahedral cations) are represented by discrete levels, while e_g^* (in octahedra) and t_2^* (in tetrahedra) form bands because of the covalency of thiospinels.

Consider the energy band schemes (see Fig. 132) of a semiconductor-ferromagnetic, $FeCr_2S_4$ (daubreelite) and of a Pauli paramagnetic with metallic type conductivity Co_3S_4 (linnaeite) [450, 481].

In daubreelite the filled levels and bands of the high-spin Fe^{2+} in tetrahedral position ($e^2 t^4$) and Cr^{3+} in octahedral position (t_{2g}^3) are separated from the unfilled bands and levels by the forbidden band (with the Fermi level within the latter).

In linnaeite the discrete levels of the low-spin Co^{3+} in the octahedral position ($t_{2g}^6 e_g^0$) and Co^{2+} in the tetrahedral position ($e^4 t_2^3$) are occupied, but the e_g^* (Co_{oct}^{3+}) bands remain empty and the t_2 (Co_{tetr}^{2+}) bands are half-filled ($e^4 t_2^3$) with the Fermi level within these overlapping $e_g^* - t_2$ bands. This leads to metallic conductivity of Co_3S_4, characteristic of metal Pauli paramagnetism.

8.3.4 States of Iron in Sulfides According to Mössbauer Spectra Parameters

Some important characteristics of the chemical bond can be obtained for those sulfides which have elements with Mössbauer nuclei (Fe^{57} and in a lesser extent Sn^{119}, Te^{125}, Au^{197}) in their composition. In some respects Mössbauer spectroscopy provides still greater possibilities for sulfides than for silicates: isomer shift variations are larger in sulfides and magnetic hyperfine splitting, practically absent in silicates, is observed in many sulfides.

The Mössbauer parameters are of diversified significance in descriptions of the chemical bond in sulfides: they allow determination of valence states, electron configurations, ionicity–covalency, high- and low-spin states, exchange interactions, and magnetic ordering.

Let us compare the isomer shift (IS) values in sulfides with those in oxides, silicates, sulfates and other compounds with oxygen ligands (Table 47).

Increasing isomer shift values reflect decreasing s electron density at the nucleus and increasing d electron density (screening s electrons and thus decreasing their density at the nucleus). Thus increasing covalency in sulfides leads to increasing s electron density at the iron nucleus and to decreasing isomer shift.

Ranges of IS value variation of Fe^{2+} in sulfides are rather large and include the values typical both for most covalent and essentially ionic compounds.

Table 47. Isomer shift relative iron in stainless steel (mm s^{-1})

Compounds	Fe^{2+}	Fe^{3+}
Sulfides	0.38–1.20	0.34–0.47
Oxydes, silicates	1.00–1.60 (usually 1.2–1.4)	0.35–0.80

Table 48. Relation of spin state, oxydation state and coordination of iron in sulfides and related compounds to Mössbauer parameters

High-spin Fe^{2+} and Fe^{3+} ($S=2$ and $(5/2)$)						Low-spin Fe^{2+} ($S=0$)			
Sulfosalts			Sulfides			Disulfides			
Stannine	Fe_4^{2+}	0.87 2.76	Sphalerite	Fe_4^{2+}	0.75 0.80	Pyrite	Fe_6^{2+}	0.42 0.63
Berthierite	Fe_6^{2+}	1.20 2.67	Pentlandite	Fe_4^{2+}	0.58 0.40	Marcasite	Fe_6^{2+}	0.38 1.52
			Troilite	Fe_6^{2+}	0.86 —	Löllingite	Fe_6^{2+}	0.39 1.68
			Pyrrhotite	Fe_6^{2}	0.76 —			
			Cubanite	Fe_4^{2+}	0.51 0.88			
			Arsenopyrite	Fe_6^{3+}	0.34 1.05			
			Chalcopyrite	Fe_4^{3+}	0.36 —			
			Bornite	Fe_4^{3+}	0.50 —			

Isomer shift (relative to stainless steal, mm s^{-1}) and quadrupole splitting (mm s^{-1}) (first and second columns respectively).

Smallest IS values are observed for low-spin Fe^{2+} in disulfides and diarsenides (pyrite, marcasite, löllingite and others) with complex radicals $[S_2]^{2-}$, $[As_2]^{2-}$ while the largest IS values occur in sulfosalts with complex radicals $[SbS_2]^{2-}$, $[SnS_4]^{4-}$.

For simple sulfides (without complex radicals) isomer shift increases, dependent on valency and coordination: $Fe_6^{2+} > Fe_4^{2+} > Fe_4^{3+}$.

Quadrupole splitting (electric-field gradient) is observed in all sulfides of Fe^{2+} (usually 0.4–0.9 mm s^{-1} at 300 K and up to 2.0 mm s^{-1} at 80 K). The largest values of quadrupole splitting have been observed in sulfosalts (2.67 mm s^{-1} at 300 K and 3.60 mm s^{-1} at 80 K in berthierite).

In magnetically ordered ferromagnetic sulfide minerals (cubanite, pyrrhotite) and antiferromagnetic sulfides (troilite, chalcopyrite), one observes magnetic hyperfine structure arising from exchange interaction. The values of the local magnetic field at the Fe^{57} nucleus are for Fe^{2+} from 225 kOe (pyrrhotite) to 332 kOe (cubanite), for Fe^{3+} 365 kOe (chalcopyrite).

States of Fe in sulfides which can be discerned from Mössbauer spectra data are indicated in Table 48.

Much more scarce are Mössbauer studies of sulfides at other nuclei; however the available data on Sn, Au, Te provide important information on the chemical bond in the compounds of these elements.

The importance of the Mössbauer parameters obtained will increase with advances in their theoretical interpretation and with their use in conjunction with other experimental data (especially with X-ray spectroscopy data).

8.3.5 Polarity and Donor–Acceptor Bonds in Sulfides and Sulfosalts of As, Sb, Bi According to NQR Data [191, 181]

In the method of nuclear quadrupole resonance (NQR), nuclei As^{75}, Sb^{121}, Sb^{123}, Bi^{209} enter the list of most convenient NQR nuclei. Natural abundance of As^{75} and Bi^{209} is 100%, that of Sb^{121} 57.25, and Sb^{125} 42.75%. These elements enter the composition of important ore minerals which belong at the same time to compound which found technical applications over the past years.

The principle parameter of NQR is the frequency of the resonance, from which one determines the constant of quadrupole splitting eQq. The characteristics of the mineral in this constant are represented by electric-field gradient q at the site of the nucleus with quadrupole moment eQ; essential but additional information can be obtained from NQR line width analysis and relaxation parameters.

It is the same characteristic which is obtained from quadrupole splitting in Mössbauer spectra and two parameters B_l^m in EPR spectra ($B_2^0 = D$ and $B_2^2 = E$). Let us recall that variation of these parameters in Mössbauer and EPR spectra are related to small distortions of octahedron or tetrahedron which can lead to sharp changes of electric-field gradient.

However, in contrast to iron group ions with usually octahedral or tetrahedral coordination, for As, Sb, and Bi pyramidal groups RX_3 are typical (see Fig. 135). The very structure of these groups suggests the noncubic crystalline field. However, it is not structure distortion, but redistribution of electronic density

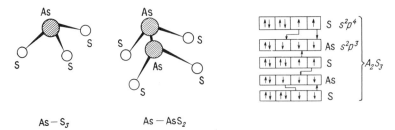

Fig. 135. Pyramidal groups As–S$_3$ and As–AsS$_2$ in R$_2$S$_3$ and R$_2$As$_3$ type compounds and the electron distribution in them [191]

which gives rise to the electric-field gradient. The angles X–R–X vary little in coordination groups of the same type: AsS$_3$–99°, SbS$_3$–96°, BiS$_3$–94°, AsO$_3$–102°; this suggests that sp hybridization is of no great importance.

Owing to this structural peculiarity of the RX$_3$ group one could establish the principle factor influencing the variations of the quadrupole-splitting constant in NQR spectra of sulfides and sulfosalts of As, Sb, and Bi as polarity of the R–X bond (i.e., As–S, Sb–S, Bi–S) [191]. Here polarity corresponds to ionicity of the bond.

Thus increasing of covalency decreases the eQq value in the series As$_2$O$_3$ (220 Mc s^{-1}) – As$_2$S$_3$ (143 Mc s^{-1}) – As$_2$Se$_3$ (118 Mc s^{-1}) – As$_2$Te$_3$ (88 Mc s^{-1}) and in the series Sb$_2$O$_3$ (541 Mc s^{-1}) – Sb$_2$S$_3$ (318 Mc s^{-1}) – Sb$_2$Se$_3$ (272 Mc s^{-1}) [191].

Comparison of the variation of eQq in different structural types enables the following aspects of electronic density distribution to be revealed: (1) ionicity-covalency of the R–X bonds (As–S, Sb–S, Bi–S) with participation of donor–acceptor bonds; (2) effects of As–As bonds in some structures; (3) influence of M–S bonds (when there are metals M = Cu, Pb, Ag, Tl) on electronic density distribution in M–S–R (with R = As, Sb, Bi).

In donor-acceptor bonds, the donor atoms with many electrons in outer p and d orbitals or with s^2 electrons not involved in the bond ("lone pairs of s^2 electrons") transfer their electrons to the acceptor atoms with a smaller number of electrons in p and d orbitals.

Thus acceptors are atoms of first elements in the transition metal rows, for example, Ti ($3d^24s^2$) having only four electrons in five d, one s and three p orbitals, while such elements as oxygen ($2s^22p^4$), sulfur ($3s^23p^4$), halogens (s^2p^5), can be donors owing to both the lone pairs of s^2 electrons and p-electrons.

Polarity of the bonds of As, Sb, Bi (s^2p^2) is accompanied by weak donor (sulfur) – acceptor (As, Sb, Bi) interaction and by a dipole moment originating from the lone pair of s^2 electrons.

An estimation of ionicity from NQR spectra can be made as follows [191]:

The electric-field gradient value is proportional to a number of nonbalanced p electrons N_p which can be determined from the ratio of the eQq_{exp} measured for a given crystal to eQq_{at} of the free atom:

$N_p = eQq_{exp}/eQq_{at}.$

However, N_p is related to the degree of sp hybridization (after Pauling) $\lambda_{sp} = \cos\vartheta/(1-\cos\vartheta)$, where ϑ is the angle X–R–X, and to the desired degree of covalency (i_{cov}):

$$N_p = 3\lambda_{sp}(1+i_{cov}).$$

From example, for proustite, Ag_3SbS_3:

$$\lambda_{sp}(\text{As–S}) = 0.19,$$
$$eQq_{exp}(\text{As}^{75}) = 132.35 \text{ Mc s}^{-1},$$
$$eQq_{at}(\text{As}^{75}) = 600 \text{ Mc s}^{-1}.$$

From $N_p = 3\lambda_{sp}(1+i_{cov})$ one determines $i_{cov} = 0.61$, i.e., the As–S bond in proustite is 61% covalent. For pyrargirite, Ag_3SbS_3:

$$\lambda_{sp} = 0.12,$$
$$eQq_{exp}(\text{Sb}^{121}) = 324.74 \text{ Mc s}^{-1},$$
$$eQq_{at}(\text{Sb}^{121}) = 2000 \text{ Mc s}^{-1},$$
$$i_{cov} = 0.55.$$

However, these estimations of covalency, as well as the concept of donor–acceptor bond, correspond to the valence bond model. In the MO theory, there is no necessity in this concept which is replaced here by the c_i coefficients in the antibonding molecular orbitals made from the metal and ligand atomic orbitals.

8.3.6 Structural Features of Sulfides and Related Compounds from the Standpoint of the Electronic Structure

The accumulation of crystal structure determinations for sulfides and related compounds illustrates that these structures differ from those of oxygen compounds and are more diverse than in oxides. In some cases there are principal differences: for example, linkage of coordination polyhedra in sulfides by edges and faces instead of linkage by corners and edges common in oxygen compounds.

It was naturally to try to understand the causes of these general differences and to explain why atoms with their electron configurations in sulfides form arrangements in the actual structure which are often rather strange and even unique [414, 442, 455, 461, 471]. Attempts at prediction of new compounds with semiconducting and more special properties from structural and valence analogy have been further stimulation for these interpretations [422, 436, 441, 456, 459, 464, 473].

It is also necessary to explain the causes of lattice parameter variation in series of isostructural compounds, peculiarities of isomorphism and, in general, causes of more limited isomorphism in sulfides, causes of stable nonstoichiometry in a number of types of crystals etc. and, as a consequence, the nature of the special electrical, magnetic, optical and mechanical properties of sulfides.

It should be noted that the difference between structural types of sulfides and oxides of the same metal is more pronounced for transition metal compounds than for compounds of nontransition metals.

Thus, for example, MgS, CaS, PbS as MgO, CaO, PbO form with the NaCl structure and both ZnS and ZnO form with the sphalerite-wurtzite structure in spite of a sharp difference in their properties. On the contrary, whereas transition metal oxides TiO, VO, CrO, MnO, FeO, CoO, NiO form with the NaCl structure, sulfides of the same metals crystallize with the NiAs structure (TiS, VS, CrS, FeS, CoS), the ZnS or NaCl structure (MnS), or with the peculiar NiS structure. Disulfides of all these metals form with the pyrite (FeS_2) structure.

Which features of the electronic structure of sulfur cause these differences between sulfides and oxides?

It is first of all the smaller valence state ionization energies of $3s$, $3p$ and $3d$ orbitals of sulfur (see Fig. 57) that lead to their greater mixing with the metal orbitals, i.e., to an increasing covalency of the bonds. Involvement of $3d$ orbitals increases the coordination possibilities of sulfur as compared with oxygen: besides coordination number 2, 3, 4 the coordinations 5, 6, 7 and more are possible for sulfur.

The amount of the S $3d$ orbitals in the bond are described in the MO method by coefficients of sulfur d atomic orbitals in the molecular orbital.

Greater covalency leads to interactions described as donor-acceptor bonds where sulfur is the donor of electrons in the presence of unfilled close orbitals of metal.

Covalent S–S bonds can lead to the formation of complex radicals $[S_2]^{2-}$ which appear under the influence of interaction with metal d orbitals (see below).

As in oxygen polymeric compounds, sulfur atoms in sulfide can be bridging and nonbridging. In complex sulfides and sulfosalts one observes rather considerable variation of the interatomic M–S distances. Sulfur (as Se and Te) is a much more "polarizable" ion than oxygen: because of their greater covalency, its bonds are more directed, and asymmetric positions of metal atoms lead to a great asymmetry of sulfur ions.

In the framework of the valence bond method, one distinguishes [462] twenty types of sulfur atoms (Fig. 136). In the basis of this division there are the three features:

1. Formal charge, i.e., difference in number of outer electrons as compared with the sulfur atom with six outer electrons ($3s^2 3p^4$):

$$S^{2+}(s^2p^2), S^+(s^2p^3), S^0(s^2p^4), S^{2-}(s^2p^6).$$

2. Types of orbitals (hybrid) in which these electrons are distributed; one distinguishes "argon" sulfur with hybrid orbitals composed only of $3s$ and $3p$ orbitals (complete filling of these orbitals corresponds to the electron configuration of argon) and "trans-argon" sulfur with hybrid orbitals composed, besides $3s$ and $3p$, also of $3d$ orbitals.

3. Types of hybridization corresponding to observed coordinations of sulfur (in Fig. 136 two points indicate two unpaired s^2 or p^2 electrons, lines correspond to hybrid orbitals and the number of lines determines the coordination number).

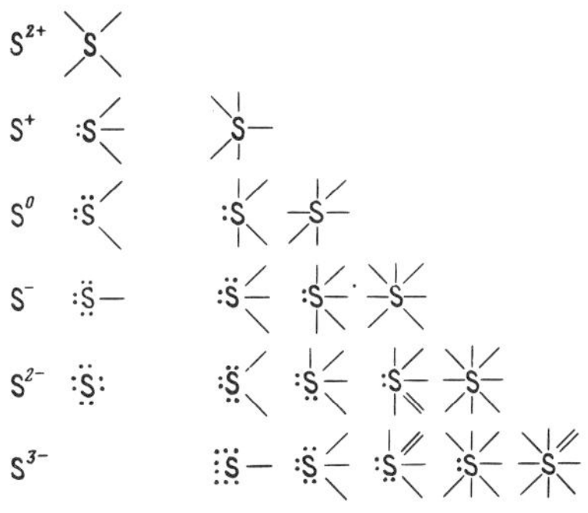

Fig. 136. Sulfur atom types. (After [462])

Thus one obtains the type listed in Table 49: unpaired electrons (shown by points in Fig. 136) are written in brackets and the further type of the hybrid orbital is written; for example, in the case of $S^-[s^2p^2]pd^2$ there are the two pairs of the unpaired s^2 and p^2 electrons and the three orbitals, each a hybrid pd^2 orbital.

Such a classification permits determination of the number and directions of hybrid orbitals for all observed coordinations of sulfur.

However in such a description (1) the final electron configuration is obtained as a set of several extreme configurations, (2) charges of metal and sulfur are determined as the mean from several resonance states of the compound, (3) the positive charge of sulfur obtained and the negative of metal have not been supported by experimental or theoretical data. For example, in ZnS with tetrahedral coordinations of Zn and S one obtains a quadricovalent argon state of S^{2+} with $s^2p^2 \to 4(sp^3)$, i.e., $Zn^{2-}S^{2+}$ with mean charges $Zn^{-0.67}S^{+0.67}$ [462].

In PbS with octahedral coordination of Pb and S, the state of sulfur is obtained as a result of the contributions of the five states: S^{2+} (12%), S^+ (50%), S (32%), S^- (6%), S^{2-} (0.2%) with mean charges $Pb^{-0.43}S^{+0.43}$.

In MoS$_2$ (with six-fold coordination of Mo with respect to S in the form of trigonal pyramide and with three-fold coordination of S with respect to Mo): S^+ (47%), S^0 (44%), S^- (9%), S^{2-} (0.3%) and Mo^{2-} (18%), Mo^- (45%), Mo^0 (30%), Mo^+ (7%) with mean charges $Mo^{-0.74}S_2^{+0.37}$.

Consider now the pecularities of bond states of metals in compounds with sulfur. These are first of all (1) greater overlapping of orbitals M–S–M correspond-

Table 49. Hybridization types of sulfur

Charge and number of electrons	Argonic sulfur	Transargonic sulfur
S^{2+} 4	sp^3	
S^+ 5	$[s^2]p^3$	sp^3d
S^0 6	$[s^2p^2]p^2$	$[s^2]p^3d\,sp^2d^2$
S^- 7	$[s^2p^4]p$	$[s^2p^2]pd^2\,[s^2]p^3d^2sp^3d^3$
S^{2-} 8	$[s^2p^6]$	$[s^2p^4]pd\,[s^2p^2]pd^3\,[s^2]p^3d^3sp^3d^4$
S^{3-} 9	—	$[s^2p^6]d\,[s^2p^4]pd^2\,[s^2p^2]n^2d^2\,[s^2]p^3d^4sp^3d^5$

ing to greater covalency, and (2) transition metal cation interaction M–M owing to overlapping of t_{2g} orbitals leading to the realization of crystal structures with contiguous edges and faces of coordination polyhedra and to the metallic component of the bond.

Thus, besides M–S bonds (which are here more covalent and with greater diversity of sulfur coordination possibilities) and S–S bonds (leading to formation of disulfides) there are M–M bonds influencing the formation of the structural type of sulfides. In arsenides and sulfoarsenides As–As bonds lead to the polymerization of the structures. Interaction M–M leads to shortened M–M distances and formation of isolated pairs (for first elements of transition metal row), for example, V–V pairs in patronite, VS_3.

Greater covalency and stronger crystal fields cause the existence of transition metals, common in sulfides low-spin (spin-paired) states, which do not occur in oxygen minerals.

Since compounds with paired spins are more stable here, the existence of low-spin electron configurations and M–M interations (also leading to pairing of unpaired electrons of the two neighboring atoms) control the formation of the structural types and limits of stability of structural types in solid solutions. These interrelations explain, for example, phase relations and structures of disulfides, diselenides, ditellurides, triarsenides and sulfoarsenides of Fe, Co, Ni [412, 444, 484] and thiospinels [481].

Consider changes in the transition metal sulfides series (MS, MS_2) from the beginning to the end of the row (from Ti, V ... to Cu) thus adding only one d electron to the electron configuration of each following metal.

With the increasing of d electron number, the energies of their levels (or narrow d bands) are lowered: each supplemented electron is affected (because of imperfect screening by d electrons) by the greater effective charge of the nucleus increasing its ionization potential. At the same time covalency and crystal field strength increase and the d electron bands are broadened.

If a number of electrons in antibonding orbitals increase, the total energy of the bond decreases, as well as the microhardness, reflectance and stability of the compound, while interatomic distances increase [421, 484].

The low-spin or high-spin state is reflected very strongly in lattice parameters. For example, in the series MS_2, the high-spin MnS_2 ($t_{2g}^3 e_g^3$) has the largest lattice parameter (6.109 Å), while the low-spin FeS_2 ($t_{2g}^6 e_g^0$) without antibonding electrons) has the shortest lattice parameter (5, 418 Å).

In several of the series, in spite of the closeness of the interatomic distances (and closeness of the supposed covalent radii), one observes limitation in isomorphous substitutions. This is connected with more specific features of different electron configurations (types of hybridization, orientations of the interacting orbitals, and relative energies of the d electron bands).

In particular, one of the causes of limitation in isomorphous substitutions can be the difference of the spin states. For example, nonexistence of solid solutions between Fe_3S_4 (greigite) and $FeNi_2S_4$ (violarite) is explained by the fact that while both form with the same thiospinel structure, Fe_{oct}^{2+} is high-spin in greigite and low-spin in violarite [481].

Fig. 137. Position of d band relative to valence band explaining partial reduction of the cation

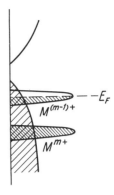

Besides nonstoichiometry related to heterovalent substitutions with charge compensation by vacancies, nonstoichiometry related to delocalization of d electrons is characteristic for many types of sulfides.

In some cases (not related to substitutions) it is accompanied by the appearance of mixed-valence states (for example, in sulfides of Cu) but in other cases it leads to the formation of vacancies (for example, in iron-deficient pyrrhotite $Fe_{1-x}S$).

In metals of the second half of the transition metal series of selenides and tellurides, the energies of d orbitals are strongly lowered and d bands, not only of M^{m+}, but also of $M^{(m-1)+}$, fall below the top of the valence band (Fig. 137). This leads to partial reduction of the cation capturing electron from valence band (i.e., from S^{2-}) in which holes form ($S^{2-} + e^+ \to S^-$; $2S^- \to S_2^{2-}$).

Low enough energy of the d band of $M^{(m-1)+}$ leads to complete reduction of the cation. For example, Cu^{2+} (d^9) reduces in sulfides to very stable diamagnetic Cu^+ (d^{10}) by capture of the electron from the valence band. At the same time, part of S^{2-} transfers into S^- and $S^- + S^- \to S_2^{2-}$ leads to the appearance of dumb-bell disulfide radicals S_2^{2-} and to deformation of the structure. If all S^{2-} ions transfer into $S^- \to S_2^{2-}$, then disulfides of the MS_2 type form.

The capture of electrons and holes in sulfides does not lead to formation of electron-hole centers because of delocalization, but strongly influences their conductivity.

8.3.7 Survey of Data on the Chemical Bond in Sulfides

Thus far the discussion of sulfides was concerned with aspects of the theoretical and experimental characteristics of general features of the chemical bond. Consider now applications of these data to concrete groups of compounds. Completeness of the descriptions of such a type depends on the availability of the experimental data and calculations for each group, i.e., first of energy band scheme calculations by different methods, especially by the semi-empirical tight-binding methods; one takes into account also the availability of simple qualitative band schemes with the information from molecular orbital and ligand field con-

Table 50. $A^N B^{8-N}$ compounds

Tetrahedral	NaCl structure	Chalcopyrite structure
A^{IV} (C, Si, Ge, Sn)	—	—
$A^{IV}B^{IV}$ (SiC)	—	—
$A^{III}B^{V}$ (GaAs, ZnSb)	—	$(A^{II}B^{IV})\,C_2^V$ (CuGaS$_2$)
$A^{II}B^{IV}$ (ZnS)	(PbS)	$(A^{I}B^{III})\,C_2^{VI}$ (CuFeS$_2$)
$A^{I}B^{VII}$ (CuCl)	(NaCl)	—

siderations, spectroscopic parameters, chemical bond characteristics from analysis of the electrical and magnetic properties.

The compounds whose electronic structure were studied in most detail and with most consistency are crystals grouped in the $A^N B^{8-N}$ type (N is the number of the periodic system group). The essence of this grouping is the fact that each heteroatomic crystal $A^N B^{8-N}$ has a homoatomic analog C^{IV} (i.e., C, Si, Ge, Sn). The covalent component of the bond in the heteroatomic $A^N B^{8-N}$ can be equated with the bond state in its homoatomic analog (see above).

There are in this group the following types of compounds (Table 50; the examples of typical compounds are indicated in parentheses). The $A^N B^{8-N}$ compounds represented as minerals are mainly $A^{II}B^{VI}$.

Tetrahedral Sulfides of the ZnS Type. These (as all tetrahedral $A^N B^{8-N}$) belong to the important class of diamond-like semiconductors [436, 471]. The bond-orbital approximation (see Chap. 7.1) was developed and then successfully applied to study energy band structure and many properties in this type of compound: dielectric constants, elastic constants, effective charges, magnetic susceptibilities, X-ray core electron shifts etc. [144, 153].

Energy band schemes for ZnS and other tetrahedral $A^N B^{8-N}$ have been calculated by different methods [161] and their reflectance spectra have been assigned on the basis of these band structures [156, 157, 161].

Methods of determination of ionicity from the spectroscopic data (using energy gap values) and the system of covalent radii [396] were developed only for the tetrahedral $A^N B^{8-N}$ solids.

By means of electron spectra inner electron level shifts correlated with ionicity were determined [483].

States of the impurity ions in ZnS determined from EPR, luminescence and optical absorption spectra are interpreted in the framework of molecular orbital and ligand field theories.

Tetrahedral Triple Sulfides $A^I B^{III} C_2$ (Chalcopyrite, CuFeS$_2$, type) and $A^{II}B^{IV}C_2$. This structural type can be obtained by substitution of one Zn atom in ZnS structure for A^I and B^{III} or A^{II} and B^{IV}. The compounds have attracted attention over the last years, and for many of them energy band schemes were calculated. For chalcopyrite data exist calculated in detail for the construction of the band structure [159].

Sulfides with PbS Structure belong as ZnS to crystals with the most extensively studied electronic structure; for PbS were performed: calculations by the semi-empirical tight-binding method [153], most rigorous calculations of energy band structure, taking into account relativistic correction and spin-orbit coupling [127, 128], assignment of peaks in the reflectance spectra [127], interpretations of electrical and thermomagnetic properties [460], superfine structure and g-factor of EPR spectra of Mn^{2+} in galena.

Sulfides and Related Compounds with NiAs Structure. A vast group of transition metal monofulfides [446] crystallizes not with the NaCl–PbS structure but with the hexagonal nickeline (NiAs) structure (see survey of the structures of monochalcogenides in [459]). Coordination polyhedra in this structural type (octahedra distorted along the L_3 axis) are joined by their faces, which leads to a decrease of Madelung energy (due to cation–cation repulsion across the adjacent faces), but this decrease is compensated by the same exchange cation–cation interaction increasing the bond energy.

The qualitative scheme of band structure (see Fig. 130) describes changes of electrical and magnetic properties with cation–cation distances and overlap integral values.

Important information is obtained from Mössbauer spectra for troilite-pyrrhotite.

Thiospinels $M^{2+}_{tetr}M^{3+}_{2(oct)}S_4$ as well as seleno- and tellurospinel comprise a vast class of compounds recently sinthesized and have attracted attention owing to the combination of semiconducting and ferromagnetic properties.

Alternative to thiospinel (with cation coordinations 4 and 6) for AB_2X_4 compounds are the Cr_3S_4 structure (NiAs type with ordered vacancies and cation coordinations 6 and 6) and chalcopyrite $CuFeS_2$-sulfogallat $CdGa_2S_4$ structure (with coordinations 4 and 4). A number of minerals belong to these groups: daubreelite $FeCr_2S_4$, violarite $FeNi_2S_4$, greigite $FeFe_2S_4$, indite $FeIn_2S_4$, linnaeite $CoCo_2S_4$, polydimite $NiNi_2S_4$ (and their solid solutions: siegenite), carrollite $CuCo_2S_4$ [481].

Cation–cation interactions considered above by means of qualitative band schemes (Fig. 132), together with their electrical and magnetic properties, are characteristic for these compounds, as for the NiAs type sulfides. Stability, structure and reflectivity can be discussed in the framework of the same band schemes [481]. Mössbauer spectra were obtained for iron-bearing thiospinels.

Transition Metal Dichalcogenides MX_2 for which more than 60 compounds are known are divided into two groups: (1) with pyrite, marcasite and similar structure types (sulfides, selenides, tellurides of the VIII B group of the periodic system for all three rows: Fe, Co, Ni; Ru, Rh, Pd; Os, Ir, Pt), and (2) with layer-type structures and octahedral, trigonal-prismatic or distorted coordinations (CdI_2–MoS_2 type and others); chalcogenides of the IVB, VB and VII B groups and of all three rows of the periodic system belong to this type (except Mn). Correlations between electronic structure and crystal structure, Mössbauer spectra, electrical, magnetic, and optical properties have been discussed in detail [412, 416, 444, 458, 484].

Layer dichalcogenides reveal a great diversity of electrical properties: from insulators as HfS_2, semiconductors as MoS_2, superconductor as $NbSe_2$, and semimetal such as WTe_2, to metallic conductivity in NbS_2 and magnetic properties:

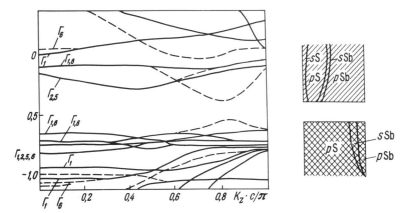

Fig. 138. Energy band scheme of Sb_2S_3 [150]

paramagnetics, ferromagnetics and antiferromagnetics, Pauli paramagnetics. The cation–anion sheets with ionic-covalent and metallic bonds within the sheets are bound to other by Van der Waals bonds [484].

Energy band schemes have been calculated for many compounds of this group. For molybdenite the semiempirical tight-binding calculations and the complete calculations of the band structure by the pseudopotential method have been carried out [143]. The measurements of optical, electrical, photoelectrical, as well as mechanical properties of molybdenite (as a solid lubricant) are correlated with its electronic structure [169, 444].

Sulfides and Sulfosalts of As, Sb, Bi. The most interesting information has been obtained by nuclear quadrupole resonance studies [191]. Peculiarities of these compounds connected with radicals RX_3 (and seldom RX_4), usual for many of them, with donor–acceptor and interlayer bonds in their polymeric variants are revealed in NQR spectra. X-ray, optical and piezoelectric spectra provide experimental parameters describing certain aspects of the chemical bonds. The energy band scheme (Fig. 138) has been calculated for antimonite, Sb_2S_3 [150].

An enormous number of chalcogenides of metal and nonmetal of all groups of the periodic system have been synthesized in the past years, including chalcogenides of alkali and alkaline-earth metals, noble metals, lantanides and actinides. Their crystal structures and properties (in many cases refractory properties) have been studied quite extensively [456]. However the data concerning the chemical bonds in these compounds are usually restricted by consideration of crystal structures and general discussions based on the valence bond method.

In the coming years one can expect extensive development of further calculations of the energy band structure of different types of sulfides and interpretations on their basis of reflectance spectra, studies and interpretations of X-ray, electron and Mössbauer spectra, further considerations and calculations of electrical and magnetic properties (including a number of special properties and their combinations), and systematic discussions of crystal structures from the standpoint of their electronic structure.

8.4 Features of Chemical Bonding in Other Classes of Mineral

If we confine ourself here to listing the most important theoretical and experimental possibilities of chemical bond descriptions and to reviewing some already available data for other classes of minerals, it will then be possible to mention briefly the vast fields of study related to certain important features of some groups of mineral or structural types. References to works on spectroscopy and properties of these compounds too numerous to be quoted here can be found in the principle monographs and reviews listed below (see also [1]).

Consider three classes of compounds: oxides, salts of oxygen acids (sulfates, carbonates etc.), and halides, as well as features of hydrogen and molecular bonding in solids.

Oxides. The most detailed theoretical calculations and measurements of spectroscopic parameters have been performed for simple (binary) oxides, as well as for complex oxides (spinels-chromites-ferrites, perovskites etc.) to which important classes of technical crystals such as most ferromagnetics and ferroelectrics belong. Detailed particular theories describing these properties and the peculiarities of their electronic structure were developed.

Monooxides form with the MgO–NaCl–PbS structure. Periclase MgO has been studied most extensively: calculation of complete energy band structure, interpretations of reflectance spectra, semi-empirical tight-binding estimations of electronic structure have been carried out (see Chaps. 4.2 and 7.1).

Semi-empirical calculations have been made also for CaO, SrO, BaO [153]. Energy band structures of a series CaO–TiO–VO–MnO–FeO–CoO–NiO have been calculated by the augmented plane wave method [154]. These energy band results are used to interpret the electrical and optical properties in this series, which are of great diversity despite their simple and identical structure: this series begins with a typical insulator CaO, the next members TiO and VO exhibit metallic-type conductivity, while MnO, FeO, CoO, NiO are antiferromagnetic insulators.

A great number of oxides with the perooskite, $CaTiO_3$, structure belong to ferro- and antiferroelectrics, pyroelectrics, semiconductors-ferroelectrics, superconductors, etc. The chemical bond characteristics enter the theories of these properties [430, 443]. Directional and nonequivalent bonds existing in these structures determine the condition of realization of ferroelectric and other states [430]. Band structure data also exist for this structural type [474].

The most important problems in a description of properties of spinels-chromites-ferrites such as ferromagnetics, antiferromagnetics etc. are localized or collective states of d electrons, exchange interactions described usually in the framework of qualitative band schemes (as for transition metal sulfides, see Chap. 8.3), and cation distribution between octahedral and tetrahedral positions, estimated often by relative values of the crystal field stabilization energies. Their diverse magnetic properties are described from the point of view of these atomic structure and chemical bonding features and are used in turn to estimate their chemical bond features.

Besides electrical and magnetic properties, spectroscopic parameters are used in descriptions of the chemical bond in oxides. Reflectance spectra in UV (studied extensively, for example, in MgO and ZnO etc.) and adjacent ultra-soft X-ray regions (studied in detail in TiO, VO, V_2O_3 etc. with the interpretation in the framework of molecular orbital and energy band schemes) are of fundamental significance. Mössbauer spectra of iron and tin oxides, optical absorption spectra of transition metal oxides, electron paramagnetic resonance of numerous impurity ions in periclase, corundum, rutile, cassiterite and, spinel; nuclear quadrupole resonance of As_2O_3, Sb_2O_3, Bi_2O_3 provide detailed characteristics of the different aspects of the chemical bond in corresponding structures.

Salts of Oxygen Acids. As in silicates, one can distinguish in salts of other oxygen acids cation polyhedra (Mg, Fe–O_6, Ca, Ba, Sr–$O_{7,8,9,12}$ etc.) and oxyanions: trigonal XO_3^{n-} (CO_3^{2-}, BO_3 etc. in carbonates, borates etc.), tetrahedral XO_4^{n-} (BO_4^{5-}, SO_4^{-2}, PO_4^{3-} in borates, sulfates, phosphates, CrO_4^5, VO_4^{3-} etc. in chromates, vanadates etc., MoO_4^{2-}, WO_4^{2-} in molybdates, wolframates, AsO_4^{3-} in arsenates etc.).

The cation polyhedra are similar, with respect to the state of the chemical bond, to corresponding polyhedra in silicates (see Chap. 8.2) and the same methods of spectroscopy of solids are used for their characterization: electron paramagnetic resonance of impurity ions, Mössbauer spectra of iron etc.

Oxyanion radicals can be divided into several series according to their local symmetry and type of electron configuration.

Tetrahedral oxyanions of the third row elements (AlO_4^{5-}, SiO_4^{4-}, PO_4^{3-}, SO_4^{2-}) are described by the same type of the molecular orbital diagram. In many cases (especially for sulfates and phosphates) they can be treated approximately as isolated clusters. For all of them calculations of the molecular orbital diagrams exist (see Chap. 3.4) and based on them, interpretations of the X-ray spectra. The formation of the free radicals (electron-hole centers) due to natural or artificial irradiation is most characteristic of these classes of minerals.

Trigonal radicals BO_3^{3-}, CO_3^{2-} (in borates and carbonates) are described because of the different local symmetry by the other molecular orbital diagrams. Calculations of the molecular orbital diagrams exist also for these radicals (see Chap. 3.4); the method of BO_3 or BO_4 identification and description has been elaborated by means of B^{11} nuclear magnetic resonance spectra.

Tetrahedral oxyanions of transition metals are described by the typical molecular orbital diagram (see Chaps. 3.2–3.4); experimental data are obtained mainly by X-ray spectra and optical absorption spectra (d–d transition and charge transfer bands, see Chap. 6.3).

Halides. Alkali metal halides M^+Hal^- represent a special group of compounds among all the classes of compounds including other groups of halides. Simple composition and structure (NaCl and CsCl types), univalence, availability of large monocrystals, and transparence, led to using them as model systems in quite numerous and different theories of solids. In particular, lattice energy calculations both in Born-Madelung and Löwdin quantum-mechanical models are closest to the experimental estimations.

Recent calculations provide complete energy band structures obtained by different methods from the most rigorous to the semiempirical (see Chaps. 4.2 and 7.1).

Different particular theories of physics of solids are nearly always applied first to alkali halides. Crystal radii of ions have been evaluated for alkali halides (see Chap. 7.4).

Extensive experimental data include reflectance spectra in UV interpreted on the basis of the energy band schemes (see Chap. 4.2), electron spectra [192], optical absorption spectra of the transition metal ions [1, 225], EPR spectra of impurity ions and of color centers etc.

Intermolecular Interactions. This is an unusual type of interaction in solids; these are interactions not between atoms and ions, but between molecules in solids within which atoms form completely saturated chemical bonds.

Because the bonds within such molecules are mutually satisfying, the intermolecular interaction cannot occur through transfer or exchange of electrons.

How then can the interaction between neutral molecules occur?

At first the nature of intermolecular forces was understood in polar molecules (with constant electric dipole): these are dipole–dipole orientation interactions (potential energy U_{pot}), plus inductive interactions between the dipole of one of molecules and the induced dipole of the other molecule (inductive energy U_{ind}). Then explanation was found for interactions between nonpolar neutral molecules: these occur mainly owing to dispersion forces between instantaneous dipoles arising due to the movement of electrons in each of the molecules (dispersion energy U_{disp}). The total energy of intermolecular attraction corresponding to these Van der Waals forces consists of the three types of interaction:

$$U_{tot} = U_{pot} + U_{ind} + U_{disp}.$$

The intermolecular forces are composed of these types of interaction plus repulsion forces.

These interactions are weak (of the order of few kcal mol^{-1}), while chemical bond energies are of the order of many tens (kcal mol^{-1}) but long-range (to 100–1000 Å).

In molecular crystals (mainly of organic compounds) and molecular liquids, their very existence is determined by the intermolecular interactions.

However among minerals there are no true molecular crystals and in aqueous solutions mainly hydrogen bonds exist.

Intermolecular (Van der Waals) interactions in minerals occur in the three cases: (1) in layer structures: between the sheets separated by an empty sheet; it is a kind of interaction between infinites in two-dimension "molecules"; (2) in chain structures: between the chains separated by an empty chain, like interaction between two infinites in one-dimension "molecules"; (3) in rare molecular crystals: between molecular groups.

Examples of intermolecular interaction can be found in layer silicates (talc, pyrophillite and others), layer sulfides (molybdenite, tetradymite and others), structures of $CdCl_2$–CdI_2, La_2O_3 types etc.

Van der Waals interactions are characteristic for the minerals of As, Sb, Bi: arsenolite A_2O_3 and senarmontite Sb_2O_3 are examples of minerals with molecular structures (As_4O_6 and Sb_4O_6); in particular, transition from the chain structures of the monoclinic As_2O_3 (claudetite), As_2S_3 (orpiment), As_2Se_3 with Van der Waals interactions to the cubic molecular crystals As_2O_3 (arsenolite), AsS (realgar), AsSe is accompanied by a change of the quadrupole interaction constants [191]. Predominant intermolecular (Van der Waals) interactions exist in these cases with contributions from other types of bonding. Distances between atoms of neighboring molecules are determined by the Van der Waals radii [113, 378].

The intermolecular interactions are measured and calculated with in the framework of the different theories [454, 469].

Hydrogen Bonds. Within the molecule H_2O, in hydroxil OH^-, oxonium H_3O^+ the chemical bonds are usual ionic–covalent bonds. For the molecule H_2O, the molecular orbital diagram is calculated [96]. However, the bonds of hydrogen entering these and related molecules and groups with other atoms in crystal or solution are unusual and differ from both the usual ionic–covalent and intermolecular bonds. These are hydrogen bonds [438, 467, 470]; in O–H...O designation the usual ionic–covalent bond is designated by O–H, while the weaker and longer hydrogen bond is shown as H...O.

What is difference in this bond from the intermolecular interactions and why does hydrogen in particular form this type of bond?

The hydrogen atom is the only atom with no inner electron shells, it is small in size and by losing its single electron become proton H^+. Thus the O–H group which forms in OH or H_2O etc. by ionic–covalent bonding of hydrogen with oxygen has unusual properties: it has polarity (represent dipole) with positive charge near the hydrogen atom; the absence of strong exchange repulsion and the small size of the hydrogen atom enable the two oxygen atoms to come nearer to each other, i.e., to the oxygen in the O–H bond (with distance 0.9–1.0 Å) and the oxygen in the hydrogen bond H...O (with distance of the order of 1.5–1.7 Å). Then the distance between the oxygens in O–H...O shortens to 2.5–2.7 Å a distance essentially shorter than the sum of the Van der Waals radii 3.2–3.5 Å.

Thus the hydrogen atom occurring between the two oxygen atoms (or between other electronegative atoms A = O, F, N and more seldom Cl, S) can form with them nonequivalent bonds: usual O–H (or in general A–H) and hydrogen H...O (or in general H...B). The energy of the hydrogen bond (of the order of 2–5 kcal mol^{-1}) is of intermediate value between those of molecular and ionic–covalent bonds.

The theories of the H...B interactions consider electrostatic interactions, charge transfer (the variant of the donor–acceptor bond where the proton is acceptor of the lone pair of oxygen electrons; one also takes into account the contribution of hydrogen-excited orbitals), and the recent calculations of the electronic structure in the framework of the molecular orbital and valence bond methods [416, 438, 448, 467].

One distinguishes in crystals the two cases: the bonds of H_2O molecules and those of hydroxyl OH^-. In crystallohydrates (for example, gypsum $CaSO_4 \cdot 2H_2O$), in the cases of zeolithe water, interlayer water in sheet silicates, and water in pores (opal and others), in one or another form the hydrogen bonds are always exhibited.

In hydroxides, the amounts of hydrogen bond involved depend on the properties of the cations. If in ROH...OR the R cation has a small charge and a long radius, its polarizing effect on the OH group is weak and this group represents an entity. It is not hydrogen but hydroxyl bonds which are exhibited, for example, in hydroxydes $Ca(OH)_2$, $Mg(OH)_2$, and in hydroxyl groups in most silicates etc.

On the contrary, the hydroxyl bonds become more and more hydrogen bonds with increasing charge and decreasing size of R cation, for example, in diaspore $Al(OH)_3$.

Because of the small scattering power of the hydrogen atom, its position in crystal structures cannot be determined by the usual X-ray methods. In hydrogen position and hydrogen bond studies, methods of neutron and electron diffraction, nuclear magnetic resonance, infrared and Raman spectra, and inelastic neutron scattering are used. The position of the hydrogen atom can be determined also by lattice energy calculations [404, 447].

References

1. Marfunin, A.S.: Spectroscopy, luminescence and radiation centers in minerals. Berlin-Heidelberg-New York: Springer 1979. (This book also contains 1087 references which complement the bibliography of the present book)

The Atom in Mineralogy and Geochemistry

2. Clark, F.W.: The data of geochemistry; see in US Geol. Surv. Bull. *770*, 1924 (1908)
3. Fersman, A.E.: Geochemistry, Vols. I–IV. Moscow (1933–1936); see in Fersman, A.E.: Selected works. Moscow: Acad. Sci. USSR, 1955–1958
4. Fyfe, W.S.: Geochemistry of solids. An introduction. New York: McGraw Hill 1964
5. Gavrusevich, B.A.: Principles of general geochemistry. Moscow: Nedra 1968
6. Goldschmidt, V.M.: Geochemische Verteilungsgesetze. Vidensk. Selskab. Skrifter Chr.-Oslo, Vols. I–VIII; see Chap. II. Beziehungen zwischen den geochemischen Verteilungsgesetzen und den Bau der Atome (1922–1927)
7. Grigoviev, D.P.: Principles of the mineral constitution. Moscow: Nedra 1967
8. Lebedev, V.I.: Principles of the energy analysis of geochemical processes. Leningrad: Univ. Leningrad 1957
9. Mason, B.: Principles of geochemistry. New York: Wiley 1967
10. Saukov, A.A.: Geochemistry. Moscow: Nauka 1966
11. Shcherbina, V.V.: Geochemistry. Moscow: Nedra 1972
12. Vernadskii, V.I.: Outlines of geochemistry. Moscow: ONTI 1927
13. Vinogradov, A.P.: Chemical evolution of the earth. Moscow: Acad. Sci. USSR 1959
14. Zawaritskii, A.N.: Introduction to the Petrochemistry of Igneous Rocks. Moscow: Acad. Sci. USSR 1950

Quantum Theory and the Structure of Atoms. Atomic Spectroscopy

15. Ahrens, L.H., Taylor, S.R.: Spectrochemical analysis. London, Paris: Pergamon 1961
16. Aller, L.H.: Atoms, stars, and nebulae. Cambridge: Harvard Univ. Press 1971
17. Angino, E.E., Billings, G.K.: Atomic absorption spectrometry in geology. Amsterdam: Elsevier 1967
18. Bohm, D.: Quantum theory. New York: Prentice Hall 1952
19. Bratzev, V.F.: Atomic wave function tables. Leningrad: Nauka 1970
20. Clementi, E.: Analytical self-consistent field functions for positive ions. J. Chem. Phys. *38*, 996–1008 (1963)
21. Condon, E., Shortly, G.: The theory of atomic spectras. Cambridge: Univ. Press 1957
22. Eljashevich, M.A.: Atomic and molecular spectroscopy. Moscow: Physmathisdat 1962
23. Frish, S.E.: Optical spectra of atoms. Moscow: Physmathisdat 1963
24. Gamow, G.: Thirty years that shook physics. The story of quantum theory. Garden City, N.Y.: Doubleday 1966
25. Green, H.: Matrix methods in quantum mechanics. New York: Barnes and Noble 1968
26. Grotrian, W.: Graphische Darstellung der Spektren von Atomen mit ein, zwei und drei Valenzelecktronen. Berlin: Springer 1928

27. Hartree, D.: The calculation of atomic structure. New York: Wiley; London: Chapman and Hall, 1957
28. Herman, F., Skillman, Sh.: Atomic structure calculations. Englewood Cliffs N.Y.: Prentice-Hall 1963
29. Herzberg, G.: Atomic spectra and atomic structure. New York: Dover, 1944
30. Hund, F.: Linienspektren und Periodisches System der Elemente. Berlin: Springer 1927
31. Kauzmann, W.: Quantum chemistry. An introduction. New York: Academic Press 1957
32. Landau, L., Lifshitz, E.: Lehrbuch der theoretischen Physik. III Quantenmechanik. Berlin: Akademie Verlag 1965
33. Lieberman, D., Waber, J.T., Cromer Don, T.: Selfconsistent-field Dirac-Slater wave functions for atoms and ions. Phys. Rev. *137*A, 27–34 (1965)
34. Lonzich, S.V., Nedler, V.V., Reichbaum, Ja.M.: Spectral analysis in ore deposit prospecting. Moscow: Nedra 1969
35. Ludwig, G.: Wave mechanics. Oxford: Pergamon 1968
36. Mandelstam, S.L.: Spectroscopy and astrophysics. Priroda *2*, 8–19 (1970)
37. Martinov, D.Ja.: A textbook of general astrophysics. Moscow: Nauka 1965
38. Moore, Ch.E.: Atomic energy levels as derived from the analysis of optical spectra. Washington: Gov. Print. Off. 1971
39. Mott, N.F., Sneddon, I.N.: Wave mechanics and its applications. New York: Dover 1963
40. Schiff, L.I.: Quantum mechanics. New York: McGraw-Hill 1955
41. Shore, B.W., Menzel, D.H.: Principles of atomic spectra. New York: Wiley 1968
42. Slater, J.C.: The quantum theory of atomic structure, Vols. I–II. New York: McGraw-Hill 1960
43. Sobelman, I.I.: Introduction to atomic spectra theory. Moscow: Physmathisdat 1963
44. Sommerfeld, A.: Atombau und Spektallinien. Braunschweig: Vieweg 1951
45. Waerden, B.L.van der (ed.): Sources of quantum mechanics. Amsterdam: North-Holland 1967
46. Theissing, H.H., Caplan, P.J.: Spectroscopic calculations for a multielectron ion. New York: Interscience 1966
47. Waber, J.T., Cromer, D.T.: Orbital radii of atoms and ions. J. Chem. Phys. *42*, 4116–4123 (1965)
48. Walker, S., Straw, H.: Spectroscopy. Vol. I. Atomic, microwave and radiofrequency spectroscopy. Vol. II. Ultraviolett, visible, infra-red and Raman spectroscopy. New York: MacMillan 1961, 1962

Crystal Field Theory

49. Ballhausen, C.J.: An introduction to ligand field theory. New York: McGraw-Hill 1962
50. Bersuker, I.B.: Inner asymmetry in complex compounds. Zh. Strukt. Khim. *2*, 350–360, 734–739; *3*, 64–69, 563–568 (1962)
51. Bethe, H.: Termaufspaltung in Kristallen. Ann. Phys. *3*, N 1–2 (1929)
52. Burns, R.G.: Mineralogical applications of crystal field theory. Cambridge Univ. Press 1970
53. Coleman, A.J.: The symmetry group made easy. Advan. Quant. Chem. *4*, 83–108 (1968)
54. Cotton, F.A.: Chemical applications of group theory. New York: Wiley 1971
55. Curie, D.: Champ cristallin et luminescence. Application de la théorie des groupes à la luminescence cristallin. Paris: Gauthier-Villars 1968
56. Dunn, T.M., McClure, D.S., Pearson, R.G.: Some aspects of crystal field theory. New York: Harper and Row 1965
57. Englman, R.: The Jahn-Teller effect in molecules and crystals. New York: Wiley 1972

58. Ferraro, J.R., Ziomek, J.S.: Introductory group theory and its application to molecular structure. New York: Plenum 1969
59. Figgis, B.N.: Introduction to ligand fields. New York: Wiley 1966
60. Griffith, J.S.: The theory of transition metal ions. Cambridge Univ. Press 1964
61. Goodenough, J.B.: Spin-orbit-coupling effects in transition-metal compounds. Phys. Rev. *171*, 466–479 (1968)
62. Herzfeld, Ch.M., Meijer, P.H.E.: Group theory and crystal field theory. Solid State Phys. *12*, 2–91 (1961)
63. Jaffe, H.H., Orchin, M.: Symmetry in chemistry. New York: Wiley 1965
64. Jong, W.F.de: Elementare Gruppentheorie in der Kristallographie. Tscherm. Mineral. Petr. Mitteil. *10*, 140–156 (1965)
65. Knox, R.S., Gold, A.: Symmetry in the solid state. New York: Benjamin 1964
66. Kristofel, N.N.: Tables of point group characters. Tartu: Univ. Press 1970
67. Leushin, A.M.: Tables of function transforming into irreducible representations of point groups. Moscow: Nauka 1968
68. Lever, A.B.P.: Inorganic electronic spectroscopy. Amsterdam: Elsevier 1968
69. Mathiak, K.: Gruppentheorie (für Chemiker, Physico-Chemiker, Mineralogen). Berlin: VEB Deut. Verlag der Wissenschaften 1968
70. Orgel, L.: An Introduction to transition-metal chemistry: ligand-field theory. London: Methuen 1966
71. Petrashen, M.I., Trifonov, E.D.: Application of group theory in quantum mechanics. Moscow: Physmathisdat 1967
72. Schläfer, H.L., Gliemann, G.: Einführung in die Liganden-Feld Theorie. Leipzig: Geest and Portig 1967
73. Sturge, M.D.: The Jahn-Teller effect in solids. Solid State Phys. *20*, 92–213 (1967)
74. Tanabe, Y., Sugano, S.: On the absorption spectra of complex ions. J. Phys. Soc. Japan *9*, 753–766, 766–780 (1954)
75. Wilson, E.B., Decius, J.C., Cross, P.C.: Molecular vibrations. The theory of infrared and raman vibrational spectra. New York: McGraw-Hill 1955

Molecular Orbital Theory; Chemical Bond Theories

(See also below: references to chemical bonding in some groups and classes of minerals)

76. Ahrens, L.H.: Ionization potentials and metal-amino acid complex formation in the sedimentary cycle. Geochim. Cosmochim. Acta *30*, 1111–1119 (1966)
77. Bader, R.F.W., Henneker, W.H., Cade, P.E.: Molecular charge distribution and chemical binding. J. Chem. Phys. *46*, 3341–3363 (1967)
78. Ballhausen, C.J., Gray, H.B.: Molecular orbital theory. New York: Benjamin 1964
79. Bash, H., Hollister, C., Moscowitz, J.W.: Sigma molecular orbital theory. New Haven: Yale Univ. Press 1970
80. Batsanov, S.S., Zvyagina, R.A.: Overlap integrals and effective charge problem. Novosibirsk: Nauka 1966
81. Batsanov, S.S.: Concept of electronegativity: results and perspectives. Usp. Khim. *37*, 778–815 (1968)
82. Bersuker, I.B.: Structure and properties of coordination compounds. Leningrad: Chimia 1975
83. Burton, P.G.: Current methods in MO theory for inorganic systems and their future development. An interpretative review. Coord Chem. Rev. *12*, 37–71 (1974)
84. Charkin, O.P., Bobikina, G.V., Dyatkina, M.E.: Orbital ionization potentials for atoms and ions in valence configurations. In: Structure of molecule and quantum chemistry. Kiev: Naukova Dumka 1970
85. Charkin, O.P.: Valence states of atoms. Zh. Strukt. Khim. *14*, 389–415 (1973)
86. Cook, D.B.: Ab initio valence calculations in Chemistry. London: Butterworths 1974
87. Coulson, C.A.: Valence. Oxford: Univ. Press 1961

88. Dahl, J.P., Ballhausen, C.J.: Molecular orbital theories of inorganic complexes. Advan. Quant. Chem. *4*, 170–226 (1968)
89. Day, M.C., Selbin, J.: Theoretical inorganic chemistry. New York: Reinhold 1969
90. Dyatkina, M.E., Rosenberg, E.L.: Quantum-mechanical calculations of transition metal compounds. Moscow: Viniti 1974
91. Dyatkina, M.E.: Principles of molecular orbital theory. Moscow: Nauka 1975
92. Golutvin, Ju.M.: Heats of formation and chemical bond types in inorganic crystals. Moscow: Nauka 1962
93. Goodenough, J.B.: Magnetism and the chemical bond. New York, London: Interscience 1963
94. Gray, H.B.: Electrons and chemical bonding. New York: Benjamin 1965
95. Gubanov, V.A., Zhukov, V.P., Litinskii, A.O.: Semi-empirical methods of molecular orbitals in quantum chemistry. Moscow: Nauka 1976
96. Herzberg, G.: The spectra and structures of simple free radicals. An introduction to molecular spectroscopy. Ithaca, London: Cornwell Univ. Press, 1971
97. Hultgren, R.: Equivalent chemical bonds formed by s, p, and d eigenfunctions. Phys. Rev. *40*, 891–907 (1932)
98. Johnson, K.H., Smith, F.C.: Chemical bonding of a molecular transition-metal ion in a crystalline environment. Phys. Rev. B *5*, 831–843 (1972)
99. Jorgensen, C.K.: Absorption spectra and chemical bonding in crystals. Oxford: Pergamon 1962
100. Kimball, G.E.: Directed valence. J. Chem. Phys. *8*, 188–198 (1940)
101. Levin, A.A., Syrkin, Ja.K., Dyatkina, M.E.: Problem of monoatomic multicharge ions and chemical bonding in inorganic crystals. Usp. Khim. *38*, 193–231 (1969)
102. Liehr, A.D.: Interaction of electromagnetic radiation with matter. I. Theory of optical rotatory power: Topic B. Diagonal dihedral compounds and compounds of lower symmetry. J. Phys. Chem. *68*, 3629–3733 (1964)
103. McDowell, Ch.A.: Ionization potentials and electron affinities. Phys. Chem. Advan. Treatise *111*, 496–537 (1969)
104. McWeeney, R., Scutliffe, B.T.: Methods of molecular quantum mechanics. Theor. Chemistry. A Series of Monographs. Vol. II. London, New York: Academic Press 1969
105. Mulliken, R.S.: Electronic population analysis on LCAO MO molecular wave functions. J. Chem. Phys. *23*, 1833–1840 (1955)
106. Murrell, J.N., Kettle, S.F.A., Tedder, J.M.: Valence theory. London: Wiley 1965
107. Murrell, J.N., Harget, A.J.: Semi-empirical self-consistent field molecular orbital theory of molecules. London: Wiley-Interscience 1972
108. Nefedov, V.I., Fomichev, V.A.: Electronic structure of third row element's tetrahedral oxyanions. Zh. Strukt. Khim. *9*, 126–132 (1968)
109. Nefedov, V.I.: Structure of molecules and the chemical bond. Moscow: Viniti 1974
110. Offenhartz, P.: Atomic and molecular orbital theory. New York: McGraw-Hill 1970
111. Orchin, M., Jaffe, H.H.: The importance of antibonding orbitals. New York: Houghton Mifflin Comp. 1967
112. Osipov, O.A., Minkin, V.I.: Handbook of dipole momenta. Moscow: Nauka 1965
113. Pauling, L.: The nature of the chemical bond. Ithaca N.Y.: Corwell Univ. Press 1960
114. Pople, J.A., Beveridge, D.L.: Approximate molecular orbital theory. New York: McGraw-Hill 1970
115. Povarennych, A.S.: Use of element electronegativity in crystallochemistry and geochemistry. Part I. Zapiski Ukrainian otd. Vses. Mineralog. obshchestva, 1962, Part II. In: Theoretical and genetical problems of mineralogy and geochemistry. Kiev: Naukova Dumka 1962, 1963
116. Roberts, J.D.: Notes on molecular orbital calculations. New York: Benjamin 1961
117. Rosenberg, E.M., Dyatkina, M.E.: Electron density distribution and charges on atoms in molecules. Zh. Strukt. Khim. *12*, 1058–1061 (1971)
118. Royer, D.J.: Bonding theory. New York: McGraw-Hill 1968
119. Shustorovich, E.M.: Chemical bond. Moscow: Nauka 1973
120. Slater, J.C.: Electronic structure of molecules. New York: McGraw-Hill 1963

121. Slater, J.C., Johnson, K.H.: Self-consistent field X_α-cluster method for polyatomic molecules and solids. Phys. Rev. B 5, 844–853 (1972)
122. Spiridonov, V.P., Tatevskii, V.M.: Concerning the concept of electronegativity of atoms. Zh. Phys. Khim. 37, 994–1232, 1233–1241, 1583–1586, 1973–1979 (1963)
123. Syrkin, Ja.K.: Effective charges and electronegativity. Usp. Khim. 31, 397–416 (1962)
124. Tatevskii, V.M.: Quantum mechanics and theory of molecule structure. Moscow Univ. Ed. 1965
125. Vechten, J.A. van: Quantum dielectric theory of electronegativity in covalent systems. Phys. Rev. 182, 891–905 (1969)
126. Wolfsberg, M., Helmholz, L.: The spectra and electronic structure of the tetrahedral ions MnO_4^-, CrO_4^-, and ClO_4^-. J. Chem. Phys. 20, 837–843 (1952)

Energy Band Theory and Reflectance Spectra of Minerals

127. Au-Yang, M.Y., Cohen, M.L.: Electronic structure and optical properties of SnS_2 and $SnSe_2$. Phys. Rev. 178, 1279–1283 (1969)
128. Bass, D.D., Parada, N.J.: Calculation of the optical constants of PbTe from augmented-plane-wave k.p energy bands. Phys. Rev. B 1, 2692–2699 (1970)
129. Boucaert, L.P., Smoluckowski, P., Wigner, E.: Theory of Brillouin zones and symmetry properties of wave functions in crystals. Phys. Rev. 50, 58–67 (1936)
130. Bube, R.H.: Electronic properties of crystalline solids. An introduction of fundamentals. New York, London: Academic 1974
131. Callaway, J.: Energy band theory. New York, London: Academic 1964
132. Cardona, M.: Modulation spectroscopy. Solid State Phys. Suppl. 11 (1969)
133. Ciraci, S.: The bond orbital model study of the valence band of cubic-tetrahedrally coordinated semiconductors. Phys. Stat. Sol. (b) 70, 689–703 (1975)
134. Cohen, M.L., Heine, V.: The fitting of pseudo-potentials to experimental data and their subsequent application. Solid State Phys. 24 (1970)
135. Cohen, M.L., Lin, P.J., Roessler, D.M., Walker, W.C.: Ultraviolet optical properties and electronic band structure of magnesium oxide. Phys. Rev. 55, 992–996 (1967)
136. Conklin, J.B., Johnson, L.E., Pratt, G.W.: Energy bands in PbTe. Phys. Rev. 137, 1282–1294 (1965)
137. Connell, G.A.N., Wilson, J.A., Yoffe, A.D.: Effects of pressure and temperature on exciton absorption and band structure of layer crystals: molybdenum disulphide. J. Phys. Chem. Solids 30, 287–296 (1969)
138. Cornwell, J.F.: Group theory and electronic energy bands in solids. Amsterdam: North-Holland 1969
139. Cotton, F.A.: Chemical applications of group theory. New York: Interscience 1963
140. Dimmock, T.O.: The calculation of electronic energy bands by the augmented plane wave method. Solid State Phys. 26, 104–274 (1971)
141. Frei, V.: The symmetry properties of the energy bands of α- and β-quartz. Czech. J. Phys. B17, 147–152, 233–248 (1967)
142. Greenaway, D.L., Harbeke, G.: Optical properties and band structure of semiconductors. Oxford: Pergamon 1968
143. Harper, P.G., Edmondson, D.R.: Electronic band structure of the layer-type crystal MoS_2 (atomic model). Phys. Stat. Sol. (b) 44, 59–69 (1971)
144. Harrison, W.A.: Bond-orbital model and the properties of tetrahedrally coordinated solids. Phys. Rev. B 8, 4487–4498 (1973)
145. Hecht, H.G.: The present status of diffuse reflectance theory. In: Modern aspects of reflectance spectroscopy. Wendlandt, W.W. (ed.). New York: Plenum 1968
146. Heine, V., Weaire, D.: Pseudopotential theory of cohesion and structure. Sol. State Phys. 24, 250–463 (1970)
147. Kahn, A.H., Leyendecker, A.J.: Electronic energy bands in strontium titanate. Phys. Rev. 135A, A1321–1315 (1964)
148. Kalikhman, V.L., Umanskii, Ja.S.: Transition metal chalcogenides with complex structure and features of their Brillonin zone filling. Usp. Phys. Nauk. 3 (1972)

149. Kittel, Ch.: Introduction to solid state physics. New York: Wiley 1966
150. Khasabov, A.G., Nikiforov, I.Ja.: Energy band structure of the ferroelectric-semiconductor Sb_2S_3. Kristallografiya *16*, 41–45 (1971)
151. Korsunskii, M.I., Mitryaeva, N.M., Muratov, E.M.: Reflectance spectra of silicon, galena, and pyrite in 0.62–5.48 eV region. Izv. Akad. Nauk Kaz. SSR, Ser. Phys.-Math. *6*, 11–22 (1969)
152. Kortüm, G.: Reflexionspektroskopie. Grundlagen, Methoden, Anwendungen. Berlin, Heidelberg, New York: Springer 1969
153. Levin, A.A.: Introduction to quantum chemistry of solids. Chemical bond and energy band structure in tetrahedral semiconductors. Moscow: Chimia 1974
154. Mattheiss, L.F.: Electronic structure of the 3 *d* transition-metal monoxides. Phys. Rev. *135*, 290–306, 306–315 (1972)
155. Motulevich, G.P.: Optical properties of nontransition metals. Proc. Lebedev Phys. Inst. *55*, 3–150 (1971)
156. Phillips, J.C.: The fundamental optical spectra of solids. Solid State Phys. *18*, 55–164 (1966)
157. Phillips, J.C.: Bonds and bands in semiconductors. New York, London: Academic 1973
158. Roberts, G.G., Lind, E.L., Davis, E.A.: Photoelectronic properties of synthetic mercury sulphide crystals. J. Phys. Chem. Sol. *30* (1969)
159. Rowe, J.E., Shay, J.L.: Extension of the quasicubic model to ternary chalcopyrite crystals. Phys. Rev. B*3*, 451–453 (1971)
160. Ruffa, A.R.: The valence bond approximation in crystals – application to an analysis of the ultraviolet spectrum in quartz. Phys. Stat. Sol. *28*, 605–616 (1968)
161. Segall, B.: Band structure. In: Physics and Chemistry of II–VI compounds. Amsterdam: North-Holland 1967
162. Slater, J.C.: Quantum theory of solids, Vol. III. Insulators, semiconductors and metals. New York: McGraw-Hill 1967
163. Sobolev, V.V., Zalevskii, B.K., Vorobjev, V.G.: Reflectance spectra of some minerals. In: Semiconductors studies. Kishinev: Nauka 1968
164. Sobolev, V.V.: Reflectance spectrum of molybdenite. Opt. i Spektroskopiya *18*, 334–336 (1965)
165. Starostin, N.V.: Energy band structure of fluorite-type crystals. Phys. Tverdogo Tela *11*, 1624–1626 (1969)
166. Vyalsov, L.H.: Optical methods of identification of ore minerals. Moscow: Nedra 1976
167. Walter, J.P., Cohen, A.L., Petroff, Y., Balkanski, M.: Calculated and measured reflectivity of ZnTe and ZnSe. Phys. Rev. B*1*, 2661–2667 (1970)
168. Wendlandt, W.W., Hecht, H.G.: Reflectance spectroscopy. New York: Interscience 1966
169. Williams, R.H., McEvoy, A.J.: Photoemission studies of MoS_2. Phys. Stat. Sol. (b) *47*, 217–224 (1971)

Spectroscopy and the Chemical Bond

(See extensive references in: Marfunin, A.S.: Spectroscopy, luminescence and radiation centers in minerals and materials [1])

170. Adler, I.: X-ray emission spectrography in geology. Amsterdam: Elsevier 1966
171. Alger, R.S.: Electron spin resonance in chemistry. Techniques and applications. New York: Interscience, 1968
172. Altshuler, S.A., Kozyrew, B.M.: Paramagnetische Elektronenresonanz. Leipzig: Teubner 1963
173. Andrew, R.: Nuclear magnetic resonance. Cambridge: Univ. Press 1956
174. Ayscough, P.B.: Electron spin resonance in chemistry. London: Methuen 1967

175. Bancroft, G.M.: Mössbauer spectroscopy: An introduction for inorganic chemists and geochemists. Maidenhead: McGraw-Hill 1974
176. Barinskii, R.L., Nefedov, V.I.: X-ray spectroscopic determination of atomic charges in molecules. Moscow: Nauka 1966
177. Carrington, A., McLachlan, A.D.: Introduction to magnetic resonance with applications to chemistry and chemical physics. New York: Harper and Row, 1967
178. Goldanskii, V.I., Herber, R.H. (eds.): Chemical applications of Mössbauer spectroscopy. New York, London: Academic 1968
179. Fedin, E.I., Semin, G.K.: Applications of nuclear quadrupole resonance in crystallochemistry. Zh. Strukt. *1* (1960)
180. Goodman, B.A., Raynor, J.B.: Electron spin resonance of transition metal complexes. Advan. Inorg. Chem. Radiochem. *13*, 135–362 (1970)
181. Jeffrey, G.A., Sakurai, T.: Applications of nuclear quadrupole resonance. Progr. Sol. State Chem. *1*, 380–416 (1964)
182. Lösche, A.: Kerninduktion. Berlin: VEB. Deut. Verlag der Wissenschaften 1957
183. Low, W.: Paramagnetic resonance in solids. Sol. State Phys. Suppl. *2* (1960)
184. Low, W.: Electron spin resonance. A tool in mineralogy and geology. Advan. Electron. Electron Phys. *24*, 51–108 (1968)
185. Malysheva, T.V.: Mössbauer effect in geochemistry and cosmochemistry. Moscow: Nauka 1975
186. Marfunin, A.S., Bershov, L.V.: Application of electron paramagnetic resonance in mineralogy. Moscow: Viniti 1964
187. Marfunin, A.S.: Nuclear magnetic and nuclear quadrupole resonance in minerals. Moscow: Viniti 1966
188. Nefedov, V.I.: Application of X-ray emission spectroscopy in chemistry. Moscow: Viniti 1973
189. Nemoshkalenko, V.V., Aleshin, V.G.: Electron spectroscopy of crystals. Kiev: Naukova Dumka 1976
190. Orton, J.W.: Electron paramagnetic resonance. An introduction to transition group ions in crystals. London: Iliffe 1968
191. Penkov, I.N.: Application of nuclear quadrupole resonance in mineral investigations. Izv. Akad. Nauk SSSR, Ser. Geol. *12*, 41–52 (1966)
192. Siegbahn, K., Nordling, C., Fahlman, A., Nordberg, R., Hamrin, K., Hedman, J., Johansson, G., Bergmark, T., Karlsson, S.-E., Lindgren, I., Lindberg, B.: ESCA: Atomic, molecular and solid state structure studied by means of electron spectroscopy. Nova Acta Regiae Soc. Sci. Upsaliensis Ser. IV, *20* (1967)
193. Turner, D.W., Baker, C., Baker, A.D., Brunde, C.R.: Molecular photoelectron spectroscopy. London: Wiley Interscience 1970
194. Zimkina, T.M., Fomichev, W.A.: Ultrasoft X-ray spectroscopy. Leningrad: Nauka 1971

Optical Absorption Spectra and Nature of Colors of Minerals

195. Abu-Eid, R.M.: Absorption spectra of transition metal bearing minerals at high pressures. In: The physics and chemistry of minerals and rocks. Strens, R.G.J.: (ed.): London: Wiley 1976
196. Agnetta, G., Garofano, T., Palma-Vittorelli, Palma, M.U.: Low-temperature optical absorption of ferrous fluosilicate crystals. Phil. Mag. *7*, 495–498 (1962)
197. Allen, G.C., Hush, N.S.: Intervalence transfer absorption. Progr. Inorg. Chem. *8*, 357–390 (1967)
198. Allen, G.C., Warren, K.D.: The electronic spectra of the hexafluoro complexes of the first transition series. Struct. Bonding *9*, 49–137 (1971)
199. Bakhtin, A.I., Minko, O.E., Vinkurov, V.M.: Isomorphous substitutions and colours of tourmaline. Izv. Akad. Nauk SSSR, Ser. Geol. *6*, 73–83 (1975)
200. Barsanov, G.P., Yakoleva, M.E.: Colours of minerals. Proc. Mineralogical Museum, Vol. IV, Moscow: Nauka 1963

201. Bates, C.H., White, W.B., Roy, R.: The solubility of transition metal oxides in zinc oxide and the reflectance spectra of Mn^{2+} and Fe^{3+} in tetrahedral fields. J. Inorg. Nucl. Chem. 28, 397 (1966)
202. Bevan, H., Dawes, S.V., Ford, R.A.: The electronic spectra of titanium dioxide. Spectrochim. Acta 13, 43–49 (1958)
203. Burns, R.G.: Partitioning of transition metals in mineral structures of the mantle. In: The physics and chemistry of minerals and rocks. Strens, R.G.J. (ed.): London: Wiley 1976
204. Burns, G., Geiss, E.A., Jenkins, B.A., Nathan, M.I.: Cr^{3+} fluorescence in garnets and other crystals. Phys. Rev. 139, 1687–1693 (1965)
205. Carlin, R.L.: Electronic structure and stereochemistry of cobalt (II). Transition Metal Chemistry, Vol. I. New York: Marcel Dekker 1965
206. Chesnokov, B.V.: Absorption curves of some minerals coloured by titanium. Dokl. Akad. Nauk SSSR 129, 1162–1163 (1959)
207. Chesnokov, B.B.: Trivalent titanium in eclogites of the South Urals. Geokhimiya 1, 68–72 (1960)
208. Clogston, A.M.: Interaction of magnetic crystals with radiation in the range 10^4–10^5 cm^{-1}. Appl. Phys. Suppl. 31, 198–205 (1960)
209. Curtis, C.D.: Application of the crystal-field theory to the inclusion of trace transition elements in minerals during magmatic crystallization. Geochim. Cosmochim. Acta 28, 389 (1964)
210. Deutschein, O.: Die linienhafte Emission and Absorption der Chromphosphore. Ann. Phys. 14, 712–728 (1932)
211. Dietz, R.E., Kamimura, H., Sturge, M.D., Yariv, A.: Electronic structure of copper impurities in ZnO. Phys. Rev. 132, 1559–1569 (1963)
212. Dionne, G.F.: Calculation of crystal field energy level splittings of the Ti^{3+} ion in $RbAl(SO_4)_2 \cdot 12H_2O$. Phys. Rev. 137, 743–748 (1965)
213. Dorschner, J.: Diffuse interstellar bands and garnet grains. Nature (Lond.) Phys. Sci. 231, 124–125 (1971)
214. Drickamer, H.G.: The effect of high pressure on the electronic structure of solids. Solid State Phys. 17, 1–133 (1965)
215. Druzhinin, V.V.: Splitting of $3d^5$ configuration terms by octahedral crystal field. Opt. i Spektroskopiya 22, 824–827 (1967)
216. Farrell, E.F., Newnham, K.E.: Crystal-field spectra of chrysoberyl, alexandrite, peridote and sinhalite. Am. Mineralogist. 50, 1971–1981 (1965)
217. Farrell, E.F., Newnham, R.E.: Electronic and vibrational absorption spectra in cordierite. Am. Mineralogist. 52, 380–388 (1967)
218. Faye, G.H.: The optical absorption spectra of certain transition metal ions in muscovite, lepidolite, and fuchsite. Can. J. Earth Sci. 5, 31–38 (1968)
219. Faye, G.H.: The optical absorption spectra of iron in six-coordinate sites in chlorite, biotite, phlogopite and vivianite. Some aspects of pleochroism in the sheet silicates. Can. Mineralogist. 9, 403–425 (1968)
220. Faye, G.H., Harris, D.C.: On the origin of colour and pleochroism in andalusite from Brazil. Can. Mineralogist. 10, 47–56 (1969)
221. Faye, G.H., Manning, P.G., Nickel, E.H.: The polarized optical absorption spectra of tourmaline, cordierite, chloritoid and vivianite: ferrous-ferric electronic interaction as a source of pleochroism. Am. Mineralogist. 53, 1174–1201 (1968)
222. Faye, G.H., Hogarth, D.D.: On the origin of reverse pleochroism of a phlogopite. Can. Mineral. 10, 25–34 (1969)
223. Faye, G.H.: The optical absorption spectrum of tetrahedrally bonded Fe^{3+} in orthoclase. Can. Mineralogist. 10, 112–117 (1969)
224. Faye, G.H.: Relationship between crystal field splitting parameter, Δ_{VI} and M_{host}–0 bond distance as an aid in the interpretation of absorption spectra of Fe^2 bearing materials. Can. Mineralogist. 11, 473–487 (1972)
225. Ferguson, J.: Spectroscopy of $3d$ complexes. Prog. Inorg. Chem. 12, 159–294 (1970)
226. Ferguson, J., Guggenheim, H.J.: Absorption of light by pairs of exchange-coupled manganese and nickel ions in cubic perovskite fluorides. J. Chem. Phys. 45, 1134–1141 (1965)

227. Fersman, A.E.: Colours of minerals. Moscow: Acad. Sci. USSR 1936
228. Ford, R.A., Kauer, E., Rabenau, A., Brown, D.A.: The electronic states of octahedral and tetrahedral Mn^{2+} in α, β and γ managanous sulphide. Ber. Bunsenges. Phys. Chem. 67, 460–465 (1963)
229. Grum-Grzhimailo, S.V., Brilliantov, N.A., Sviridov, D.T.: Absorption spectra of crystals coloured by Fe^{3+} at temperatures down to 1.7° K. Opt. i Spektroskopiya 14, 228–233 (1963)
230. Grum-Grzhimailo, S. V., Boksha, O.N., Varina, T.M.: Absorption spectrum of olivine. Kristallografiya 14, 339–342 (1969)
231. Gumlich, H.-E., Schulz, H.J.: Optical transitions in ZnS-type crystals containing cobalt. J. Phys. Chem. Solids 27, 187–195 (1966)
232. Ham, F.S., Slack, G.A.: Infrared absorption and luminescence spectra of Fe^{2+} in cubic ZnS: role of the Jahn-Teller coupling. Phys. Rev. B4, 777–798 (1971)
233. Hush, N.S.: Intervalence transfer absorption. Progr. Inorg. Chem. 8, 391–444 (1968)
234. Hush, N.S., Hobbs, R.J.M.: Absorption spectra of crystals containing transition metal ions. Progr. Inorg. Chem. 10, 259–486 (1968)
235. Iorgensen, Chr.K.: Electron transfer spectra. Progr. Inorg. Chem. 12, 101–158 (1970)
236. Judd, D.B.: Measurements and specification of color. In: Analytical Absorption Spectroscopy. Mellon, G. (ed.), pp. 515–600. New York: Wiley; London: Chapman and Hall 1960
237. Keester, K.L., White, W.B.: Crystal-field spectra and chemical bonding in manganese minerals. Papers and Proc. 5th Gen. Meet. Cambridge, England: London: Mineral. Soc. 1968
238. Kleber, W., Bautsch, H.J., Adam, J.: Absorptionsspektren natürlicher Granate im Bereich von 200 bis 1500 nm und ihre Bedeutung. Kristall u. Techn. 4, 537–550 (1969)
239. Kolbe, E.: Über die Färbung von Mineralien durch Mn, Cr und Fe. Neues Jahrb. Mineral. B 69, Abt. A, 183–254 (1935)
240. Kortüm, G.: Diffuse reflectance spectra of mercuric jodide on different adsorbents. Trans. Faraday Soc. 58, 1624–1613 (1962)
241. Krebs, J.J., Maisch, W.G.: Exchange effects in the optical absorption spectrum of Fe^{3+} in Al_2O_3. Phys. Rev. 4, 757–764 (1971)
242. Krishnamurthy, R., Schaap, W.B., Perumareddi, J.R.: Interpretation of the spectra of cyanoaquo complexes of chromium (III) by the theory of non-cubic ligand fields. Inorg. Chem. 6, 1338–1352 (1967)
243. Krishnamurthy, R., Schaap, W.B.: A general procedure for obtaining ligand field potentials and relative orbital energies for $d^1(d^9)$ ions various ligand fields. Progr. Coord. Chem. Amsterdam: Elsevier 1967
244. Kubelka, P., Munk, F.: Ein Beitrag zur Optik der Farbanstriche. Z. Tech. Phys. 11a, 593–601 (1931)
245. Kubelka, P.: New contributions to the optics of intensely light-scattering materials. J. Opt. Soc. Am. 38, 448–457 (1948)
246. Lehmann, G., Harder, H.: Optical spectra of di- and trivalent iron in corundum. Am. Mineralogist 55, 98–105 (1970)
247. Low, W., Rosengarten, G.: The optical spectrum and ground state splitting of Mn^{2+} in the crystal of cubic symmetry. Mol. Spectrosc. 12, 319–346 (1964)
248. Ludi, A., Feitknecht, W.: Lichtabsorption und Struktur von Kristallverbindungen der Übergangsmetalle. Helv. Chim. Acta 46, 2226–2238 (Co, Ni), 2238–2248 (Cu) (1963)
249. Macfarlane, R.M.: Optical and magnetic properties of trivalent vanadium complexes. J. Chem. Phys. 40, 373–377 (1964)
250. Mandarino, J.A.: Absorption and pleochroism: Two much-neglected optical properties of crystals. Am. Mineralogist. 44, 65–77 (1959)
251. Manning, P.G.: Absorption spectra of the manganese-bearing chain silicates pyroxmangite, rhodonite, bustamite and serandite. Can. Mineralogist. 9, 348–357 (1968)
252. Manning, P.G., Harris, D.C.: Optical-absorption and electron-microprobe studies of some high-Ti andradites. Can. Mineral. 10, 260–271 (1970)
253. Manning, P.G., Nickel, E.H.: A spectral study of the origin of colour and pleochroism of a titanaugite from Kaiserstuhl and of a riebeckite from St. Peter's Dome, Colorado. Can. Mineralogist. 10, 57–70 (1969)

254. Mao, H.K., Bell, P.M.: Crystal-field effects in spinel: oxidation states of iron and chromium. Geochim. Cosmochim. Acta 39, 865–874 (1975)
255. Mao, H.K.: Charge-transfer processes at high pressure. In: The Physics and Chemistry of Minerals and Rocks. Strens, R.G.J. (ed.): London: Wiley 1976
256. McClure, D.S.: Optical spectra of transition-metal ions in corundum. J. Chem. Phys. 36, 2757–2779 (1962)
257. McClure, D.S.: Comparison of the crystal fields and optical spectra of Cr_2O_3 and ruby. J. Chem. Phys. 38, 2284–2294 (1963)
258. McClure, D.S.: Optical spectra of exchange coupled Mn^{2+} ion pairs in ZnS:MnS. Chem. Phys. 39, 2850 (1963)
259. Marfunin, A.S., Platonov, A.N., Fedorov, V.E.: Absorption spectrum of Fe^{2+} in sphalerite. Phys. Tverdogo Tela 9, 3616–3618 (1967)
260. Marfunin, A.S., Mineeva, R.M., Mkrtchan, A.R.: Optical and Mössbauer spectroscopy of iron in rock-forming silicates. Izv. Akad. Nauk SSSR, Ser. Geol. 10, 86–102 (1967)
261. Marfunin, A.S., Mkrtchan, A.R., Nadzharyan, G.N.: Optical and Mössbauer spectra of iron in some layer silicates. Izv. Akad. Nauk SSSR, ser. Geol. 7, 87–93 (1971)
262. Melamed, N.T.: Optical properties of powders. Appl. Phys. 34, 560–568 (1963)
263. Nelson, E.D., Wong, J.Y., Schawlow, A.L.: Far infrared spectra of $Al_2O_3:Cr^{3+}$ and $Al_2O_3:Ti^{3+}$. Phys. Rev. 256, 298–308 (1967)
264. Neuhaus, A.: Über die Ionenfarbe der Kristalle und Minerale am Beispiel der Chromfärbungen. Z. Kristallogr. 113, 195–233 (1960)
265. Newnham, R.E., Santoro, R.P.: Magnetic and optical properties of dioptase. Phys. Stat. Sol. 19, K87–90 (1967)
266. Nussik, Ja.M.: Optical absorption spectra of nickel-bearing minerals. Izv. Akad. Nauk SSSR, Ser. Geol. 3, 108–112 (1969)
267. Pappalardo, R.: Absorption spectra of Cu^{2+} in different crystal coordinations. J. Mol. Spectr. 6, 554–571 (1961)
268. Pappalardo, R., Wood, D.L., Linares, R.C.: Optical absorption spectra of Ni-doped oxide systems. J. Chem. Phys. 35, 1460–1478 (1961)
269. Pappalardo, R., Wood, D.L., Onares, R.C.: Optical absorption study of Co-doped oxide systems. Chem. Phys. 35, 2041 (1961)
270. Pechkova, T.A.: Colour classifications (A review). Moscow: VNII TechEst 1969
271. Perumareddi, J.R.: Ligand field theory of d^3 and d^7 electronic configurations in noncubic fields. Phys. Chem. 71, 3144–3154, 3155–3165 (1967)
272. Piller, H.: Colour measurements in ore-microscopy. Mineral. Deposita 1 (1966)
273. Pisarev, R.V.: Absorption spectrum of $YFeO_3$. Phys. Tverdogo Tela 6, 2545–2547 (1964)
274. Pisarev, R.V.: Absorption spectrum of the antiferromagnetic $NaNiF_3$. Phys. Tverdogo Tela 7, 1382–1388 (1965)
275. Pitt, G.D., Tozer, D.C.: Optical absorption measurements on natural and synthetic ferromagnesium minerals subjected to high pressures. Phys. Earth Planet. Inter. 2, 179–188 (1970)
276. Platonov, A.N.: The nature of the colours of minerals. Kiev: Naukova Dumka 1976 (Extensive references to optical absorption spectra of minerals
277. Poole, C.P., Jr.: The optical spectra and color of chromium-containing solids. J. Phys. Chem. Solids 25, 1169–1182 (1964)
278. Pryce, M.H.L., Agnetta, G., Garofano, T., Palma-Vittorelli, M.B., Palma, M.U.: Low-temperature optical absorption of nickel fluosilicate crystals. Philos. Mag. 10, 477–496 (1964)
279. Ralph, J.E., Townsend, M.G.: Near-infrared fluorescence and absorption spectra of Co^{2+} and Ni^{2+} in MgO. J. Chem. Phys. 48, 149–154 (1968)
280. Reinen, D.: Die Lichtabsorption des Co^{2+} und Ni^{2+} in oxidischen Festkörpern mit Granatstruktur. Z. Anorg. Allg. Chem. B 337, 238 (1964)
281. Reinen, D.: Farbe und Konstitution bei anorganischen Feststoffen. II. Mitt. Die Lichtabsorption des oktaedrisch koordinierten Co^{2+} Ions in der Mischkristallreiche $Mg_{1-x}Co_xO$ und anderen oxidischen Wirtsgittern. Monatsh. Chem. 96, 730–739 (1965)

282. Rösch, S.: Darstellung der Farbenlehre für die Zwecke des Mineralogen. Fortschr. Mineral. *13*, 73–230 (1929)
283. Rösch, S.: Messung der Absorptionsfarben von Mineralen und ihre Auswertung. Optica Acta *11*, 267–279 (1964)
284. Samoilovich, M.I., Cinober, L.I., Dunin-Barkovskii, R.L.: Origin of colour in beryl with iron impurities. Kristallografiya *16*, 186–189 (1971)
285. Shankland, T.J.: Pressure shift of infrared absorption bands in minerals and the effect on radiative heat transport. J. Geophys. Res. *75*, 409–413 (1970)
286. Slack, G.A., Ham, F.S., Chrenko, R.M.: Optical absorption of tetrahedral Fe^{2+} ($3d^6$) in cubic ZnS, CdTe, and $MgAl_2O_4$. Phys. Rev. *152*, 376–402 (1966)
287. Slack, G.A., Roberts, S., Ham, F.S.: Far-infrared optical absorption of Fe^{2+} in ZnS. Phys. Rev. *155*, 170–177 (1967)
288. Slack, G.A., Chrenko, R.M.: Optical absorption of natural garnets from 1000 to 30,000 wavenumbers. J. Opt. Soc. Am. *61*, 1325–1329 (1971)
289. Smith, G., Strens, R.G.J.: Intervalence-transfer absorption in some silicate, oxide and phosphate minerals. In: The physics and chemistry of minerals and rocks. Strens, R.G.J. (ed.). London: Wiley 1976
290. Smitz-Dumont, O., Friebel, G.: Farbe und Konstitution bei anorganischen Feststoffen. 15. Die Lichtabsorption des zweiwertigen Kobalts in Silikaten vom Olivintypus. Monatsh. Chem. *98/4*, 1583–1602 (1967)
291. Stevenson, R.: Absorption spectrum of $MnCO_3$. J. Appl. Phys. *39*, 1143–1145 (1968)
292. Sviridov, D.T., Sviridova, R.K.: Calculation of garnet group spectra with Fe^{3+} ions. Kristallografiya *17*, 221–223 (1972)
293. Syono, Y., Tokonami, M., Matsui, Y.: Crystal field effect on the olivine-spinel transformation. Phys. Earth Planet. Inter. *4*, 347–352
294. Tippins, H.H.: Charge-transfer spectra of transition-metal ions in corundum. Phys. Rev. B*1*, 126–135 (1970)
295. Vargin, V.V.: Colours in titanium-bearing glasses. Dokl. Akad. Nauk SSSR *103*, 105–106 (1955)
296. Veinberg, T.I.: Absorption spectra of vanadium in phosphate glasses. Optiko-mech. Prom. *9*, 46–51 (1958)
297. Walker, I.M., Carlin, R.L.: Electronic structure of hexaque metal ions. J. Chem. Phys. *46*, 3931–3936 (1967)
298. Weakliem, H.A.: Optical spectra of Ni^{2+}, Co^{2+}, and Cu^{2+} in tetrahedral sites in crystals. J. Chem. Phys. *36*, 2117–2140 (1962)
299. White, W.B., Keester, K.L.: Optical absorption spectra of iron in the rockforming silicates. Am. Mineralogist. *51*, 774–791 (1966)
300. White, W.B., Keester, K.L.: Selection rules and assignments for the spectra of ferrous iron in pyroxenes. Am. Mineralogist. *52*, 1508–1514 (1967)
301. White, W.B., Roy, R., McCrighton, J.: The "alexandrite effect", an optical study. Am. Mineralogist. *52*, 867–871 (1967)
302. White, W.B., McCarthy, G.J., Scheetz, B.E.: Optical spectra of chromium, nickel, and cobalt-containing pyroxenes. Am. Mineralogist. *56*, 72–89 (1971)
303. Wickersheim, K.A., Lefever, R.A.: Absorption spectra of ferric iron-containing oxides. J. Chem. Phys. *36*, 844–850 (1962)
304. Wilkins, R.W.T., Farrell, E.F., Naiman, C.S.: The crystal field spectra and dichroism of tourmaline. J. Phys. Chem. Solids *30*, 43–56 (1969)
305. Wood, D.J., Ferguson, J., Knox, K., Dillon, J.F.: Crystal field spectra of d^3, d^7 ions. III. Spectrum of Cr^{3+} in various octahedral crystal fields. J. Chem. Phys. *39*, 890–898 (1965)
306. Wood, D.L.: Absorption, fluorescence, and Zeeman effect in emerald. J. Chem. Phys. *42*, 3404–3410 (1965)
307. Wood, D.L., Remeika, J.P.: Optical absorption of tetrahedral Co^{3+} and Co^{2+} in garnets. J. Chem. Phys. *46*, 3595–3602 (1967)
308. Wood, D.L., Ballman, A.A.: Blue synthetic quartz. Am. Mineralogist. *51*, 216–2 (1966)
309. Wood, D.L., Imbusch, G.F., Macfarlane, R.M., Kisliuk, P., Larkin, D.M.: Optical spectrum of Cr^{3+} ions in spinels. J. Chem. Phys. *48*, 5255–5263 (1968)

310. Wood, D.L., Nassau, K.: The characterization of beryl and emerald by visible and infrared absorption spectroscopy. Am. Mineralogist. *53*, 777–800 (1968)
311. Wright, W.D.: The measurement of colour. London: Hilger 1969

Lattice Energies, Crystal Potentials, Intracrystalline Distribution

312. Avilov, V.V.: Calculation of Madelung constant of crystals. Phys. tverdogo tela *14*, 2550–2554 (1972)
313. Amoros, J.L.: La energie de Madelung en los minerales. Los polimorfos de la silice. Estud. Geol. *22*, 135–141 (1966)
314. Artman, J.O., Murphy, J.C.: Lattice sum evaluations of ruby spectral parameters. Phys. Rev. *135*, 1622–1639 (1964)
315. Bachinski, D.J.: Bond strength and sulfur isotopic fractionation in coexisting sulfides. Econ. Geol. *64*, 56–65 (1969)
316. Boeyens, J.C., Gafner, G.: Direct summation of Madelung energies. Acta Crystallogr. A*25*, 411–414 (1969)
317. Borisov, Ju.A., Bulgakov, N.N.: Semi-empirical method of bond energy calculation. Zh. Strukt. Khim. *13*, 103–110 (1972)
318. Born, J., Zemann, I.: Gitterenergetische Berechnungen an Tonerde-Granaten. Acta Crystallogr. *16*, 1064–1065 (1963)
319. Burns, R.G.: Site preferences of transition metal ions in silicate crystal structures. Chem. Geol. *5*, 275–283 (1970)
320. Calais, J.-L., Mansikka, K., Pettersson, G., Vallen, J.: A calculation of the cohesive energy and elastic constants of MgO. Arkiv for Fysik B *34*, 361–366 (1967)
321. Damon, P.E.: Behavior of some elements during magmatic crystallisation. Geochim. Cosmochim. Acta *32*, 564–567 (1968)
322. Darwent, B.de: Bond dissociation energies in simple molecules. U.S. Dep. Comm. Nat. Bur. Stand. *31*, (1970)
323. Dunitz, J.D., Orgel, L.E.: Electronic properties of transition metal oxides. Cation distribution amongst octahedral and tetrahedral sites. J. Phys. Chem. Solids *3*, 318–333 (1957)
324. Fyfe, W.S.: Lattice energies, phase transformations and volatiles in the mantle. Phys. Earth Planet. Interiors *3*, 196–200 (1970)
325. Giese, R.F.,Jr., Weller, S., Datta,P.: Electrostatic energy calculations of diaspore (α-AlOOH), goethite (α-FeOOH) and groutite (α-MnOOH). Z. Kristallogr. *134*, 275–284 (1971)
326. Grebenshchikov, R.G.: Lattice energies of silicates and heats of formation of silicon-oxygen anion radicals. Zh. Neorg. Khim. *9*, 1038–1048 (1964)
327. Hafner, St., Raymond, M.: Self-consistent ionic potentials, fields, and field gradients at the lattice sites of corundum (α-Al_2O_3). J. Chem. Phys. *49*, 3570–3579 (1968)
328. Hafner, St., Raymond, M.: Effect of oxygen octupoles on the electric field gradient at the Al^{3+} sites in corundum (α-Al_2O_3). J. Chem. Phys. *52*, 279–281 (1970)
329. Henderson, P., Dale, I.M.: The partitioning of selected transition element ions between olivine and groundmass of oceanic basalts. Chem. Geol. *5*, 267–274 (1970)
330. Hutching, M.T.: Point-charge calculations of energy levels of magnetic ions in crystalline electric fields. Sol. State Phys. *16*, 227–273 (1964)
331. Jäger, E., Perthel, R.: About the crystal field theory of tetrahedral sites in spinel lattices. Phys. Stat. Sol. *38*, 735–746 (1970)
332. Keller, W.D.: The bonding energies of some silicate minerals. Am. Mineralogist. *39*, 783–793 (1954)
333. Löwdin, P.-O.: Quantum theory of cohesive properites of solides. Advan. Phys. *5*, 1–172 (1956)
334. Löwdin, P.-O.: A theoretical investigation into some properties of ionic crystals. A quantum mechanical treatment of the cohesive energy, the interionic distance, the elastic constants, and the compression at high pressures with numerical applications to some alkali halides. Uppsala. Inaug. Diss. 1948

335. Marakushev, A.A., Bezmen, N.I.: Thermodynamics of sulfides and oxides with relation to ore formation problems. Part I. Lattice energies of sulfides and oxides. Moscow: Nauka 1972
336. Matusi, Y., Banno, Sh.: Partition of divalent transition metals between coexisting ferromagnesian minerals. Chem. Geol. 5, 259–265 (1970)
337. Mineeva, R.M.: Potentials of non-equivalent sites in rock-forming silicates. Dokl. Akad. Nauk SSSR 210, 1443–1446 (1973)
338. Mineeva, R.M.: Cation ordering interpretation in olivines according to energy calculations. In: Constitution and properties of minerals. Kiev: Nauka 1975
339. Mineeva, R.M.: Relationship between Mössbauer spectra and defect structure in biotites from electric-field gradient calculations. Phys. Chem. Minerals 2, 267–278 (1978)
340. Navrotsky, A.: The intracrystalline cation distribution and the thermodynamics of solid solution formation in the system $FeSiO_3$–$MgSiO_3$. Am. Mineralogist. 56, 201–211 (1971)
341. Nickel, E.H.: Bond strength and sulfur isotopic fractionation in coexisting sulfides. Econ. Geol. 64, 934–936 (1969)
342. Pettersson, G., Vallen, J., Calais, J.-L., Mansikka, K.: Experimental and theoretical determination of the elastic constants of NaCl. Arkiv Fysik B34, 371–376 (1967)
343. Pistor, W.: Optical absorption spectra of V^{3+} and Cr^{3+} in $ScCl_3$ and $InCl_3$. Phys. Stat. Sol. 40, 581–592 (1970)
344. Raymond, M.: Electric-field-gradient calculations in the aluminium silicates (Al_2SiO_5). Phys. Rev. B3, 3692–3701 (1971)
345. Raymond, M.: Electric-field-gradient calculations in spodumene ($LiAlSi_2O_6$). Carnegie Inst. Ann. Rept. Dir. Geophys. Lab. 1970–1971. Washington, 227–229 (1971)
346. Raymond, M.: Madelung constants for several silicates. Carnegie Inst. Ann. Rept. Dir. Geophys. Lab. 1970–1971. Washington, 225–227 (1971)
347. Sahl, K., Zemann, J.: Gitterenergetische Berechnungen an Zirkon. Ein Beitrag zur Ladungsverteilung in der Silikatgruppe. Tscherm. mineral. Petr. Mitteil. 10, 97–114 (1965)
348. Sakamoto, Y.: Madelung's coefficient of sulvanite, Cu_3VS_4. J. Sci. Hiroshima Univ. Ser. A, Div. 2, 27, 111–124 (1964)
349. Saltzman, M.N., Schor, R.: Madelung energy of perovskite structures. J. Chem. Phys. 42, 3698–3700 (1965)
350. Saxena, S.K.: Thermodynamics of rock-forming crystalline solutions. Berlin, Heidelberg, New York: Springer 1973
351. Sengupta, D., Artman, J.O., Sawatzky, G.A.: "Overlap" contributions to the electric-field-gradient components at the Fe^{3+} site in FeOCl. Phys. Rev. B4, 1481–1486 (1971)
352. Sharma, R.R.: Nuclear quadrupole moment of Al^{27} in Al_2O_3. Phys. Rev. Letters 25, 1622–1623 (1970)
353. Sherman, J.: Crystal energies of ionic compounds and thermochemical applications. Chem. Rev. 11, 93–170 (1932)
354. Slaughter, M.: Chemical binding in silicate minerals. Geochim. Cosmochim. Acta 30, 299–314, 323–339 (1966)
355. Schwartz, H.P.: The effect of crystal field stabilization on the distribution of transition metals between metamorphic minerals. Geochim. Cosmochim. Acta 31, 503–517 (1967)
356. Smith, J.V., Ribbe, P.H.: Atomic movements in plagioclase feldspars: kinetic interpretation. Contrib. Mineral. Petrol. 21, 157–202 (1969)
357. Smyth, J.R., Smith, J.V.: Electrostatic energy for ion clustering in intermediate plagioclase feldspar. Mineral. Mag. 37, 181–184 (1969)
358. Tosi, M.P.: Cohesion of ionic solids in the Born model. Sol. State Phys. 16, 1–120 (1964)
359. Urusov, V.S.: Energy crystallochemistry. Moscow: Nauka 1975
360. Vallen, J., Pettersson, G., Calais, J.L., Mansikka, K.: Calculation of elastic constants of NaF. Arkiv Fysik B 34, 199–213 (1967)
361. Waddington, T.C.: Lattice energies and their significance in inorganic chemistry. Advan. Inorgan. Chem. Radiochem. 1, 158–218 (1959)

362. Whittaker, E.J.W.: Madelung energies and site preference in amphiboles. Am. Mineralogist. *56*, 980–996 (1971)
363. Wood, R.: Madelung constants for the calcium carbide and pyrite crystal structures. J. Chem. Phys. *87*, 598–600 (1962)
364. Zemann, J.: Der Verlauf der Madelungschen Zahl für den Wurtzit-Typ bei Änderung des Achsenverhältnisses. Tscherm. Mineral. Petrogr. Mitt. *12*, 439–442 (1968)
365. Zoltai, T., Buerger, M.J.: Relative energies of rings of tetrahedra. Z. Krist. *114*, 1–8 (1960)

Ionic and Atomic, Orbital and Mean Radii

366. Ahrens, L.H.: The use of ionization potentials. I. Ionic radii of the elements. Geochim. Cosmochim. Acta *2*, 155–169 (1952)
367. Batsanov, S.S.: System of atomic radii. Zh. Strukt. Khim. *3*, 616–628 (1963)
368. Biggar, G.M.: The ionic radius of nickel. Mineral. Mag. *37*, 299–300 (1969)
369. Bokij, G.B.: Crystallochemistry. Moscow: Nauka 1971
370. Brill, R.: Determination of electron distribution in crystals by means of X-rays. Sol. State Phys. *20*, 1–35 (1967)
371. Cho, S.J.: Spin-polarized electronic energy-band structure in EuS. Phys. Rev. *157*, 632–640 (1967)
372. Fumi, F.G., Tosi, M.P.: Ionic sizes and Born repulsive parameters in the NaCl-type alkali halides. Phys. Chem. Sol. *25* (1964)
373. Järvinen, M., Inkinen, O.: An X-ray diffraction study of rubidium chloride. Phys. Status Solidi *21*, 127–135 (1967)
374. Kordes, E.: Berechnung der Ionenradien allein den Ionenabständen. Z. Krist. *115*, 169–184 (1961)
375. Kordes, E.: Direkte Berechnung der Ionenradien in Alkalihalogeniden aus der Lichtbrechung bzw. Molrefraktion und dem Ionenabstand. Tscherm. Mineral. Petr. Mitt. *8*, 13–23 (1962)
376. Krasnov, K.S.: Ionic radii in inorganic molecules. Zh. Strukt. Khim. *4*, 884–891 (1963)
377. Kristofel, N.N., Turkson, E.E.: Calculation of wave function of ions in fluorite crystal. Proc. Inst. Phys. Astron. Akad. Nauk Est. SSR, *23*, 216–218 (1963)
378. Kruglyak, Ju.A.: Solvated ions radii. Zh. Phys. Khim. *41*, 867–869 (1967)
379. Ladd, M.F.C.: The radii of spherical ions. Theoret. Chim Acta *12*, 333–336 (1968)
380. Lebedev, V.I.: Ionic-atomic radii and their significance in geochemistry and chemistry. Leningrad: Univ. Ed. 1969
381. Linkoaho, M.V.: The deformations of the ions in NaCl and AgCl crystals and the temperature parameters of ions in some alkali halides. Acta Cryst. A*25*, 450–455 (1969)
382. Margolina, A.F., Buntar, A.G.: Electron diffraction study of NaF. Kristallografiya *13*, 892–894 (1968)
383. Maslen, V.W.: The electron density in fluorite. An analysis of some X-ray diffraction data. Proc. Phys. Soc. *91*, 466–474 (1967)
384. Maslen, V.W.: Crystal ionic radii. Proc. Phys. Soc. *91*, 259–260 (1967)
385. Meisalo, V., Inkinen, O.: An X-ray diffraction analysis of potassium bromide. Acta Cryst. *22* (1967)
386. Morris, D.F.: Ionic radii and enthalpies of hydration of ions. Struct. Bonding *4*, 63–82 (1968)
387. Petrashen, M.I., Abarenkov, I.V., Kristofel, N.N.: Approximate wave functions of free ions and ions in crystal. Vestn. Leningr. Univ., Ser. Phys.-Math. *16*, 3 (1960)
388. Schoknecht, G.: Röntgen-Kristallstrukturanalyse mit Faltungsintegralen. Meßverfahren und Bestimmung der Elektronendichte in NaCl. Naturforscher *12*a, 983–995 (1957)
389. Schomaker, V., Stevenson, D.P.: Some revisions of the covalent radii and the additivity rule for the lengths of partially ionic single covalent bonds. J. Am. Chem. Soc. *63*, 37–40 (1941)
390. Shannon, R.D., Prewitt, C.T.: Effective ionic radii in oxides and fluorides. Acta Cryst. B*25*, 925–946 (1969)

391. Shannon, R.D., Prewitt, C.T.: Revised values of effective ionic radii. Acta Cryst. B25, 946–960 (1969)
392. Slater, J.C.: Introduction to Chemical Physics. New York, London: McGraw-Hill, 1939
393. Slater, J.C.: Atomic radii in crystals. J. Chem. Phys. 41, 3199–3204 (1964)
394. Sutton, L.E. (ed.): Tables of interatomic distances and configuration in molecules and ions. London: Chem. Soc. 1958
395. Tosi, M.P.: Cohesion of ionic solids in the Born model. Sol. State Phys. 16, 1–120 (1964)
396. Van Vechten, J.A., Phillips, J.C.: New set of tetrahedral covalent radii. Phys. Rev. B2, 2160–2167 (1970)
397. Wasastjerna, J.A.: On the radii of ions. Soc. Sci. Fennica, Comment. Phys.-Math. 1, 38 (1923)
398. Weiss, R.J.: X-ray determination of electron distribution. Amsterdam: North-Holland; New York: Wiley 1966
399. Whittaker, E.J.W., Muntus, R.: Ionic radii for use in geochemistry. Geochim. Cosmochim. Acta 34, 945–956 (1966)
400. Witte, H., Wölfel, E.: Electron distribution in NaCl, LiF, CaF_2 and Al. Rev. Mod. Phys. 30, 1 (1958)

Chemical Bonds in Some Classes and Groups of Minerals and Inorganic Materials

401. Adler, D.: Insulating and metallic states in transition metal oxides. Sol. State Phys. 21, 1–115 (1968)
402. Adler, D.: Mechanisms for metal–nonmetal transitions in transition-metal oxides and sulfides. Rev. Mod. Phys. 40, 714–736 (1968)
403. Albers, W., Haas, C.: Band structure and the mechanism of electrical conduction in transition metal compounds. Phys. Letters 8, 300–302 (1964)
404. Baur, W.H.: On hydrogen bonds in crystalline hydrate. Acta Cryst. 19, 909–916 (1965)
405. Baur, W.H.: The prediction of bond length variations in silicon-oxygen bonds. Am. Mineralogist. 56, 1573–1599 (1971)
406. Baur, W.H.: Bond length variation and distorted coordination polyhedra in inorganic crystals. Trans. Am. Cryst. Ass. 6, 129–155 (1970)
407. Belov, N.V., Pobedimskaya, E.A.: Characteristic features of crystallochemistry of sulfides, sulfosalts and related compounds. Lvov Univ. Mineral. Sbornik 20, 326–340 (1966)
408. Belov, N.V.: Outlines of structural mineralogy. Moscow: Nedra 1976
409. Berghoff, G., Paeslack, J.: Sauerstoff als Koordinations-Zentrum in Kristallstrukturen. Z. Krist. 126, 112–123 (1968)
410. Bershov, L.V., Marfunin, A.S.: Estimations of the chemical bond from Mn^{2+} EPR spectra superfine structure. Dokl. Akad. Nauk SSSR 155, 632–635 (1964)
411. Biornacki, S.W.: Molecular orbital approximation study of the Co^{2+} impurity in ZnSe – ligand hyperfine structure. Phys. Stat. Sol. 51, 829–840 (1972)
412. Bither, T.A., Bouchard, R.J., Cloud, W.H., Donchue, P.C., Siemons, W.J.: Transition metal pyrite dichalcogenides. Inorgan. Chem. 7, 2208–2220 (1968)
413. Bloch, A.M.: Structure of water and geological processes. Moscow: Nedra 1969
414. Bokij, G.B.: Crystal structures of arsenides, sulfides, arsenosulfides and related compounds. Novosibirsk: Akad. Nauk Ed. 1964
415. Breeze, A., Perkins, P.G.: Energy band structure of silica. J. Chem. Soc. Faraday Trans. 1 169, 1237–1242 (1973)
416. Bratoz, S.: Electronic theories of hydrogen bonding. Advan. Quant. Chem. 3, 209–239 (1967)
417. Brostigen, G., Kjekshus, A.: Bonding schemes for compounds with the pyrite, marcasite, and arsenopyrite type structures. Acta Chem. Scand. 24, 2993–3012 (1970)
418. Brown, G.E., Gibbs, G.V.: Oxygen coordination and the Si–O bond. Am. Mineralogist. 54, 1528–1539 (1969)

419. Brown, G.E., Gibbs, G.V., Ribbe, P.H.: The nature and the variation in length of the Si–O and Al–O bonds in framework silicates. Am. Mineralogist. 54, 1044–1061 (1969)
420. Brown, G.E., Gibbs, G.V.: Stereochemistry and ordering in the tetrahedral portion of silicates. Am. Mineralogist. 55, 1587–1607 (1970)
421. Burns, R.G., Vaughan, D.J.: Interpretation of the reflectivity behavior of ore minerals. Am. Mineralogist. 55, 1578–1586 (1970)
422. Busch, G., Hulliger, F.: Mineralien als Vorbilder für neue Halbleiterverbindungen. Helv. Phys. Acta 33, 657 (1960)
423. Collins, G.A.G., Cruickshank, D.W.J., Breeze, A.: Ab initio calculations of the silicate ion, orthosilicic acid and their $L_{2,3}$ X-ray spectra. J. Chem. Soc. Faraday Trans. II 7, 1189–1195 (1972)
424. Chiari, G., Gibbs, G.V.: Bond length, valence angle and bond overlap population variations for arsenate tetrahedral oxyanions. Atti Acad. Sci. Turin 108, 879–901 (1974)
425. Cruickshank, D.W.J.: The role of 3d-orbitals in π-bonds between a) silicon, phosphorus, or chlorine and b) oxygen or nitrogen. J. Chem. Soc. 5486–5504 (1961)
426. Dikov, Ju.P., Debolsky, E.I., Romashenko, Yu.N., Dolin, S.P., Levin, A.A.: Molecular orbitals of $Si_2O_7^{6-}$, $Si_3O_{10}^{8-}$ etc., and mixed (P, Al, P, Si) applied to clusters and X-ray spectroscopy data of silicates. Phys. Chem. Miner. 1, 27–42 (1977)
427. Dolin, S.P., Dyatkine, M.E.: Electron structure of the ions PO_4^{3-}, SO_4^{2-}, and ClO_4^{-}. Zh. Strukt. Khim. 13, 966–968 (1972)
428. Dolin, S.P., Shegolev, W.F., Dyatkina, M.E.: Electron structure of the ions SiO_4^{4-}, GeO_4^{6-}, and AlO_4^{5-}. Zh. Strukt. Khim. 13, 964–965 (1972)
429. Drickamer, H.G., Frank, C.W.: Electronic transitions and the high pressure chemistry and physics of solids. London: Chapman and Hall 1973
430. Fesenko, E.G.: Perovskite-type compounds and ferroelectricity. Moscow: Atomisdat 1972
431. Gibbs, G.V., Hamil, M.M., Louisnathan, S.J., Bartell, L.S., Yow, H.: Correlations between Si–O bond lengths, Si–O–Si angle and bond overlap populations calculated using extended Hückel molecular orbital theory. Am. Mineralogist. 57, 1578–1613 (1972)
432. Gibbs, G.V., Louisnathan, S.J., Ribbe, P.H., Fillips, M.W.: Semi-empirical molecular orbital calculations for the atoms of the tetrahedral framework in anorthite, low albite, maximum microcline and reedmergnerite. In: The Feldspars. Zussman, I. (eds.). Manchester Press 1974
433. Giese, R.F.: Interlayer bonding in kaolinite, dickite and nacrite. Clays Clay Miner. 21, 145–149 (1973)
434. Giese, R.F.: Surface energy calculations for muscovite. Nature Phys. Sci. 248, 580–581 (1974)
435. Giese, R.F.: Interlayer bonding in talc and pyrophyllite. Clays Clay Miner. 23, 165–166 (1975)
436. Gorunova, N.A.: Complex diamond-like semiconductors. Moscow: Sovetskoe radio 1968
437. Goodenough, J.B.: Descriptions of outer d electrons in thiospinels. J. Phys. Chem. Solids 30, 261–280 (1969)
438. Hamilton, W.C., Ibers, J.A.: Hydrogen bonding in solids. New York, Amsterdam: Benjamin 1968
439. Harrison, W.A.: Metallic Bonds. Physical Chemistry. An Advanced Treatise. Vol. V. Valency, Eyring, H. (ed.). New York: Academic 1970
440. Hirschfelder, J.O., Curtiss, Ch.F., Bird, R.B.: Molecular theory of gases and liquids. New York: John Wiley 1960
441. Hulliger, F., Mooser, E.: Semiconductivity in pyrite, marcasite and arsenopyrite phases. J. Phys. Chem. Solids 26, 429–433 (1965)
442. Hulliger, F.: Crystal chemistry of the chalcogenides and pnictides of the transition elements. Struct. Bonding 4, 83–299 (1968)
443. Jaynes, E.T.: Ferroelectricity. Princeton: University Press 1953

444. Jellinek, F.: Sulphides. In: Inorganic Sulphur Chemistry. Nickless, G. (ed.). Amsterdam: Elsevier 1968
445. Kjekshus, A., Nicholson, D.G.: The significance of back-bonding in compounds with pyrite, marcasite, and arsenopyrite-type structures. Acta Chem. Scand. 25, 866–876 (1971)
446. Kjekshus, A., Pearson, W.B.: Phases with the nickel arsenide and closely related structures. Prog. Solid State Chem. 1, 83–174 (1964)
447. Ladd, M.F.C.: The location of hydrogen atoms in crystalline ionic hydrates. Z. Krist. 26, 147–152 (1968)
448. Lin, Sh.H.: Hydrogen Bonding. Physical Chemistry, Advanced Treatise, Vol. V. Valency. Ehryng, H. (ed.). New York: Academic 1970
449. Lip, K.L., Fowler, W.B.: Electronic structure of SiO_2. Phys. Rev. B10, 1391–1399, 1400–1408 (1975)
450. Lotgering, F.K., Van Stapele, R.P., Van der Steen, G.H.A.M., Van Wieringen, J.S.: Magnetic properties, conductivity and ionic ordering in R_{1-x}, Cu_x, Cr_2S_4. J. Phys. Chem. Solids 30, 799–804 (1969)
451. Louisnathan, S.J., Gibbs, G.V.: Bond length variation in TO_4^n tetrahedral oxyanions of the third row elements: T–Al, Si, P, S and Cl. Mater. Res. Bull. 7, 1281–1292 (1972)
452. Louisnathan, S.J., Hill, R.J., Gibbs, G.V.: Tetrahedral bond length variations in sulfates. Phys. Chem. Miner. 1, 53–69 (1977)
453. Marfunin, A.S., Mkrtchan, A.R.: Mössbauer spectra of Fe^{57} in sulfide minerals. Geochimia 10, 1094–1103 (1967)
454. Margenau, H., Kestner, N.R.: Theory of intermolecular forces. London: Pergamon 1971
455. Mooser, E., Pearson, W.B.: The chemical bond in semiconductors. Prog. Semicond. 5, 103–139 (1960)
456. Medvedeva, Z.S.: Chalcogenides of the elements of III B subgroup of the periodic system. Moscow: Nauka 1968
457. Nefedov, V.I., Urusov, V.S., Kakhana, M.M.: ESCA study of the chemical bond in Na, Mg, Al, and Si minerals. Geochimia 1, 11–19 (1977)
458. Nickel, E.H.: The application of ligand field concepts to an understanding of the structural stabilities and solid-solution limits of sulphides and related minerals. Chem. Geol. 5, 233–241 (1970)
459. Obolonchik, V.A.: Selenides. Moscow: Metallurgia 1972
460. Pantelides, S.T., Harrison, W.A.: Electronic structure, spectra, and properties of 4:2-coordinated materials. I. Crystalline and amorphous SiO_2 and GeO_2. Phys. Rev. B13, 2667–2691 (1976)
461. Pauling, L.: The nature of the chemical bond in sulvanite, Cu_3VS_4. Tscherm. Mineral. Petr. Mitt. 10, 379–384 (1964)
462. Pauling, L.: Crystallography and chemical bonding of sulfide minerals. Mineral. Soc. Am. Spec. Pap. 3, 125–134 (1970)
463. Pauwels, L.J.: Energy level diagrams of the "NiAs"-, "pyrite"- and "spinel"-type sulfides of Fe, Co and Ni. Bull. Soc. Chim. Belges 79, 549–566 (1970)
464. Pearson, W.B.: The crystal structures of semiconductors and a general valence rule. Acta Cryst. 17, 1–15 (1964)
465. Phillips, J.C.: Ionicity of the chemical bond in crystals. Rev. Mod. Phys. 42, 317–356 (1970)
466. Phillips, J.C., Van Vechten, J.A.: Spectroscopic analysis of cohesive energies and heats of formation of tetrahedrally coordinated semiconductors. Phys. Rev. B2, 2147–2160 (1970)
467. Pimentel, G.Cl., McClellan, A.L.: The hydrogen bond. San-Francisco, London: Freeman 1960
468. Prewitt, C.: Crystal structures of pyroxenes at high temperatures. In: Physics and Chemistry of minerals and rocks. Strens, R.G.J. (ed.). New York: Niley-Interscience 1976
469. Rouweler, G.C.J.: Measurements of Van der Waals forces. Utrecht: Diss., 1972

470. Ryskin, Ja.I., Stavitskaya, G.P.: Hydrogen bond and structure of hydrosilicates. Moscow: Nauka 1972
471. Samsonov, G.V.: Classification of chalcogenides. In: Chalcogenides. Kiev: Naukova Dumka 1967
472. Scanlon, W.W.: The physical properties of semiconducting sulfides, selenides and tellurides. Intern. Mineral. Ass. Papers Proc. 3rd Gen. Meet. Washington *1*, 135–143 (1962)
473. Smith, D.G.W., O'Nions, R.K.: Investigations of bonding in some oxide minerals by oxygen K emission spectroscopy. Chem. Geol. *9*, 29–43 (1977)
474. Soules, Th.F., Kelly, E.J., Vaught, D.M., Richardson, J.W.: Energy-band structure of $SrTiO_3$ from a self-consistent-field tight-binding calculation. Phys. Rev. B*6*, 1519–1532 (1972)
475. Suchet, J.P., Bailly, F.: La liaison chimique dans les cristaux minéraux. Annales Chim. *10*, 517–532 (1965)
476. Taylor, D.: The relationship between SiO distances and Si–O–Si bond angles in the silica polymorphs. Mineral. Mag. *38*, 629–631 (1972)
477. Tossell, J.A., Vaughan, D.J., Johnson, K.H.: X-ray photoelectron, X-ray emission and UV spectra of SiO_2 calculated by the SCF X_α scattered wave method. Chem. Phys. Letters *20*, 329–334 (1973)
478. Tossell, J.A.: The electronic structures of silicon, aluminum and magnesium in tetrahedral coordination with oxygen from SCF–MO calculations. J. Am. Chem. Soc. *97*, 17, 4840–4844 (1974)
479. Tossell, J.A.: The electron structure of Mg, Al and Si in octahedral coordination with oxygen from SCF X_αMO calculations. J. Phys. Chem. Solids *36*, 1273–1280 (1975)
480. Tossell, J.A., Gibbs, G.V.: Molecular orbital studies of spectra and geometries of mineral and inorganic compounds. Phys. Chem. Miner. *2*, 21–58 (1977)
481. Vaughan, D.J., Burns, R.G., Burns, V.M.: Geochemistry and bonding of thiospinel minerals. Geochim. Cosmochim. Acta *85*, 365–381 (1971)
482. Verhoogen, J.: Physical properties and bond type in Mg–Al oxides and silicates. Am. Mineralogist. *43*, 552–579 (1958)
483. Vesely, C.J., Langer, D.W.: Electronic core levels of the IIB–VIA compounds. Phys. Rev. B*4*, 451–462 (1971)
484. Wilson, J.A., Yoffe, A.D.: The transition metal dichalcogenides. Discussion and interpretation of the observed optical, electrical and structural properties. Advan. Phys. *8*, 193–335 (1969)

Subject Index

Actinides
 multiplets and terms, weak crystal field 75
 optical transition types and colors 238
Additive systems of ionic and atomic radii 268
Affinity, electron 142, 151
Altaite, PbTe
 energy band structure 173, 174
 reflectance spectra 175, 176
Analysis
 atomic absorption 54
 spectrographic 54
 X-ray spectrum 55
Andalusite
 lattice sum calculations 263
 nuclear magnetic resonance of Al^{27} 285
Anorthite, silicium effective charge 280
Antibonding molecular orbitals 99, 135, 148
Arsenolite, As_2O_3
 intermolecular interactions 314
 nuclear quadrupole resonance of As^{75} 302
Arsenopyrite, FeAsS, isomer shift, quadrupole splitting 300
APW (Augmented Plane Wave) method 168
Atom
 in energy band model 156
 in geochemistry and non-additive of properties 94, 137
 in molecular orbital scheme 137
 quantum-mechanical model 18
 Rutherford-Bohr model 3, 14
 vector model 47
Atomic orbitals, s, p, d, f types 33
Atomization energy 248
Azurite, optical absorption spectra, Cu^{2+} energy levels 235

Berthierite, $FeSb_2S_4$, isomer shift, quadrupole splitting 300
Beryl
 of Cr^{3+} 212
 effective charges 280
 of Mn^{3+} 213

 nuclear magnetic resonance of Be^9 285
 optical absorption spectra of V^{3+} 207
Bond orbital approximation 154, 242, 280
Bornite, Cu_5FeS_4, isomer shift, quadrupole splitting 300
Brillouin zone 159

Calcite
 EPR of Mn^{2+} 282
 optical absorption spectrum of Ni^{2+} 233
Cattierite, CoS_2
 energy band scheme 295
 exchange interactions and properties 299
Chalcopyrite
 antiferromagnetic 301
 chemical bond and energy band structure 308
 Mössbauer spectra 300
Characters 64, 67
 for C_{2h} group 64, 65
 for C_{2v} group 73
 for C_{4v} 65, 67
 tables 72
Charge transfer spectra 186, 198, 199
Chemical bond
 contemporary state 94
 and crystallochemistry 94, 275, 286
 in crystals 239
 selfconsistency of atomic properties 137
 stages of development 95
Chromites, cation distribution, exchange interactions, properties 311
Cinnabar, HgS
 absorption edge and color 237
 energy band structure 177
Cobalt, energy levels in crystal fields, optical absorption spectra 227–231
Colours of minerals 236
Coordinate systems, relation between rectangular and polar 23, 256
Copper, energy level diagram, optical absorption spectra 233
Corundum, electric field gradient and lattice sums 263
 energy level diagrams and optical absorption spectra of
  Cr^{3+} 211–213

Corundum, electric field gradient and lattice sums
    EPR of Mn^{2+} 282
    Fe^{3+} 219, 220
    Mn^{3+} 214, 215
    Ni^{2+} 232
    Ti^{3+} 206
    V^{2+} 209
    V^{3+} 208
Coulomb integrals 113, 136
    in bond orbital method 241, 243
Crystal field 56
    medium (iron group) 76
    origin of crystal field model 60
    parameters 81
    potential 255
    strong (Pd and Pt groups) 76
    weak (rare earths, actinides) 75
Crystal field stabilization energy 263, 283
Crystal field strength, Dq 60
    determination from optical absorption spectra 84
    Fe^{2+} 223
    Fe^{3+} 221
    lattice sum calculation 255
    Ni^{2+} 231
Crystal orbitals 156, 240
    symmetry transformations 163
    with spin-orbit interaction 164

Daubreelite, $FeCr_2S_4$
    antiferromagnetism 299
    energy band scheme 295
Degenerate states 58
Diamond
    bond orbital calculation 242
    electron density 271
    energy band structure 177
Diaspore, hydroxyl bonds 315
Dichalcogenides, chemical bond and properties 309
Diffuse reflectance spectra 197
Dipole momenta in molecules 147
Disthene
    lattice sums 263
    nuclear magnetic resonance of Al^{27} 285
Double symmetry groups 164

Effective charges 143, 147, 148, 241
    and dipole momenta 147
    in silicates and aluminosilicates 279
Electron configurations
    free atoms 45
    iron group ions 77
    molecules 101
Electronegativity 136
Energy band calculation methods 167

Energy band schemes
    altaite PbTe 173, 174
    cristobalite 242, 281
    galena 173
    NaCl 172
    other minerals 177
    periclase 171
    quartz 242
    silicates 280
    sphalerite 176
    sulfides 293
Energy band theory and mineragraphy 169
Energy gap 292, 299
Energy level diagram for Fe^{2+} in tetrahedral field 222
Energy level diagrams for ions in octahedral field
    d^1 204, 205
    d^2 209
    d^3 210, 212
    d^4 214
    d^5 215, 219
    d^6 222
    d^7 228
    d^8 232
    d^9 234
EPR (Electron Paramagnetic Resonance) 179, 180, 182, 184, 188, 282
ESCA (Electron Spectroscopy for Chemical Analysis) 182, 279
Exchange integral 113
Exchange interactions 294, 311

Field gradient 181, 189, 262
Fine structure in atomic spectra 9, 10
Fluorite
    energy band scheme 177
    wave functions of atoms in crystal 272
Forbidden transitions 11, 71, 200, 202
Fractionation of sulfur isotopes and lattice energy 254

Galena, PbS
    $A^N B^{8-N}$ type 308
    chemical bond, theory and calculation 309
    energy band scheme 173
    reflectance spectrum 175
    spin-orbit splitting 164
Gallite, $CuGaS_2$, energy band scheme 177
Garnets
    of Co^{2+} in octahedron, tetrahedron, dodecahedron 231
    lattice sums 263
    of Ni^{2+} 233
    optical absorption spectra of Fe^{3+} 219
    Y-Fe-Gr, concentration dependence in spectra 220

Gold, energy band scheme 177
Greenockite, CdS
  absorption edge and color 237
  of Co^{2+} 230
  effective charges and chemical bond 308
  optical absorption spectra of V^{3+} 209
Greigite, Fe_3S_4
  chemical bond and properties 309
  high-spin state 306
Group orbitals
  in octahedra 116, 117
  in tetrahedra 128
  oxygen in CO_2^- 104, 107
Group overlap integral 126, 129
Group theory 60, 61
Grotrian diagrams 11
Gypsum, hydrogen bonds 314

Halides, alcali, atomic orbital and total electron density superposition 265
  $A^N B^{8-N}$ type 308
  chemical bond 312
  crystal radii 272
  energy band scheme 172
  experimental ionic radii 272
  lattice energy 248
  superfine structure of Mn^{2+} EPR spectra 282
Hamiltonian operator
  concept of 21
  in energy band theory 156
  free atom 28, 42, 52
  for hydrogen atom, molecule and molecular ion 109
  for ion in crystal field 81
Hematite, electric field gradient 263
High-spin states 86
Hybrid atomic orbitals 132, 185
  in MO schemes 134
Hydrogen atom
  electronic structure and spectrum 6
  energy level diagram 7
  orbital radii and energies calculation 5
  quantum-mechanical calculation 109
Hydrogen bonding 314

Identity operation 63, 67
Indite, $FeIn_2S_4$, physical properties 309
Integrals in quantum chemistry 113
Interatomic distances
  critical distances in oxides and sulfides 297
  in crystals and molecules 269
  and dipole momenta 147
  lattice energy calculation 249
  lattice sums calculation 255
  low-spin states 298

  molecular orbital schemes 267
  self-consistence 137
  semiempirical tight-binding method 241
  Si-O distances variation in silicates and aluminosilicates 286
Interband transitions 166, 174, 292
Intercrystalline distribution 285
Intermolecular forces 313
Intervalence-transfer absorption 199
Ionicity-covalency of the chemical bonding
  in energy band schemes 294
  in minerals from EPR parameters 282
  in molecular orbital schemes 150
  and spectra 185
  in sulfides and sulfosalts from NQR spectra 301
Ionization potential
  of atoms 8
  and lattice energy 247
  molecules and complexes in MO schemes 150
  simple molecules and radicals 152
  VSIE 137
Iron, energy level diagrams and absorption spectra 218–227
Irreducible representations 63
  designations 68
  table for different symmetry groups 69

Jahn-Teller effect 87
Jaipurite, CoS
  energy band scheme 295
  exchange interactions and properties 298

Kramers-Kronig relation 170
Kubelka-Munk function 197

Laporte selection rule 200
Lattice energy 246
  geochemical applications 253
  NaCl 248, 250
  simplified formulas 252
  TiO_2 247
Lattice sums 255
LCAO (Linear Combination of Atomic Orbitals) 98
  in bond orbital method 154, 168
Ligands 56
Linnaeite, Co_3S_4
  energy band scheme 295
  Pauli paramagnetism 299
Low-spin states 86
  in iron sulfides 306
  and isomorphous substitutions 306
  Mn^{3+} 214

Madelung constant 241, 250, 262
Malachite, energy levels and absorption spectra of Cu^{2+} 235
Manganese
  energy level diagram and spectra 213–217
  superfine structure in EPR spectra 282
Marcasite
  low-spin state 306
  Mössbauer parameters 300
Millerite, NiS, exchange interactions and properties 297
Molecular orbitals
  antibonding 99, 135, 148
  designations 102
  formation of MO in octahedral complex 116
  molecule ion H_2^+ 108
  molecule of O_2 type 103
  non-bonding 99, 101, 102
  σ and π orbitals 99
  population analysis 143
  radicals of CO_2^- type 105
Molecular orbital schemes
  complex CrF_6^{3-} 120
  with hybrid atomic orbitals 134
  molecular ion CO_2^- 105
  molecule CO_2 104
  molecule O_2 103
  SiO_4^{4-} 277, 278
  third row elements in octahedron and tetrahedron 119
  transition metal in octahedron and tetrahedron 119
Molybdenite, MoS_2
  energy band structure 177
  intermolecular interactions 313
  properties and electronic structure 309
Monticellite, absorption spectra of Co^{2+} in two sites 230
Mössbauer spectroscopy 179–181, 285, 300

Nickel, energy level diagram and absorption spectra 231
Nickeline, NiAs
  chemical bond 309
  energy band scheme and properties 297
  structure type 304
Normalization constant 25, 98, 111, 256
Nuclear Magnetic Resonance (NMR) 179, 180, 183, 187, 285
Nuclear Quadrupole Resonance (NQR) 179, 180, 187, 301, 310

Olivine
  absorption spectrum of Co-analog 228
  effective charges in forsterite 280
  EPR of Mn^{2+} 282

Opale
  effective charges 280
  hydrogen bonds 315
Operators 21
OPW (Orthogonalize Plane Wave) method 168
Orpiment, As_2S_3
  absorption edge and color 237
  nuclear quadrupole resonance of As^{75} 302
Orthoclase, effective charges 280
Overlap integrals 114
  of cations in sulfides and oxides 297
  group 126, 129

Pauli exclusion principle 43, 46
Pauli paramagnetism 299
Periclase, MgO
  chemical bond and properties 311
  Co^{2+} 229
  Cr^{3+} 211
  energy band structure 171
  EPR of Mn^{2+} 282
  lattice energy 252
  Ni^{2+} 232
  optical absorption spectra of V^{2+} 209
  reflectance spectra 166
Periodic system of elements, construction 42
Perovskite structure type, energy bands, chemical bonds, properties 311
Planck constant 4
Polydymite, Ni_3S_4, chemical bond and properties 309
PP (pseudopotential) method in energy band theory 168
Proustite, bond polarity by NQR 303
Pyrargyrite, bond polarity by NQR 303
Pyrite, FeS_2
  chemical bond and properties 309
  energy band scheme 295
  low-spin state and lattice parameters 299, 306
  Mössbauer parameters 300
Pyrophyllite, intermolecular interactions 313
Pyrrhotite
  energy band scheme and Fe deficit 307
  Mössbauer parameters 300

Quantum numbers 9, 25, 28, 48
Quartz
  effective charges 280
  energy band scheme 177, 281
  molecular orbital diagram 277
  optical absorption spectrum of Co^{2+} 231
  X-ray and ESCA spectra 278

Racah parameters 81
Radial distribution function 38
  Gd 40
  Na, Ti 39
Radicals, N_2^-, O_2^-, S_2^-, F_2^-, Cl_2^-, O_2^{3-}, S_2^{3-},
  molecular orbitals and terms 104
Radii
  covalent 270
  crystal 272
  free atom orbital 38
  hydrogen atom 5
  ionic 264
  mean 264
  orbital 270
Rare earths
  multiplets and terms 74
  optical transitions and color 238
  weak crystal field 75
Realgar, AsS
  nuclear quadrupole resonance of As^{75} 302
  Van der Waals interactions 314
Reflectance spectra
  alkali halides 172
  altaite PbTe 173–176
  galena PbS 173–176
  periclase MgO 166
  sphalerite ZnS 176, 177
  sulfides 291, 292
  wurtzite ZnS 176, 177
Ruby, optical absorption spectra 212, 213
Rutile
  lattice energy 247
  optical absorption spectra 203, 206

Secular equation 110
Selection rules 70
  and optical transitions 200–202
  tables for different symmetry groups 73, 74, 92, 93
Selfconsistent field 2, 37, 137
Senarmontite, Sb_2O_3
  nuclear quadrupole resonance 302
  van der Waals interactions 314
Sillimanite
  lattice sums 263
  nuclear magnetic resonance 285
Solid state spectroscopy 178
  and chemical bond parameters 183
Spectrochemical series 84
Spectroscopy and astrophysics 54
Sphalerite, ZnS
  bond orbital calculations 241, 245
  chemical bond features 308
  covalent radii 270
  energy band structure 176
  EPR of Mn^{2+} 282

ESCA and core electron level shifts 308
  homopolar energy estimation 293
  ionicity in energy band scheme 294
  Mössbauer parameters 300
  optical absorption spectra of
    Co^{2+} 230
    Cr^{2+} 213
    Fe^{2+} 226, 227
    Ni^{2+} 233
    V^{3+} 209
  reflectance spectra 176, 177
Spherical harmonic 257
Spin 40
  in EPR 182
Spinels
  cation distribution and properties 311
  optical absorption spectra of
    Co^{2+} 230
    Cr^{3+} 211, 213
    Fe^{2+} 227
    Ni^{2+} 233
Spin-forbidden transitions 201
Spin-orbit interaction 87
  in energy band schemes (double groups) 164
  parameters 88
  T term splitting 89
Splitting of d orbitals
  in cubal field 60
  in octahedral field 58
  in tetrahedral field 59
Spodumene
  lattice sums 263
  nuclear magnetic resonance of Li^7, Al^{27} 285
Sulfosalts
  chemical bond features 310
  Mössbauer spectra 300
  nuclear quadrupole resonance 301
Symmetry groups, designations 63

Talc, intermolecular interactions 313
Tanabe-Sugano diagrams 81
Terms
  derivation 50
  designations 49
  iron group ions 77
  molecules 102
  numerical value estimation from free atom spectra 53
  for ns and np configurations 51
  transition metal atoms 52
Thiospinels
  chemical bond features 309
  ferromagnetics-semiconductors 289
  simplified energy band schemes 295

Titanium, energy level diagram and optical absorption spectra 203–207
Tourmaline
 Mn^{3+} 213
 optical absorption spectra of V^{3+} 208
Troilite, FeS
 energy band scheme 295
 exchange interactions and properties 297
 high-spin state 298
 Mössbauer parameters 300
 semiconductor-antiferromagnetic 298

Uncertainty principle 16
Units of measurements
 absorption intensity 195
 optical transition energy 191

Vaesite, NiS_2
 energy band scheme 295
 exchange interactions and metallic conductivity 299
Valence 148
Valence bond method 130
Valence State Ionization Energy (VSIE) 122, 124, 136, 138

Vanadium, energy level diagram and optical absorption spectra 207–209
Van der Waals interactions 313
Violarite, $FeNi_2S_4$
 chemical bond and properties 309
 high-spin state 306
Vivianite, absorption spectrum 224

Wave function 17, 24, 98
 angular part 24, 26, 32
 calculations 35
 in crystals 156
 in Molecular Orbital and Valence Bond methods 130
 radial part 24, 30
 representation by complex numbers 157
Wave vector 156, 157

X-ray and ESCA spectroscopy 181, 182
 energy levels 180, 186
  SiO_4^{4-} molecular orbitals by X-ray and ESCA spectroscopy 279

Zircon
 effective charges 280
 lattice sums 263

Minerals and Rocks

Editor-in-Chief: P. J. Wyllie
Editors: W. v. Engelhardt, T. Hahn

Springer-Verlag
Berlin
Heidelberg
New York

Volume 1
W. G. Ernst
Amphiboles
Crystal Chemistry, Phase Relations and Occurrence
1968. 59 figures. X, 125 pages
ISBN 3-540-04267-9

Volume 2
E. Hansen
Strain Facies
1971. 78 figures, 21 plates
X, 208 pages
ISBN 3-540-05204-6

Volume 3
B. R. Doe
Lead Isotopes
1970. 24 figures. IX, 137 pages
ISBN 3-540-05205-4

Volume 4
O. Braitsch
Salt Deposits, Their Origin and Composition
Translators: P. J. Burek, A. E. M. Nairn
In Consultation with A. G. Herrmann, R. Evans
1971. 47 figures. XIV, 297 pages
(German edition s. Mineralogie and Petrographie 3)
ISBN 3-540-05206-2

Volume 5
G. Faure, J. L. Powell
Strontium Isotope Geology
1972. 51 figures. IX, 188 pages
ISBN 3-540-05784-6

Volume 6
F. Lippmann
Sedimentary Carbonate Minerals
1973. 54 figures. VI, 228 pages
ISBN 3-540-06011-1

Volume 7
A. Rittmann
Stable Mineral Assemblages of Igneous Rocks
A Method of Calculation

With contributions by
V. Gittini, W. Hewers, H. Pichler, R. Stengelin
1973. 85 figures. XIV, 262 pages
ISBN 3-540-06030-8

Volume 8
S. K. Saxena
Thermodynamics of Rock – Forming Crystalline Solutions
1973. 67 figures. XII, 188 pages
ISBN 3-40-06175-4

Volume 9
J. Hoefs
Stable Isotope Geochemistry
1973. 37 figures. IX, 140 pages
ISBN 3-540-06176-2

Volume 10
J. T. Wasson
Meteorites
Classification and Properties
1974. 70 figures. X, 316 pages
ISBN 3-540-06744-2

Volume 11
W. Smykatz-Kloss
Differential Thermal Analysis
Application and Results in Mineralogy
1974. 82 figures, 36 tables. XIV, 185 pages
ISBN 3-540-06906-2

Volume 12
R. G. Coleman
Ophiolites
Ancient Oceanic Lithosphere
1977. 72 figures, 18 tables. IX, 229 pages
ISBN 3-540-08276-X

Volume 13
M. S. Paterson
Experimental Rock Deformation: The Brittle Field
1978. 56 figures, 5 tables.
Approx. 260 pages
ISBN 3-540-08835-0

A Springer Journal

Physics and Chemistry of Minerals

In Cooperation with the International Mineralogical Association (I.M.A.)

Editors

S. S. Hafner, University of Marburg, Institute of Mineralogy, Lahnberge, 3550 Marburg, Federal Republic of Germany

A. S. Marfunin, IGEM, Academy of Sciences of the USSR, Staromonetnyi 35, Moscow 109017, USSR

C. T. Prewitt, State University of New York at Stony Brook, Department of Earth and Space Sciences, Stony Brook, New York 11794, USA

Advisory Board

T. J. Ahrens, Pasadena; A. Authier, Paris; P. M. Bell, Washington; G. B. Bokij, Moscow; V. Gabis, Orléans la Source; T. Hahn, Aachen; H. Jagodzinski, Munich; J. C. Jamieson, Chicago; N. Kato, Nagoya; R. C. Liebermann, Stony Brook; J. D. C. McConnell, Cambridge, U.K.; A. C. McLaren, Clayton; N. Morimoto, Osaka; A. Navrotsky, Tempe; R. E. Newnham, University Park; A. F. Reid, Port Melbourne; R. D. Shannon, Wilmington; I. Sunagawa, Sendai; D. W. Strangway, Toronto; R. G. J. Strens, Newcastle upon Tyne; V. M. Vinokurov, Kazan; E. J. W. Whittaker, Oxford; B. J. Wuensch, Cambridge, MA.

Physics and Chemistry of Minerals is an international journal that publishes articles and short communications about physical and chemical studies on minerals, as well as solids related to minerals.

The journal supports competent interdisciplinary work in mineralogy as it relates to physics or chemistry. Emphasis is placed on the application of modern techniques and theories, and the use of models to interpret atomic structures and physical or chemical properties of minerals.

Subjects of interest include:

Relationships between atomic structure and crystalline state (structures of various states, crystal energies, crystal growth, thermodynamic studies, phase transformations, solid solution, exsolution phenomena, etc.)

General solid state spectroscopy (ultraviolet, visible, infrared, Raman, ESCA, luminescence, X-ray, electron paramagnetic resonance, nuclear magnetic resonance, gamma ray resonance, etc.)

Experimental and theoretical analysis of chemical bonding in minerals (application of crystal field, molecular orbital, band theories, etc.)

Physical properties (magnetic, mechanical, electric, optical, thermodynamic, etc.)

Relations between thermal expansion, compressibility, elastic constants, and fundamental properties of atomic structure, particularly as applied to geophysical problems.

Electron microscopy in support of physical and chemical studies.

Sample copies and subscription information upon request.

Springer-Verlag
Berlin
Heidelberg
New York

Date Due

UML 735